W9-ACM-750

CLASSICAL MECHANICS

This book is in the
ADDISON-WESLEY SERIES IN ADVANCED PHYSICS

CLASSICAL MECHANICS

By

HERBERT GOLDSTEIN, Ph.D.

Harvard University

ADDISON-WESLEY PUBLISHING COMPANY, INC.

READING, MASSACHUSETTS · PALO ALTO · LONDON · DALLAS · ATLANTA

Printed in the United States of America

Library of Congress Catalog No. 50–7669

Seventh printing—April 1965

To the Memory

of

LOUIS J. KLEIN

Teacher and Friend

CONTENTS

PREFACE

An advanced course in classical mechanics has long been a time-honored part of the graduate physics curriculum. The present-day function of such a course, however, might well be questioned. It introduces no new physical concepts to the graduate student. It does not lead him directly into current physics research. Nor does it aid him, to any appreciable extent, in solving the practical mechanics problems he encounters in the laboratory.

Despite this arraignment, classical mechanics remains an indispensable part of the physicist's education. It has a twofold role in preparing the student for the study of modern physics. First, classical mechanics, in one or another of its advanced formulations, serves as the springboard for the various branches of modern physics. Thus, the technique of action-angle variables is needed for the older quantum mechanics, the Hamilton-Jacobi equation and the principle of least action provide the transition to wave mechanics, while Poisson brackets and canonical transformations are invaluable in formulating the newer quantum mechanics. Secondly, classical mechanics affords the student an opportunity to master many of the mathematical techniques necessary for quantum mechanics while still working in terms of the familiar concepts of classical physics.

Of course, with these objectives in mind, the traditional treatment of the subject, which was in large measure fixed some fifty years ago, is no longer adequate. The present book is an attempt at an exposition of classical mechanics which does fulfill the new requirements. Those formulations which are of importance for modern physics have received emphasis, and mathematical techniques usually associated with quantum mechanics have been introduced wherever they result in increased elegance and compactness. For example, the discussion of central force motion has been broadened to include the kinematics of scattering and the classical solution of scattering problems. Considerable space has been devoted to canonical transformations, Poisson bracket formulations, Hamilton-Jacobi theory, and action-angle variables. An introduction has been provided to the variational principle formulation of continuous systems and fields. As an illustration of the application of new mathematical techniques, rigid body rotations are treated from the standpoint of matrix transformations. The familiar Euler's theorem on the motion of a rigid body can then be presented in terms of the eigenvalue problem for an orthogonal matrix. As a consequence, such diverse topics as the inertia tensor, Lorentz transformations in Minkowski space, and resonant frequencies of small oscillations become capable of a unified mathematical treatment. Also, by this technique it becomes possible to include at an early stage the difficult concepts of reflection operations and pseudotensor quantities, so important in modern quantum mechanics. A further advantage of matrix methods is that "spinors" can be introduced in connection with the properties of Cayley-Klein parameters.

Several additional departures have been unhesitatingly made. All too often, special relativity receives no connected development except as part of a highly specialized course which also covers general relativity. However, its vital importance in modern physics requires that the student be exposed to special relativity at an early stage in his education. Accordingly, Chapter 6 has been devoted to the subject. Another innovation has been the inclusion of velocity-dependent forces. Historically, classical mechanics developed with the emphasis on static forces dependent on position only, such as gravitational forces. On the other hand, the velocity-dependent electromagnetic force is constantly encountered in modern physics. To enable the student to handle such forces as early as possible, velocity-dependent potentials have been included in the structure of mechanics from the outset, and have been consistently developed throughout the text.

Still another new element has been the treatment of the mechanics of continuous systems and fields in Chapter 11, and some comment on the choice of material is in order. Strictly interpreted, the subject could include all of elasticity, hydrodynamics, and acoustics, but these topics lie outside the prescribed scope of the book, and adequate treatises have been written for most of them. In contrast, no connected account is available on the classical foundations of the variational principle formulation of continuous systems, despite its growing importance in the field theory of elementary particles. The theory of fields can be carried to considerable length and complexity before it is necessary to introduce quantization. For example, it is perfectly feasible to discuss the stress-energy tensor, microscopic equations of continuity, momentum space representations, etc., entirely within the domain of classical physics. It was felt, however, that an adequate discussion of these subjects would require a sophistication beyond what could naturally be expected of the student. Hence it was decided, for this edition at least, to limit Chapter 11 to an elementary description of the Lagrangian and Hamiltonian formulation of fields.

The course for which this text is designed normally carries with it a prerequisite of an intermediate course in mechanics. For both the inadequately prepared graduate student (an all too frequent occurrence) and the ambitious senior who desires to omit the intermediate step, an effort was made to keep the book self-contained. Much of Chapters 1 and 3 is therefore devoted to material usually covered in the preliminary courses.

With few exceptions, no more mathematical background is required of the student than the customary undergraduate courses in advanced calculus and vector analysis. Hence considerable space is given to developing the more complicated mathematical tools as they are needed. An elementary acquaintance with Maxwell's equations and their simpler consequences is necessary for understanding the sections on electromagnetic forces. Most entering graduate students have had at least one term's exposure to modern physics, and frequent advantage has been taken of this circumstance to indicate briefly the relation between a classical development and its quantum continuation.

A large store of exercises is available in the literature on mechanics, easily accessible to all, and there consequently seemed little point to reproducing an extensive collection of such problems. The exercises appended to each chapter therefore have been limited, in the main, to those which serve as extensions of the text, illustrating some particular point or proving variant theorems. Pedantic museum pieces have been studiously avoided.

The question of notation is always a vexing one. It is impossible to achieve a completely consistent and unambiguous system of notation that is not at the same time impracticable and cumbersome. The customary convention has been followed of indicating vectors by bold face Roman letters. In addition, matrix quantities of whatever rank, and tensors other than vectors, are designated by bold face sans serif characters, thus: A. An index of symbols is appended at the end of the book, listing the initial appearance of each meaning of the important symbols. Minor characters, appearing only once, are not included.

References have been listed at the end of each chapter, for elaboration of the material discussed or for treatment of points not touched on. The evaluations accompanying these references are purely personal, of course, but it was felt necessary to provide the student with some guide to the bewildering maze of literature on mechanics. These references, along with many more, are also listed at the end of the book. The list is not intended to be in any way complete, many of the older books being deliberately omitted. By and large, the list contains the references used in writing this book, and must therefore serve also as an acknowledgement of my debt to these sources.

The present text has evolved from a course of lectures on classical mechanics that I gave at Harvard University, and I am grateful to Professor

J. H. Van Vleck, then Chairman of the Physics Department, for many personal and official encouragements. To Professor J. Schwinger, and other colleagues I am indebted for many valuable suggestions. I also wish to record my deep gratitude to the students in my courses, whose favorable reaction and active interest provided the continuing impetus for this work. תושלב״ע

HERBERT GOLDSTEIN

Cambridge, Mass.
March 1950

CHAPTER 1

SURVEY OF THE ELEMENTARY PRINCIPLES

The motion of material bodies formed the subject of some of the earliest researches pursued by the pioneers of physics. From their efforts there has evolved a vast field known as analytical mechanics or dynamics, or simply, mechanics. In the present century the term "classical mechanics" has come into wide use to denote this branch of physics in contradistinction to the newer physical theories, especially quantum mechanics. We shall follow this usage, interpreting the name to include the type of mechanics arising out of the special theory of relativity. It is the purpose of this book to develop the structure of classical mechanics and to outline some of its applications of present-day interest in pure physics.

Basic to any presentation of mechanics are a number of fundamental physical concepts, such as space, time, simultaneity, mass, and force. In discussing the special theory of relativity the notions of simultaneity and of time and length scales will be examined briefly. For the most part, however, these concepts will not be analyzed critically here; rather, they will be assumed as undefined terms whose meanings are familiar to the reader.

1–1 Mechanics of a particle. The essential physics involved in the mechanics of a particle is contained in *Newton's Second Law of Motion*, which may be considered equivalently as a fundamental postulate or as a definition of force and mass. For a single particle, the correct form of the law is:

$$\mathbf{F} = \frac{d\mathbf{p}}{dt}, \tag{1–1}$$

where $\mathbf{F}$ is the total force acting on the particle and $\mathbf{p}$ is the *linear momentum* of the particle defined as follows: Let s be the curve traced by the particle in its motion, and $\mathbf{r}$ the radius vector from the origin to the particle. The vector velocity can then be defined formally by the equation:

$$\mathbf{v} = \frac{d\mathbf{r}}{dt}, \tag{1–2}$$

where the derivative is evaluated by the usual limiting process (cf. Fig. 1–1):

$$\frac{d\mathbf{r}}{dt} = \underset{\Delta t \to 0}{\mathrm{L}} \frac{\mathbf{r}_2 - \mathbf{r}_1}{\Delta t} = \underset{\Delta t \to 0}{\mathrm{L}} \frac{\Delta \mathbf{s}}{\Delta t} = \frac{d\mathbf{s}}{dt}.$$

(This last form for the derivative explicitly indicates that **v** is tangent to the curve.) Then the linear momentum **p** is defined in terms of the velocity as

$$\mathbf{p} = m\mathbf{v}, \tag{1–3}$$

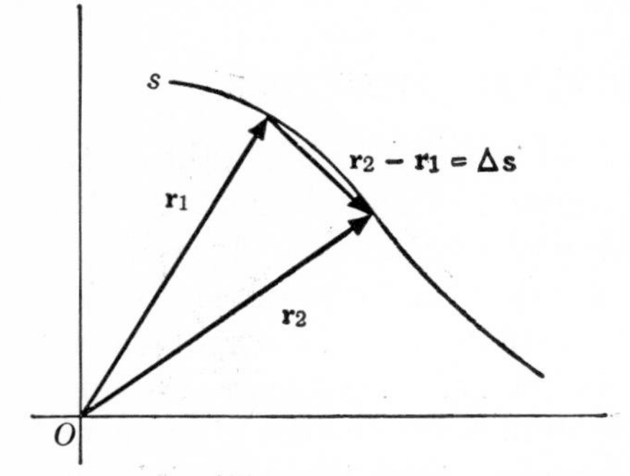

FIG. 1–1. Motion of a particle in space, illustrating the definition of velocity.

so that (1–1) can be written

$$\mathbf{F} = \frac{d}{dt}(m\mathbf{v}). \tag{1–4}$$

In most cases the mass of the particle is constant and Eq. (1–1) reduces to:

$$\mathbf{F} = m\frac{d\mathbf{v}}{dt} = m\mathbf{a}, \tag{1–5}$$

where **a** is called the acceleration of the particle and is defined by

$$\mathbf{a} = \frac{d^2\mathbf{r}}{dt^2}. \tag{1–6}$$

Many of the important conclusions of mechanics can be expressed in the form of conservation theorems, which indicate under what conditions various mechanical quantities are constant in time. Eq. (1–1) directly furnishes the first of these, the

Conservation Theorem for the Linear Momentum of a Particle: If the total force, **F**, *is zero then* $\dot{\mathbf{p}} = 0$ *and the linear momentum,* **p**, *is conserved.*

The angular momentum of the particle about point O, denoted by **L**, is defined as

$$\mathbf{L} = \mathbf{r} \times \mathbf{p}, \tag{1–7}$$

where **r** is the radius vector from O to the particle. Notice that the order of the factors is important. We now define the *moment of force* or *torque* about O as

$$\mathbf{N} = \mathbf{r} \times \mathbf{F}. \tag{1–8}$$

The equation analogous to (1–1) for **N** is obtained by forming the cross product of **r** with Eq. (1–4):

$$\mathbf{r} \times \mathbf{F} = \mathbf{N} = \mathbf{r} \times \frac{d}{dt}(m\mathbf{v}). \tag{1–9}$$

Eq. (1–9) can be written in a different form by using the vector identity:

$$\frac{d}{dt}(\mathbf{r} \times m\mathbf{v}) = \mathbf{v} \times m\mathbf{v} + \mathbf{r} \times \frac{d}{dt}(m\mathbf{v}),$$

where the first term on the right obviously vanishes. In consequence of this identity Eq. (1–9) takes the form

$$\mathbf{N} = \frac{d}{dt}(\mathbf{r} \times m\mathbf{v}) = \frac{d\mathbf{L}}{dt}. \tag{1–10}$$

Note that both **N** and **L** depend upon the point O, about which the moments are taken.

As was the case for Eq. (1–1), the torque equation, (1–10), also yields an immediate conservation theorem, this time the

Conservation Theorem for the Angular Momentum of a Particle: If the total torque, **N**, *is zero then* $\dot{\mathbf{L}} = 0$, *and the angular momentum* **L** *is conserved.*

Next consider the work done by the external force **F** upon the particle in going from point 1 to point 2. By definition this work is

$$W_{12} = \int_1^2 \mathbf{F} \cdot d\mathbf{s}. \tag{1–11}$$

For constant mass (as will be assumed from now on unless otherwise specified), the integral in Eq. (1–11) reduces to

$$\int \mathbf{F} \cdot d\mathbf{s} = m \int \frac{d\mathbf{v}}{dt} \cdot \mathbf{v}\, dt = \frac{m}{2} \int \frac{d}{dt} (v^2)\, dt,$$

and therefore

$$W_{12} = \frac{m}{2} (v_2^2 - v_1^2). \tag{1–12}$$

The scalar quantity $mv^2/2$ is called the kinetic energy of the particle and is denoted by T, so that the work done is equal to the change in the kinetic energy:

$$W_{12} = T_2 - T_1. \tag{1–13}$$

If the force field is such that the work W done around a closed orbit is zero, i.e.,

$$\oint \mathbf{F} \cdot d\mathbf{s} = 0, \tag{1–14}$$

then the force (and the system) is said to be *conservative.* Physically it is clear that a system cannot be conservative if friction or other dissipation forces are present, for $\mathbf{F} \cdot d\mathbf{s}$ due to friction is always positive and the integral cannot vanish. By Stokes' Theorem, the condition for conservative forces, Eq. (1–14), can be written:

$$\nabla \times \mathbf{F} = 0,$$

and since the curl of a gradient always vanishes **F** must therefore be the gradient of some scalar:

$$\mathbf{F} = -\nabla V, \tag{1–15}$$

where V is called the *potential,* or *potential energy.* The existence of V can be established without the use of theorems of vector calculus. If Eq. (1–14) holds, the work W_{12} must be independent of the path of integration between end points 1 and 2. It follows then that it must be possible to express W_{12} as the change in a quantity which depends only upon the positions of the end points. This quantity may be designated by $-V$, so that for a differential path length we have the relation:

$$\mathbf{F} \cdot d\mathbf{s} = -dV$$

or

$$F_s = -\frac{\partial V}{\partial s},$$

which is equivalent to Eq. (1–15). Note that in Eq. (1–15) we can add to V any quantity constant in space, without affecting the results. Hence, *the zero level of V is arbitrary.*

For a conservative system the work done by the forces is

$$W_{12} = V_1 - V_2. \tag{1–16}$$

Combining Eq. (1–16) with Eq. (1–13) we have the result

$$T_1 + V_1 = T_2 + V_2, \tag{1–17}$$

which states in symbols the

Energy Conservation Theorem for a Particle: If the forces acting on a particle are conservative, then the total energy of the particle, $T + V$, is conserved.

1–2 Mechanics of a system of particles. In generalizing the ideas of the previous section to systems of many particles, we must distinguish between the *external forces* acting on the particles due to sources outside the system, and *internal forces* on, say, some particle i due to all other particles in the system. Thus the equation of motion (Newton's Second Law) for the ith particle is to be written:

$$\sum_j \mathbf{F}_{ji} + \mathbf{F}_i^{(e)} = \dot{\mathbf{p}}_i, \tag{1–18}$$

where $\mathbf{F}_i^{(e)}$ stands for an external force, and $\mathbf{F}_{ji}$ is the internal force on the ith particle due to the jth particle ($\mathbf{F}_{ii}$, naturally, is zero). We shall assume that the $\mathbf{F}_{ij}$ (like the $\mathbf{F}_i^{(e)}$) obey Newton's third law of action and reaction: that the forces two particles exert on each other are equal and opposite, *and lie along the line joining them.* There are some important systems in which the forces do not follow this law, notably the electromagnetic forces between moving particles. The theorems derived below must be applied to such systems with due caution.

Summed over all particles Eq. (1–18) takes the form:

$$\frac{d^2}{dt^2}\sum_i m_i\mathbf{r}_i = \sum_i \mathbf{F}_i^{(e)} + \sum_{\substack{i,j \\ i\neq j}} \mathbf{F}_{ji} \tag{1–19}$$

The first sum on the right is simply the total external force $\mathbf{F}^{(e)}$, while the second term vanishes, since the law of action and reaction states that each pair $\mathbf{F}_{ij} + \mathbf{F}_{ji}$ is zero. To reduce the left-hand side we define a vector $\mathbf{R}$ as the average of the radii vectors of the particles, weighted in proportion to their mass, i.e.,*

$$\mathbf{R} = \frac{\Sigma m_i \mathbf{r}_i}{\Sigma m_i} = \frac{\Sigma m_i \mathbf{r}_i}{M}. \tag{1-20}$$

The vector $\mathbf{R}$ defines a point known as the *center of mass*, or more loosely as the center of gravity, of the system. With this definition (1-19) reduces to

$$M \frac{d^2\mathbf{R}}{dt^2} = \sum_i \mathbf{F}_i^{(e)} \equiv \mathbf{F}^{(e)}, \tag{1-19a}$$

which states that the center of mass moves as if the total external force were acting on the entire mass of the system concentrated at the center of mass. Purely internal forces therefore have no effect on the motion of the center of mass. An oft-quoted example is the motion of an exploding shell; the center of mass of the fragments traveling as if the shell were still in a single piece (neglecting air resistance). The same principle is involved in jet and rocket propulsion. In order that the motion of the center of mass be unaffected, the ejection of the exhaust gases at high velocity must be counterbalanced by the forward motion of the vehicle.

By Eq. (1-20) the total linear momentum of the system,

$$\mathbf{P} = \Sigma m_i \frac{d\mathbf{r}_i}{dt} = M \frac{d\mathbf{R}}{dt}, \tag{1-21}$$

is the total mass of the system times the velocity of the center of mass. Consequently, the equation of motion for the center of mass, (1-21), can be restated as the

Conservation Theorem for the Linear Momentum of a System of Particles: If the total external force is zero, the total linear momentum is conserved.

We obtain the total angular momentum of the system by forming the cross product $\mathbf{r}_i \times \mathbf{p}_i$ and summing over i. If this operation is performed in Eq. (1-18) there results

$$\sum_i (\mathbf{r}_i \times \dot{\mathbf{p}}_i) = \sum_i \frac{d}{dt} (\mathbf{r}_i \times \mathbf{p}_i) = \dot{\mathbf{L}} = \sum_i \mathbf{r}_i \times \mathbf{F}_i^{(e)} + \sum_{i,j} \mathbf{r}_i \times \mathbf{F}_{ji}. \tag{1-22}$$

* This definition may be more familiar if Eq. (1-20) is written in terms of the cartesian coordinates

$$X = \frac{\Sigma m_i x_i}{\Sigma m_i}, \qquad Y = \frac{\Sigma m_i y_i}{\Sigma m_i}, \qquad Z = \frac{\Sigma m_i z_i}{\Sigma m_i}.$$

The last term on the right in (1–22) can be considered a sum of the pairs of the form

$$\mathbf{r}_i \times \mathbf{F}_{ji} + \mathbf{r}_j \times \mathbf{F}_{ij} = (\mathbf{r}_i - \mathbf{r}_j) \times \mathbf{F}_{ji}, \tag{1–23}$$

using the equality of action and reaction. But $\mathbf{r}_i - \mathbf{r}_j$ is identical with the vector $\mathbf{r}_{ij}$ from j to i, and the law of action and reaction further states that

$$\mathbf{r}_{ij} \times \mathbf{F}_{ji} = 0,$$

since $\mathbf{F}_{ji}$ is along the line between the two particles. Hence this sum vanishes and (1–22) may be written

$$\frac{d\mathbf{L}}{dt} = \mathbf{N}^{(e)}. \tag{1–24}$$

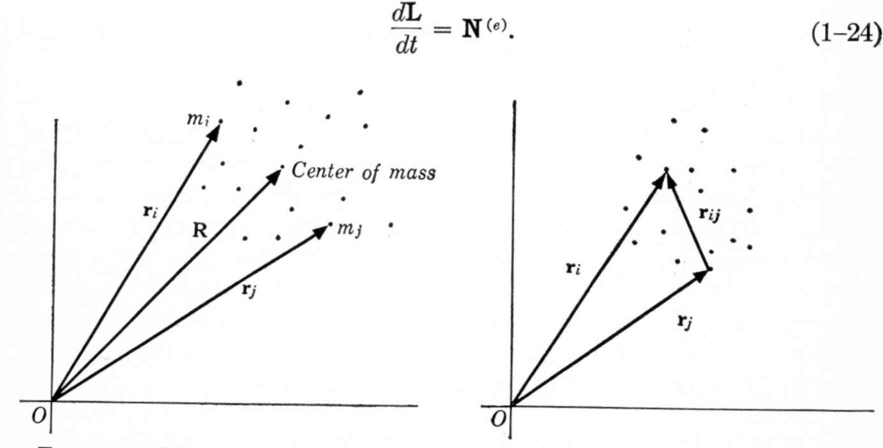

FIG. 1–2. The center of mass of a system of particles.

FIG. 1–3. The vector $\mathbf{r}_{ij}$ between the ith and jth particles.

The time derivative of the total angular momentum is thus equal to the moment of the external force about the given point. Corresponding to Eq. (1–24) is the

Conservation Theorem for Total Angular Momentum: **L** *is constant in time if the applied (external) torque is zero.*

(It is perhaps worthwhile to emphasize that this is a *vector* theorem, i.e., L_z will be conserved if $N_z^{(e)}$ is zero, even if $N_x^{(e)}$ and $N_y^{(e)}$ are not zero.)

Note that the conservation of angular momentum of a system in the absence of applied torques holds only if the law of action and reaction is valid. In a system involving moving charges, where this law is violated, it is not the total mechanical angular momentum which is conserved, but rather the sum of the mechanical and the electromagnetic "angular momentum" of the field.

Eq. (1–21) states that the total linear momentum of the system is the same as if the entire mass were concentrated at the center of mass and moving with it. The analogous theorem for angular momentum is more complicated. With the origin O as reference point the total angular momentum of the system is

$$\mathbf{L} = \sum_i \mathbf{r}_i \times \mathbf{p}_i.$$

Let $\mathbf{R}$ be the radius vector from O to the center of mass, and let $\mathbf{r}'_i$ be the radius vector from the center of mass to the ith particle. Then we have (cf. Fig. 1–4):

$$\mathbf{r}_i = \mathbf{r}'_i + \mathbf{R} \qquad (1\text{–}25)$$

and

$$\mathbf{v}_i = \mathbf{v}'_i + \mathbf{v},$$

where

$$\mathbf{v} = \frac{d\mathbf{R}}{dt},$$

is the velocity of the center of mass relative to O, and

$$\mathbf{v}' = \frac{d\mathbf{r}'}{dt},$$

is the velocity of the ith particle relative to the center of mass of the system. Using Eq. (1–25), the total angular momentum takes on the form

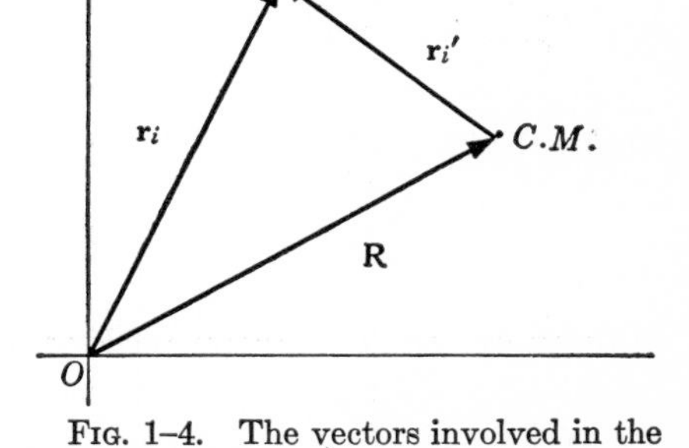

Fig. 1–4. The vectors involved in the shift of reference point for the angular momentum.

$$\mathbf{L} = \sum_i \mathbf{R} \times m_i\mathbf{v} + \sum_i \mathbf{r}'_i \times m_i\mathbf{v}'_i + \Big(\sum_i m_i\mathbf{r}'_i\Big) \times \mathbf{v} + \mathbf{R} \times \frac{d}{dt}\sum_i m_i\mathbf{r}'_i.$$

The last two terms in this expression vanish, for both contain the factor $\Sigma m_i\mathbf{r}'_i$, which, it will be recognized, defines the radius vector of the center of mass in the very coordinate system whose origin is the center of mass, and is therefore a null vector. Rewriting the remaining terms, the total angular momentum about O is:

$$\mathbf{L} = \mathbf{R} \times M\mathbf{v} + \sum_i \mathbf{r}'_i \times \mathbf{p}'_i. \qquad (1\text{–}26)$$

In words, Eq. (1–26) says that the total angular momentum about a point O is the angular momentum of the system concentrated at the center of mass, plus the angular momentum of motion about the center of mass. The form of Eq. (1–26) emphasizes that in general $\mathbf{L}$ depends on the origin O, through the vector $\mathbf{R}$. Only if the center of mass is at rest with respect to O will the angular momentum be independent of the point of reference. In this case the first term in (1–26) vanishes, and $\mathbf{L}$ always reduces to the angular momentum taken about the center of mass.

Finally, let us consider the energy equation. As in the case of a single particle, we calculate the work done by all forces in moving the system from an initial configuration 1, to a final configuration 2:

$$W_{12} = \sum_i \int_1^2 \mathbf{F}_i \cdot d\mathbf{s}_i = \sum_i \int_1^2 \mathbf{F}_i^{(e)} \cdot d\mathbf{s}_i + \sum_{i \neq j} \int_1^2 \mathbf{F}_{ji} \cdot d\mathbf{s}_i. \tag{1-27}$$

Again, the equations of motion can be used to reduce the integrals to

$$\sum_i \int_1^2 \mathbf{F}_i \cdot d\mathbf{s}_i = \sum_i \int_1^2 m_i \dot{\mathbf{v}}_i \cdot \mathbf{v}_i \, dt = \sum_i \int_1^2 d(\tfrac{1}{2} m_i v_i^2).$$

Hence the work done can still be written as the difference of the final and initial kinetic energies:

$$W_{12} = T_2 - T_1,$$

where T, the total kinetic energy of the system, is

$$T = \frac{1}{2} \sum_i m_i v_i^2. \tag{1-28}$$

Making use of the transformations to center of mass coordinates, given in Eq. (1-25), we may write T also as

$$\begin{aligned} T &= \frac{1}{2} \sum_i m_i (\mathbf{v} + \mathbf{v}_i') \cdot (\mathbf{v} + \mathbf{v}_i') \\ &= \frac{1}{2} \sum_i m_i v^2 + \frac{1}{2} \sum_i m_i v_i'^2 + \mathbf{v} \cdot \frac{d}{dt} \left(\sum_i m_i \mathbf{r}_i' \right), \end{aligned}$$

and by the reasoning already employed in calculating the angular momentum, the last term vanishes, leaving

$$T = \frac{1}{2} M v^2 + \frac{1}{2} \sum_i m_i v_i'^2. \tag{1-29}$$

The kinetic energy, like the angular momentum, thus also consists of two parts: the kinetic energy obtained if all the mass were concentrated at the center of mass, plus the kinetic energy of motion about the center of mass.

Consider now the right-hand side of Eq. (1-27). In the special case that the external forces are derivable from a potential the first term can be written as

$$\sum_i \int_1^2 \mathbf{F}_i^{(e)} \cdot d\mathbf{s}_i = - \sum_i \int_1^2 \nabla_i V_i \cdot d\mathbf{s}_i = - \sum_i V_i \Big|_1^2$$

where the subscript i on the del operator indicates that the derivatives are with respect to the components of $\mathbf{r}_i$. If the internal forces are also conservative then the mutual forces between the ith and jth particles, $\mathbf{F}_{ij}$ and $\mathbf{F}_{ji}$, can be obtained from a potential function V_{ij}. To satisfy the law of action and reaction V_{ij} can be a function only of the distance between the particles:

$$V_{ij} = V_{ij}(|\mathbf{r}_i - \mathbf{r}_j|). \tag{1–30}$$

The two forces are then automatically equal and opposite:

$$\mathbf{F}_{ji} = -\nabla_i V_{ij} = +\nabla_j V_{ij} = -\mathbf{F}_{ij}, \tag{1–31}$$

and lie along the line joining the two particles:

$$\nabla V_{ij}(|\mathbf{r}_i - \mathbf{r}_j|) = (\mathbf{r}_i - \mathbf{r}_j)f, \tag{1–32}$$

where f is some scalar function. If V_{ij} were also a function of the difference of some other pair of vectors associated with the particles, such as their velocities or (to step into the domain of modern physics) their intrinsic "spin" angular momenta, then the forces would still be equal and opposite, but would not lie along the direction between the particles.

When the forces are all conservative the second term in Eq. (1–27) can be rewritten as a sum over *pairs* of particles, the terms for each pair being of the form

$$-\int_1^2 (\nabla_i V_{ij} \cdot d\mathbf{s}_i + \nabla_j V_{ij} \cdot d\mathbf{s}_j).$$

If the difference vector $\mathbf{r}_i - \mathbf{r}_j$ be denoted by $\mathbf{r}_{ij}$, and if ∇_{ij} stands for the gradient with respect to $\mathbf{r}_{ij}$, then

$$\nabla_i V_{ij} = \nabla_{ij} V_{ij} = -\nabla_j V_{ij},$$

and

$$d\mathbf{s}_i - d\mathbf{s}_j = d\mathbf{r}_i - d\mathbf{r}_j = d\mathbf{r}_{ij},$$

so that the term for the ij pair has the form

$$-\int \nabla_{ij} V_{ij} \cdot d\mathbf{r}_{ij}.$$

The total work arising from internal forces then reduces to:

$$-\frac{1}{2}\sum_{\substack{i,j\\ i\neq j}} \int_1^2 \nabla_{ij} V_{ij} \cdot d\mathbf{r}_{ij} = -\frac{1}{2}\sum_{\substack{i,j\\ i\neq j}} V_{ij}\Big|_1^2 \tag{1–33}$$

The factor $\frac{1}{2}$ appears in Eq. (1–33) because in summing over *both* i and j each member of a given pair is included twice, first in the i summation and then in the j summation.

From these considerations it is clear that if the external and internal forces are both derivable from potentials it is possible to define a *total potential energy* of the system, V:

$$V = \sum_i V_i + \frac{1}{2} \sum_{\substack{i,j \\ i \neq j}} V_{ij}, \tag{1-34}$$

such that the total energy $T + V$ is conserved, the analog of the conservation theorem (1–17) for a single particle.

The second term on the right in Eq. (1–34) will be called the internal potential energy of the system. In general, it need not be zero and, more important, it may vary as the system changes with time. Only for the particular class of systems known as *rigid bodies* will the internal potential always be constant. Formally, a rigid body can be defined as a system of particles in which the distances r_{ij} are fixed and cannot vary with time. In such case the vectors $d\mathbf{r}_{ij}$ can only be perpendicular to the corresponding $\mathbf{r}_{ij}$, and therefore to the $\mathbf{F}_{ij}$. Therefore, in a rigid body the *internal forces do no work*, and the internal potential must remain constant. Since the total potential is in any case uncertain to within an additive constant, an unvarying internal potential can be completely disregarded in discussing the motion of the system.

1–3 Constraints. From the previous sections one might obtain the impression that all problems in mechanics have been reduced to solving the set of differential equations (1–18):

$$m_i \ddot{\mathbf{r}}_i = \sum_j \mathbf{F}_{ji} + \mathbf{F}_i^{(e)}.$$

One merely substitutes the various forces acting upon the particles of the system, turns the mathematical crank, and grinds out the answers! Even from a purely physical standpoint, however, this view is an oversimplification of the case. For example, it may be necessary to take into account the *constraints* which limit the motion of the system. We have already met one type of system involving constraints, namely rigid bodies, where the constraints on the motions of the particles are such that the distances r_{ij} remain unchanged. Other examples of constrained systems can easily be furnished. The beads of an abacus are constrained to one-dimensional motion by the supporting wires. Gas molecules within a container are constrained by the walls of the vessel to move only *inside* the container. A particle placed on the surface of a solid sphere is restricted by the constraint so that it can only move on the surface or in the region exterior to the sphere.

Constraints may be classified in various ways and we shall use the following system. If the conditions of constraint can be expressed as equations connecting the coordinates of the particles (and the time) having the form

$$f(\mathbf{r}_1, \mathbf{r}_2, \mathbf{r}_3, \ldots t) = 0, \tag{1-35}$$

then the constraints are said to be *holonomic.* Perhaps the simplest example of holonomic constraints is the rigid body, where the constraints are expressed by equations of the form:

$$(\mathbf{r}_i - \mathbf{r}_j)^2 - c_{ij}^2 = 0.$$

A particle constrained to move along a curve or on a surface is another obvious example of a holonomic constraint.

Constraints not expressible in this fashion are called nonholonomic. The walls of a gas container constitute a nonholonomic constraint. The constraint involved in the example of a particle placed on the surface of a sphere is also nonholonomic, for it can be expressed as an inequality

$$r^2 - a^2 \geqq 0$$

(where a is the radius of the sphere), which is not in the form of (1-35). Thus, in a gravitational field a particle placed on the top of the sphere will roll down the surface part of the way but will eventually fall off.

Constraints are further classified according as they are independent of time (scleronomous) or contain time explicitly (rheonomous). An example of the latter kind would be a bead sliding on a moving wire.

Constraints introduce two types of difficulties in the solution of mechanical problems. First, the coordinates r_i are no longer all independent, since they are connected by the equations of constraint; hence the equations of motion (1-18) are not all independent. Second, the forces of constraint, e.g., the force which the wire exerts on the bead, or the wall on the gas particle, are not furnished a priori. They are among the unknowns of the problem and must be obtained from the solution we seek. Indeed, imposing constraints on the system is simply another method of stating that there are forces present in the problem that cannot be specified directly but are known rather in terms of their effect on the motion of the system.

In the case of holonomic constraints, the first difficulty is solved by the introduction of *generalized coordinates.* So far we have been thinking implicitly in terms of cartesian coordinates. A system of N particles, free from constraints, has $3N$ independent coordinates or *degrees of freedom.* If there exist holonomic constraints, expressed in k equations in the form (1-35), then we may use these equations to eliminate k of the $3N$ coordinates, and we are left with $3N - k$ independent coordinates, and the system is said to have $3N - k$ degrees of freedom. This elimination of the dependent coordinates can be expressed in another way, by the introduction of new, $3N - k$, independent variables $q_1, q_2, \ldots q_{3N-k}$ in terms of which the old coordinates $\mathbf{r}_1, \mathbf{r}_2, \ldots \mathbf{r}_N$ are expressed by equations of the form

$$\begin{aligned} \mathbf{r}_1 &= \mathbf{r}_1(q_1, q_2, \ldots q_{3N-k}, t) \\ &\vdots \\ \mathbf{r}_N &= \mathbf{r}_N(q_1, q_2, \ldots q_{3N-k}, t) \end{aligned} \tag{1–36}$$

containing the constraints in them implicitly. These are *transformation* equations from the set of $(\mathbf{r}_l)$ variables to the (q_l) set, or alternatively Eqs. (1–36) can be considered as parametric representations of the $(\mathbf{r}_l)$ variables.

Usually the generalized coordinates, q_l, unlike the cartesian coordinates, will not divide into convenient groups of three which can be associated together to form vectors. Thus, in the case of a particle constrained to move *on* the surface of a sphere, the generalized coordinates are obviously two angles expressing position on the sphere, say latitude and longitude. Or, in the example of a double pendulum moving in a plane (two particles connected by an inextensible light rod and suspended by a similar rod fastened to one of the particles), the generalized coordinates are the two angles θ_1, θ_2. (Cf. Fig. 1–5.) Generalized coordinates, in the sense of coordinates other than cartesian, are often useful in systems without constraints. Thus, in the problem of a particle moving in an external central force field ($V = V(r)$) there is no constraint involved, but it is clearly more convenient to use spherical polar coordinates than cartesian coordinates. One must not, however, think of generalized coordinates in terms of conventional orthogonal position coordinates. All sorts of quantities may be impressed to serve as generalized coordinates. Thus the amplitudes in a Fourier expansion of $\mathbf{r}_j$ may be used as generalized coordinates, or we may find it convenient to employ quantities with the dimensions of energy or angular momentum.

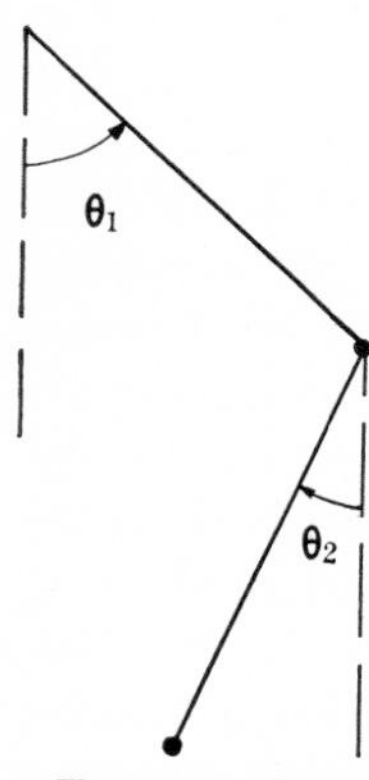

FIG. 1–5. Double pendulum.

If the constraint is nonholonomic the equations expressing the constraint cannot be used to eliminate the dependent coordinates. An oft-quoted example of a nonholonomic constraint is that of an object rolling on a rough surface without slipping. The coordinates used to describe the system will generally involve angular coordinates to specify the orientation of the body, plus a set of coordinates describing the location of the body on the surface. The constraint of "rolling" connects these two sets of coordinates; they are not independent. A change in the orientation of the body inevitably means a change in its location. Yet we cannot reduce the number of coordinates, for the "rolling" condition is not expressible as an equation between the coordinates, in the manner of (1–35). Rather, it is a condition on the *velocities* (i.e., the point of contact is stationary), a differential condition which can be given in an integrated form only *after* the problem is solved.

A simple case will illustrate the point. Consider a disk rolling on the horizontal xy plane constrained to move so that the plane of the disk is always vertical. (The disk could be one of two wheels on a common axle.) The coordinates used to describe the motion might be the x, y coordinates of the center of the disk, an angle of rotation ϕ about the axis of the disk, and an angle θ between the axis of the disk and, say, the x-axis (cf. Fig. 1–6). As a result of the constraint the magnitude of the velocity of the center of the disk is proportional to $\dot{\phi}$:

$$v = a\dot{\phi},$$

where a is the radius of the disk, and its direction is perpendicular to the axis of the disk:

$$\dot{x} = v \sin \theta,$$
$$\dot{y} = -v \cos \theta.$$

Combining these conditions, we have two *differential* equations of constraint

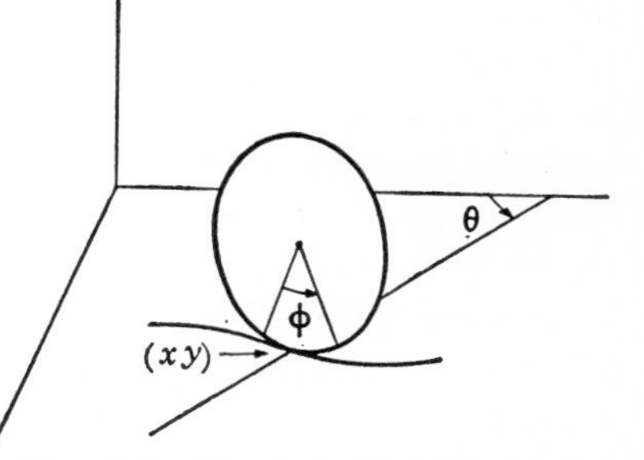

FIG. 1–6. Vertical disk rolling on a horizontal plane.

$$\begin{aligned} dx - a \sin \theta \, d\phi &= 0, \\ dy + a \cos \theta \, d\phi &= 0, \end{aligned} \tag{1–37}$$

and they cannot be integrated without in fact solving the entire problem. Such *nonintegrable* constraints are only special cases of nonholonomic constraints; the conditions of constraint may also appear in the form of inequalities, as we have seen.

Partly because the dependent coordinates can be eliminated, problems involving holonomic constraints are always amenable to a formal solution. But there is no general way of attacking nonholonomic examples. True, if the constraint is nonintegrable, the differential equations of constraint can be introduced into the problem along with the differential equations of motion, and the dependent equations eliminated, in effect, by the method of Lagrange multipliers. We shall return to this method at a later point.

However, the more vicious cases of nonholonomic constraint must be tackled individually, and consequently in the development of the more formal aspects of classical mechanics it is almost invariably assumed that any constraint, if present, is holonomic. This restriction does not greatly limit the applicability of the theory, despite the fact that many of the constraints encountered in everyday life are nonholonomic. The reason is that the entire concept of constraints imposed in the system through the medium of wires or surfaces or walls is particularly appropriate only in macroscopic or large-scale problems. But the physicist today is primarily interested in atomic problems. On this scale all objects, both in and out of the system, consist alike of molecules, atoms or smaller particles, exerting definite forces, and the notion of constraint becomes artificial and rarely appears. Constraints are then used only as mathematical idealizations to the actual physical case or as classical approximations to a quantum-mechanical property — e.g., rigid body rotations for "spin." Such constraints are always holonomic and fit smoothly into the framework of the theory.

To surmount the second difficulty, viz., that the forces of constraint are unknown a priori, we should like to so formulate the mechanics that the forces of constraint disappear. We need then deal only with the known applied forces. A hint as to the procedure to be followed is provided by the fact that in a particular system with constraints, i.e., a rigid body, the work done by internal forces (which are here the forces of constraint) vanishes. We shall follow up this clue in the ensuing sections and generalize the ideas contained in it.

1–4 D'Alembert's principle and Lagrange's equations. A virtual (infinitesimal) displacement of a system refers to a change in the configuration of the system as the result of any arbitrary infinitesimal change of the coordinates $\delta \mathbf{r}_i$, *consistent with the forces and constraints imposed on the system at the given instant t.* The displacement is called virtual to distinguish it from an actual displacement of the system occurring in a time interval dt, during which the forces and constraints may be changing. Suppose the system is in equilibrium, i.e., the total force on each particle vanishes, $\mathbf{F}_i = 0$. Then clearly the dot product $\mathbf{F}_i \cdot \delta \mathbf{r}_i$, which is the virtual work of the force $\mathbf{F}_i$ in the displacement $\delta \mathbf{r}_i$, also vanishes. The sum of these vanishing products over all particles must likewise be zero:

$$\sum_i \mathbf{F}_i \cdot \delta \mathbf{r}_i = 0. \tag{1–38}$$

As yet nothing has been said which has any new physical content. Divide $\mathbf{F}_i$ into the applied force, $\mathbf{F}_i^{(a)}$, and the force of constraint, $\mathbf{f}_i$,

$$\mathbf{F}_i = \mathbf{F}_i^{(a)} + \mathbf{f}_i, \tag{1–39}$$

so that Eq. (1–38) becomes

$$\sum_i \mathbf{F}_i^{(a)} \cdot \delta\mathbf{r}_i + \sum_i \mathbf{f}_i \cdot \delta\mathbf{r}_i = 0. \tag{1–40}$$

We now restrict ourselves to systems for which *the virtual work of the forces of constraint is zero.* We have seen that this condition holds true for rigid bodies and it is valid for a large number of other constraints. Thus, if a particle is constrained to move on a surface, the force of constraint is perpendicular to the surface, while the virtual displacement must be tangent to it, and hence the virtual work vanishes. This is no longer true if friction forces are present, and we must exclude such systems from our formulation. The restriction is not unduly hampering, since the friction is essentially a macroscopic phenomenon. We therefore have as the condition for equilibrium of a system that the virtual work of the *applied forces* vanishes:

$$\sum_i \mathbf{F}_i^{(a)} \cdot \delta\mathbf{r}_i = 0. \tag{1–41}$$

Eq. (1–41) is often called the *principle of virtual work.* Notice that the coefficients of $\delta\mathbf{r}_i$ can no longer be set equal to zero, i.e., in general $\mathbf{F}_i^{(a)} \neq 0$. Essentially this is because the $\delta\mathbf{r}_i$ are not completely independent, but are connected by the constraints. In order to equate the coefficients to zero, one must transform the principle into a form involving the virtual displacements of the q_i, which are independent. Eq. (1–41) satisfies our needs in that it does not contain the $\mathbf{f}_i$, but it deals only with statics; we want a condition involving the general motion of the system.

To obtain such a principle we use a device first thought of by James Bernoulli, and developed by D'Alembert. The equations of motion,

$$\mathbf{F}_i = \dot{\mathbf{p}}_i,$$

can be written as

$$\mathbf{F}_i - \dot{\mathbf{p}}_i = 0,$$

which states that the particles in the system will be in equilibrium under a force equal to the actual force plus a "reversed effective force" $-\dot{\mathbf{p}}_i$. Looked at in this way, dynamics reduces to statics. Instead of (1–38) we can immediately write

$$\sum_i (\mathbf{F}_i - \dot{\mathbf{p}}_i) \cdot \delta\mathbf{r}_i = 0, \tag{1–41$'$}$$

and, making the same resolution into applied forces and forces of constraint there results

$$\sum_i (\mathbf{F}_i^{(a)} - \dot{\mathbf{p}}_i) \cdot \delta\mathbf{r}_i + \sum_i \mathbf{f}_i \cdot \delta\mathbf{r}_i = 0.$$

We again restrict ourselves to systems for which the virtual work of the forces of constraint vanishes and therefore obtain:

$$\sum_i (\mathbf{F}_i^{(a)} - \dot{\mathbf{p}}_i) \cdot \delta\mathbf{r}_i = 0, \tag{1–42}$$

which is often called *D'Alembert's principle.* We have achieved our aim, in that the forces of constraint no longer appear, and the superscript $^{(a)}$ can now be dropped without ambiguity. It is still not in a useful form to furnish equations of motion for the system. We must now transform the principle into an expression involving virtual displacements of the generalized coordinates, which are then independent of each other (for holonomic constraints), so that the coefficients of the δq_i can be set separately equal to zero.

The translation from $\mathbf{r}_i$ to q_j language starts from the transformation equations (1–36):

$$\mathbf{r}_i = \mathbf{r}_i(q_1, q_2, \ldots q_n, t)$$

(assuming n independent coordinates) and is carried out by means of the usual rules of the calculus of partial differentiation. Thus $\mathbf{v}_i \equiv \dot{\mathbf{r}}_i$ is expressed in terms of the $\dot{q}_j$ by the formula

$$\mathbf{v}_i = \sum_j \frac{\partial \mathbf{r}_i}{\partial q_j} \dot{q}_j + \frac{\partial \mathbf{r}_i}{\partial t}. \tag{1–43}$$

Similarly, the arbitrary virtual displacement $\delta\mathbf{r}_i$ can be connected with the virtual displacements δq_j by:

$$\delta\mathbf{r}_i = \sum_j \frac{\partial \mathbf{r}_i}{\partial q_j} \delta q_j. \tag{1–44}$$

Notice that no variation of time, δt, is involved here, since a virtual displacement by definition considers only displacements of the coordinates. (Only then is the virtual displacement perpendicular to the force of constraint if the constraint itself is changing in time.)

In terms of the generalized coordinates the virtual work of the $\mathbf{F}_i$ becomes:

$$\sum_i \mathbf{F}_i \cdot \delta\mathbf{r}_i = \sum_{i,j} \mathbf{F}_i \cdot \frac{\partial \mathbf{r}_i}{\partial q_j} \delta q_j$$

$$= \sum_j Q_j \, \delta q_j, \tag{1–45}$$

where the Q_j are called the components of the *generalized force*, defined as

$$Q_j = \sum_i \mathbf{F}_i \cdot \frac{\partial \mathbf{r}_i}{\partial q_j}. \tag{1-46}$$

Note that just as the q's need not have the dimensions of length, so the Q's do not necessarily have the dimensions of force, but $Q_j\,\delta q_j$ must always have the dimensions of work.

We turn now to the other term involved in Eq. (1-42), which may be written as

$$\sum_i \dot{\mathbf{p}}_i \cdot \delta \mathbf{r}_i = \sum_i m_i \ddot{\mathbf{r}}_i \cdot \delta \mathbf{r}_i.$$

Expressing $\delta \mathbf{r}_i$ by (1-44) this becomes

$$\sum_{i,j} m_i \ddot{\mathbf{r}}_i \cdot \frac{\partial \mathbf{r}_i}{\partial q_j}\,\delta q_j.$$

Consider now the relation

$$\sum_i m_i \ddot{\mathbf{r}}_i \cdot \frac{\partial \mathbf{r}_i}{\partial q_j} = \sum_i \left\{ \frac{d}{dt}\left(m_i \dot{\mathbf{r}}_i \cdot \frac{\partial \mathbf{r}_i}{\partial q_j}\right) - m_i \dot{\mathbf{r}}_i \cdot \frac{d}{dt}\left(\frac{\partial \mathbf{r}_i}{\partial q_j}\right)\right\}. \tag{1-47}$$

In the last term of Eq. (1-47) we can interchange the differentiation with respect to t and q_j, for, in analogy to (1-43),

$$\frac{d}{dt}\left(\frac{\partial \mathbf{r}_i}{\partial q_j}\right) = \sum_k \frac{\partial^2 \mathbf{r}_i}{\partial q_j\,\partial q_k}\,\dot{q}_k + \frac{\partial^2 \mathbf{r}_i}{\partial q_j\,\partial t},$$

which by (1-43) is just $\partial \mathbf{v}_i/\partial q_j$. Further, from (1-43) we also see that

$$\frac{\partial \mathbf{v}_i}{\partial \dot{q}_j} = \frac{\partial \mathbf{r}_i}{\partial q_j}. \tag{1-48}$$

Substituting these changes in (1-47) there results

$$\sum_i m_i \ddot{\mathbf{r}}_i \cdot \frac{\partial \mathbf{r}_i}{\partial q_j} = \sum_i \left\{ \frac{d}{dt}\left(m_i \mathbf{v}_i \cdot \frac{\partial \mathbf{v}_i}{\partial \dot{q}_j}\right) - m_i \mathbf{v}_i \cdot \frac{\partial \mathbf{v}_i}{\partial q_j}\right\}$$

and the desired term of Eq. (1-42) expands into:

$$\sum_j \left\{ \frac{d}{dt}\left(\frac{\partial}{\partial \dot{q}_j}\left(\sum_i \frac{1}{2} m_i v_i^2\right)\right) - \frac{\partial}{\partial q_j}\left(\sum_i \frac{1}{2} m_i v_i^2\right)\right\}\delta q_j.$$

Identifying $\sum_i \frac{1}{2} m_i v_i^2$ with the system kinetic energy T, D'Alembert's principle becomes

$$\sum_j \left[\left\{ \frac{d}{dt}\left(\frac{\partial T}{\partial \dot{q}_j}\right) - \frac{\partial T}{\partial q_j}\right\} - Q_j\right]\delta q_j = 0. \tag{1-49}$$

E is not the gradient of a scalar function since $\nabla \times \mathbf{E} \neq 0$, but from $\nabla \cdot \mathbf{B} = 0$ it follows that **B** can be represented as the curl of a vector:

$$\mathbf{B} = \nabla \times \mathbf{A}, \tag{1–57}$$

where **A** is called the magnetic vector potential. Then the curl **E** equation becomes

$$\nabla \times \mathbf{E} + \frac{1}{c}\frac{\partial}{\partial t}(\nabla \times \mathbf{A}) = \nabla \times \left(\mathbf{E} + \frac{1}{c}\frac{\partial \mathbf{A}}{\partial t}\right) = 0.$$

Hence we can set

$$\mathbf{E} + \frac{1}{c}\frac{\partial \mathbf{A}}{\partial t} = -\nabla\phi$$

or

$$\mathbf{E} = -\nabla\phi - \frac{1}{c}\frac{\partial \mathbf{A}}{\partial t}. \tag{1–58}$$

In terms of the potentials ϕ and **A**, the so-called Lorentz force (1–56) becomes

$$\mathbf{F} = q\left\{-\nabla\phi - \frac{1}{c}\frac{\partial \mathbf{A}}{\partial t} + \frac{1}{c}(\mathbf{v} \times \nabla \times \mathbf{A})\right\}. \tag{1–59}$$

The terms of Eq. (1–59) can be rewritten in a more convenient form. As an example consider the x component

$$(\nabla\phi)_x = \frac{\partial\phi}{\partial x}$$

and

$$\begin{aligned}(\mathbf{v} \times \nabla \times \mathbf{A})_x &= v_y\left(\frac{\partial A_y}{\partial x} - \frac{\partial A_x}{\partial y}\right) - v_z\left(\frac{\partial A_x}{\partial z} - \frac{\partial A_z}{\partial x}\right)\\ &= v_y\frac{\partial A_y}{\partial x} + v_z\frac{\partial A_z}{\partial x} + v_x\frac{\partial A_x}{\partial x} - v_y\frac{\partial A_x}{\partial y} - v_z\frac{\partial A_x}{\partial z} - v_x\frac{\partial A_x}{\partial x},\end{aligned}$$

where we have added and subtracted the term

$$v_x\frac{\partial A_x}{\partial x}.$$

Now, the total time derivative of A_x is

$$\frac{dA_x}{dt} = \frac{\partial A_x}{\partial t} + \left(v_x\frac{\partial A_x}{\partial x} + v_y\frac{\partial A_x}{\partial y} + v_z\frac{\partial A_x}{\partial z}\right),$$

where the first term arises from the explicit variation of A_x with time, and the second term results from the motion of the particle with time, which changes the spatial point at which A_x is evaluated. The x component of $\mathbf{v} \times \nabla \times \mathbf{A}$ can therefore be written as:

$$(\mathbf{v} \times \nabla \times \mathbf{A})_x = \frac{\partial}{\partial x}(\mathbf{v} \cdot \mathbf{A}) - \frac{dA_x}{dt} + \frac{\partial A_x}{\partial t}.$$

With these substitutions (1–59) becomes

$$F_x = q\left\{-\frac{\partial}{\partial x}\left(\phi - \frac{1}{c}\mathbf{v} \cdot \mathbf{A}\right) - \frac{1}{c}\frac{d}{dt}\left(\frac{\partial}{\partial v_x}(\mathbf{A} \cdot \mathbf{v})\right)\right\}.$$

Since the scalar potential is independent of velocity this expression is equivalent to:

$$F_x = -\frac{\partial U}{\partial x} + \frac{d}{dt}\frac{\partial U}{\partial v_x}$$

where

$$U = q\phi - \frac{q}{c}\mathbf{A} \cdot \mathbf{v}. \tag{1–60}$$

U is a generalized potential in the sense of Eq. (1–54), and the Lagrangian for a charged particle in an electromagnetic field can be written

$$L = T - q\phi + \frac{q}{c}\mathbf{A} \cdot \mathbf{v}. \tag{1–61}$$

It should be noted that if only some of the forces acting on the system are derivable from a potential then Lagrange's equations can always be written in the form

$$\frac{d}{dt}\left(\frac{\partial L}{\partial \dot{q}_j}\right) - \frac{\partial L}{\partial q_j} = Q_j,$$

where L contains the potential of the conservative forces as before, and Q_j represents the forces *not* arising from a potential. Such a situation often occurs when there are frictional forces present. It frequently happens that the frictional force is proportional to the velocity of the particle, so that its x component has the form:

$$F_{fx} = -k_x v_x.$$

Frictional forces of this type may be derived in terms of a function $\mathfrak{F}$, known as *Rayleigh's dissipation function*, and defined as

$$\mathfrak{F} = \frac{1}{2}\sum_i (k_x v_{ix}^2 + k_y v_{iy}^2 + k_z v_{iz}^2),$$

where the summation is over the particles of the system. From this definition it is clear that

$$F_{fx} = -\frac{\partial \mathfrak{F}}{\partial v_x}$$

or, symbolically:

$$\mathbf{F}_f = -\nabla_v \mathfrak{F}.$$

One can also give a physical interpretation to the dissipation function. The work done *by* the system *against* friction is

$$dW_f = -\mathbf{F}_f \cdot d\mathbf{r} = -\mathbf{F}_f \cdot \mathbf{v}\, dt = (k_x v_x^2 + k_y v_y^2 + k_z v_z^2)\, dt.$$

Hence $2\mathfrak{F}$ is the rate of energy dissipation due to friction. The component of the generalized force resulting from the force of friction is then given by

$$\begin{aligned} Q_j = \sum_i \mathbf{F}_{if} \cdot \frac{\partial \mathbf{r}_i}{\partial q_j} &= -\sum \nabla_v \mathfrak{F} \cdot \frac{\partial \mathbf{r}_i}{\partial q_j} \\ &= -\sum \nabla_v \mathfrak{F} \cdot \frac{\partial \dot{\mathbf{r}}_i}{\partial \dot{q}_j}, \text{ by (1–48)}, \\ &= -\frac{\partial \mathfrak{F}}{\partial \dot{q}_j}. \end{aligned}$$

The Lagrange equations now become

$$\frac{d}{dt}\left(\frac{\partial L}{\partial \dot{q}_j}\right) - \frac{\partial L}{\partial q_j} + \frac{\partial \mathfrak{F}}{\partial \dot{q}_j} = 0,$$

so that two scalar functions, L and $\mathfrak{F}$, must be specified to obtain the equations of motion.

1–6 Simple applications of the Lagrangian formulation. The previous sections show that for systems where one can define a Lagrangian, i.e., holonomic systems with applied forces derivable from an ordinary or generalized potential, we have a very convenient way of setting up the equations of motion. We were led to the Lagrangian formulation by the desire to eliminate the forces of constraint from the equations of motion, and in achieving this goal we have obtained many other benefits. In setting up the original form of the equations of motion, Eqs. (1–18), it is necessary to work with many *vector* forces and accelerations. With the Lagrangian method one has only to deal with two *scalar* functions, T and V, which greatly simplifies the problem. A straightforward routine procedure can now be established for all problems of mechanics to which the Lagrangian formulation is applicable. One has only to write T and V in generalized coordinates, form L from them, and substitute in (1–53) to obtain the equations of motion. The needed transformation of T and V from cartesian coordinates to generalized coordinates is obtained by applying the transformation equations (1–36) and (1–43). Thus T is given in general by

$$T = \sum_i \frac{1}{2} m_i v_i^2 = \sum_i \frac{1}{2} m_i \left(\sum_j \frac{\partial \mathbf{r}_i}{\partial q_j} \dot{q}_j + \frac{\partial \mathbf{r}_i}{\partial t} \right)^2.$$

It is clear that on carrying out the expansion, the expression for T in generalized coordinates will have the form

$$T = a + \sum_j a_j \dot{q}_j + \sum_{j,k} a_{jk} \dot{q}_j \dot{q}_k \tag{1–62}$$

where a, a_j, a_{jk} are definite functions of the $\mathbf{r}$'s and t and hence of the q's and t. In fact, a comparison shows that

$$a = \sum_i \frac{1}{2} m_i \left(\frac{\partial \mathbf{r}_i}{\partial t} \right)^2,$$

$$a_j = \sum_i m_i \frac{\partial \mathbf{r}_i}{\partial t} \cdot \frac{\partial \mathbf{r}_i}{\partial q_j},$$

and

$$a_{jk} = \sum_i \frac{1}{2} m_i \frac{\partial \mathbf{r}_i}{\partial q_j} \cdot \frac{\partial \mathbf{r}_i}{\partial q_k}.$$

If the transformation equations do not contain the time explicitly, i.e., constraints independent of time (scleronomous) then only the last term in (1–62) is nonvanishing and T is always a homogeneous quadratic form in the generalized velocities.

Let us now consider some simple examples of this procedure:

1. Single particle in space
   a. Cartesian coordinates
   b. Plane polar coordinates
2. Atwood's machine
3. Time-dependent constraint—bead sliding on rotating wire

1. *Motion of one particle: using cartesian coordinates.* The generalized forces needed in Eq. (1–50) are obviously F_x, F_y, and F_z. Then

$$T = \tfrac{1}{2} m(\dot{x}^2 + \dot{y}^2 + \dot{z}^2),$$

$$\frac{\partial T}{\partial x} = \frac{\partial T}{\partial y} = \frac{\partial T}{\partial z} = 0,$$

$$\frac{\partial T}{\partial \dot{x}} = m\dot{x}, \qquad \frac{\partial T}{\partial \dot{y}} = m\dot{y}, \qquad \frac{\partial T}{\partial \dot{z}} = m\dot{z},$$

and the equations of motion are

$$\frac{d}{dt}(m\dot{x}) = F_x, \qquad \frac{d}{dt}(m\dot{y}) = F_y, \qquad \frac{d}{dt}(m\dot{z}) = F_z. \tag{1-63}$$

We are thus led back to the original Newton's equations of motion.

Motion of one particle: using plane polar coordinates. Here we must express T in terms of $\dot{r}$ and $\dot{\theta}$. The equations of transformation, i.e., the Eqs. (1–36), in this case are simply:

$$x = r\cos\theta,$$
$$y = r\sin\theta.$$

In analogy to (1–43), the velocities are given by:

$$\dot{x} = \dot{r}\cos\theta - r\dot{\theta}\sin\theta,$$
$$\dot{y} = \dot{r}\sin\theta + r\dot{\theta}\cos\theta.$$

The kinetic energy $T = \frac{1}{2}m(\dot{x}^2 + \dot{y}^2)$ then reduces formally to

$$T = \tfrac{1}{2}m(\dot{r}^2 + (r\dot{\theta})^2). \tag{1-64}$$

An alternative derivation of Eq. (1–64) is obtained by recognizing that the plane polar components of the velocity are $\dot{r}$ along $\mathbf{r}$, and $r\dot{\theta}$ along the direction perpendicular to r, denoted by the unit vector $\mathbf{n}$. Hence the square of the velocity expressed in polar coordinates is simply $\dot{r}^2 + (r\dot{\theta})^2$. The components of the generalized force can be obtained from the definition, Eq. (1–46),

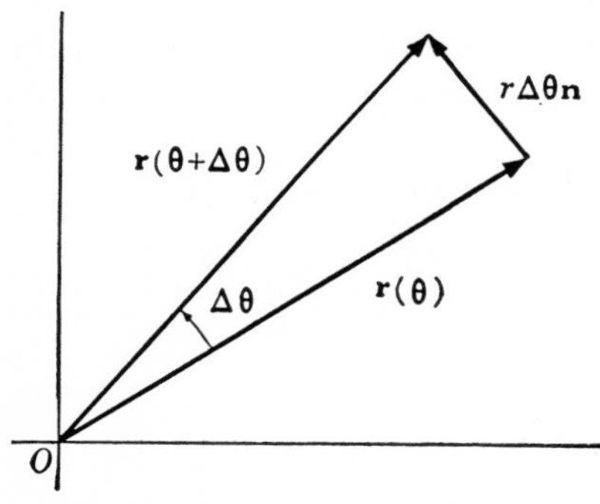

FIG. 1–7. Derivative of $\mathbf{r}$ with respect to θ.

$$Q_r = \mathbf{F}\cdot\frac{\partial\mathbf{r}}{\partial r} = \mathbf{F}\cdot\frac{\mathbf{r}}{r} = F_r,$$

$$Q_\theta = \mathbf{F}\cdot\frac{\partial\mathbf{r}}{\partial\theta} = \mathbf{F}\cdot r\mathbf{n} = rF_\theta,$$

since the derivative of $\mathbf{r}$ with respect to θ is, by definition of a derivative, a vector in the direction of $\mathbf{n}$, cf. Fig. 1–7. There are two generalized coordinates, and therefore two Lagrange equations. The derivatives occurring in the r equation are

$$\frac{\partial T}{\partial r} = mr\dot{\theta}^2, \qquad \frac{\partial T}{\partial \dot{r}} = m\dot{r}, \qquad \frac{d}{dt}\left(\frac{\partial T}{\partial \dot{r}}\right) = m\ddot{r},$$

and the equation itself is:

$$m\ddot{r} - mr\dot{\theta}^2 = F_r,$$

the second term being the centripetal acceleration term. For the θ equation we have the derivatives

$$\frac{\partial T}{\partial \theta} = 0, \qquad \frac{\partial T}{\partial \dot{\theta}} = mr^2\dot{\theta}, \qquad \frac{d}{dt}(mr^2\dot{\theta}) = mr^2\ddot{\theta} + 2mr\dot{r}\dot{\theta},$$

so that the equation becomes

$$\frac{d}{dt}(mr^2\dot{\theta}) = mr^2\ddot{\theta} + 2mr\dot{r}\dot{\theta} = rF_\theta.$$

Note that the left is just the time derivative of the angular momentum, and the right is exactly the applied torque, so that we have simply rederived the torque equation (1–24).

2. *The Atwood's machine* — an example of a conservative system with holonomic, scleronomous constraint (the pulley is assumed frictionless). Clearly there is only one independent coordinate x, the position of the other weight being determined by the constraint that the length of the rope between them is l. The potential energy is

$$V = -M_1gx - M_2g(l - x),$$

while the kinetic energy is

$$T = \tfrac{1}{2}(M_1 + M_2)\dot{x}^2.$$

Combining the two, the Lagrangian has the form:

$$L = T - V = \tfrac{1}{2}(M_1 + M_2)\dot{x}^2 + M_1gx + M_2g(l - x).$$

There is only one equation of motion, involving the derivatives

$$\frac{\partial L}{\partial x} = (M_1 - M_2)g,$$

$$\frac{\partial L}{\partial \dot{x}} = (M_1 + M_2)\dot{x},$$

so that we have

$$(M_1 + M_2)\ddot{x} = (M_1 - M_2)g,$$

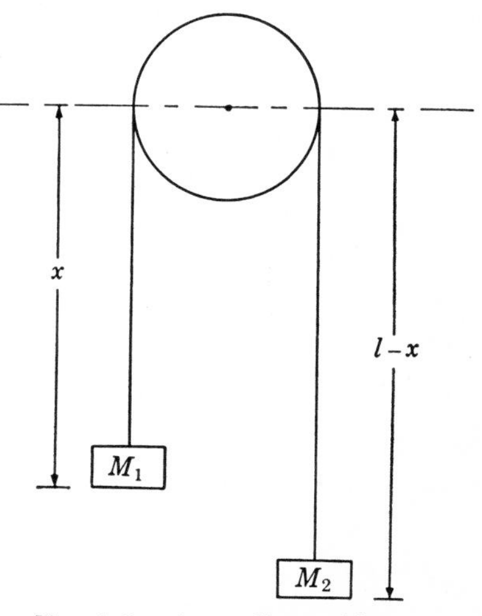

FIG. 1–8. Atwood's machine.

or

$$\ddot{x} = \frac{M_1 - M_2}{M_1 + M_2} g,$$

which is the familiar result obtained by more elementary means. This trivial problem emphasizes that the forces of constraint — here the tension in the rope — appear nowhere in the Lagrangian formulation. By the same token neither can the tension in the rope be found directly by the Lagrangian method.

3. *A bead sliding on a uniformly rotating wire in a force-free space.* This example has been chosen as a simple illustration of a constraint being time dependent, with the transformation equations therefore containing the time explicitly:

$$x = r \cos \omega t, \qquad \omega = \text{angular velocity of rotation.}$$
$$y = r \sin \omega t.$$

While one could then find T (here the same as L) by the same procedure used to obtain (1–62), it is simpler to take over (1–64) directly, expressing the constraint by the relation $\dot{\theta} = \omega$:

$$T = \tfrac{1}{2}m(\dot{r}^2 + r^2\omega^2).$$

Notice that T is not a homogeneous quadratic function of the generalized velocities, since there is now an additional term not involving $\dot{r}$. The equation of motion is then

$$m\ddot{r} - mr\omega^2 = 0$$

or

$$\ddot{r} = r\omega^2,$$

which is the well-known result that the bead moves outward because of the centripetal acceleration. Again, the method cannot furnish the force of constraint keeping the bead on the wire.

SUGGESTED REFERENCES *

J. L. Synge and B. A. Griffith, *Principles of Mechanics.* An excellent textbook of mechanics on the intermediate level, which can be read with much profit as a preliminary to the type of graduate course for which the present text is designed.

C. J. Coe, *Theoretical Mechanics.* Another text on the intermediate level, with emphasis on vector treatment. Some of the later chapters are quite advanced. The book includes a summary of vector analysis.

---

*For convenience the references at the end of each chapter are listed only by short title. Full bibliographical description will be found in the bibliography at the end of the book.

W. F. Osgood, *Mechanics.* The first five chapters of this book form an elementary introduction into the subject that is delightfully flavored by the author's long pedagogic experience. In this regard the reader's attention is directed especially to p. 102!

G. Joos, *Theoretical Physics.* Portions of Chapters V and VI are perhaps the closest of any reference to the treatment given in this chapter.

E. A. Milne, *Vectorial Mechanics.* A formidable treatise which often manages somehow to make the elegant simplicity of vector and tensor methods appear quite complicated and repellant. Nevertheless, in Part III there can be found many interesting vector theorems on the general motion of a particle and systems of particles which are deduced from first principles.

E. Mach, *The Science of Mechanics.* A classic analysis and criticism of the fundamental concepts of classical mechanics. In its earlier editions this book did much to clear the way philosophically for relativity theory.

R. B. Lindsay and H. Margenau, *Foundations of Physics.* Chapter 3 contains a clear discussion of the foundations of classical mechanics. This book, together with Mach's work, can serve as an excellent point of departure for further reading on the nature of the basic ideas involved in mechanics.

G. Hamel, *Die Axiome der Mechanik* (Vol. V, *Handbuch der Physik*). An attempt to put the axiomatic formulation of mechanics on a mathematically rigorous basis. Nordheim's article in the same volume contains a brief discussion of velocity-dependent potentials, Section 10, Chapter 2.

E. T. Whittaker, *Analytical Dynamics.* A well-known treatise which presents an exhaustive treatment of analytical mechanics from the older viewpoints. The development is marked, regrettably, by an apparent dislike of diagrams (of which there are only four in the entire book) and of vector notation, and by a fondness for the type of pedantic mechanics problems made famous by the Cambridge Tripos examinations. It remains, however, a practically unique source for the discussion of many specialized topics. For the present chapter reference should be made principally to Chapter II, especially Section 31, which discusses velocity-dependent potentials. Sections 92–94 of Chapter VIII are concerned with the dissipation function.

Lord Rayleigh, *The Theory of Sound.* The dissipation function is introduced in Chapter IV, Vol. I of this classic treatise.

EXERCISES

1. A nucleus, originally at rest, decays radioactively by emitting an electron of momentum 1.73 Mev/c, and at right angles to the direction of the electron a neutrino with momentum 1.00 Mev/c. (The Mev (million electron volt) is a unit of energy, used in modern physics, equal to 1.59×10^{-6} erg. Correspondingly, Mev/c is a unit of linear momentum equal to 5.33×10^{-17} gm-cm/sec.) In what direction does the nucleus recoil? What is its momentum in Mev/c? If the mass of the residual nucleus is 3.90×10^{-22} gm, what is its kinetic energy, in electron volts?

2. The *escape velocity* of a particle on the earth is the minimum velocity required at the surface of the earth in order that the particle can escape from the earth's gravitational field. Neglecting the resistance of the atmosphere, the system is conservative. From the conservation theorem for potential plus kinetic energy show that the escape velocity for the earth, ignoring the presence of the moon, is 6.95 mi/sec.

3. Rockets are propelled by the momentum reaction of the exhaust gases expelled from the tail. Since these gases arise from the reaction of the fuels carried in the rocket the mass of the rocket is not constant, but decreases as the fuel is expended. Show that the equation of motion for a rocket projected vertically upward in a uniform gravitational field, neglecting atmospheric resistance, is

$$m\frac{dv}{dt} = -v'\frac{dm}{dt} - mg,$$

where m is the mass of the rocket and v' is the velocity of the escaping gases relative to the rocket. Integrate this equation to obtain v as a function of m, assuming a constant time rate of loss of mass. Show, for a rocket starting initially from rest, with v' equal to 6800 ft/sec and a mass loss per second equal to 1/60th of the initial mass (values appropriate to the V–2), that in order to reach the escape velocity the ratio of the weight of the fuel to the weight of the empty rocket must be almost 300!

4. A particle moves in a plane under the influence of a force, acting toward a center of force, whose magnitude is

$$F = \frac{1}{r^2}\left(1 - \frac{\dot{r}^2 - 2\ddot{r}r}{c^2}\right),$$

where r is the distance of the particle to the center of force. Find the generalized potential which will result in such a force, and from that the Lagrangian for the motion in a plane. (The expression for F represents the force between two charges in Weber's electrodynamics.)

5. Obtain the equation of motion for a particle falling vertically under the influence of gravity when frictional forces obtainable from a dissipation function $\frac{1}{2}kv^2$ are present. Integrate the equation to obtain the velocity as a function of time and show that the maximum possible velocity for fall from rest is $v = mg/k$.

6. Two points of mass m are joined by a rigid weightless rod of length l, the center of which is constrained to move on a circle of radius a. Set up the kinetic energy in generalized coordinates.

7. Obtain the Lagrange equations of motion for a spherical pendulum, i.e., a mass point suspended by a rigid weightless rod.

8. A system is composed of three particles of equal mass m. Between any two of them there are forces derivable from a potential

$$V = -ge^{-\mu r},$$

where r is the distance between the two particles. In addition, two of the particles each exert a force on the third which can be obtained from a generalized potential of the form

$$U = -f\mathbf{v}\cdot\mathbf{r},$$

$\mathbf{v}$ being the relative velocity of the interacting particles and f a constant. Set up the Lagrangian for the system, using as coordinates the radius vector $\mathbf{R}$ of the center of mass and the two vectors

$$\boldsymbol{\rho}_1 = \mathbf{r}_1 - \mathbf{r}_3,$$
$$\boldsymbol{\rho}_2 = \mathbf{r}_2 - \mathbf{r}_3.$$

Is the total angular momentum of the system conserved?

9. Two mass points of mass m_1 and m_2 are connected by a string passing through a hole in a smooth table so that m_1 rests on the table surface and m_2 hangs suspended. Assuming m_2 moves only in a vertical line, what are the generalized coordinates for the system? Write down the Lagrange equations for the system and, if possible, discuss the physical significance any of them might have. Reduce the problem to a single second-order differential equation and obtain a first integral of the equation. What is its physical significance? (Consider the motion only so long as neither m_1 nor m_2 passes through the hole.)

10. Obtain the Lagrangian and equations of motion for the double pendulum illustrated in Fig. 1–5, where the lengths of the pendula are l_1 and l_2 with corresponding masses m_1 and m_2.

CHAPTER 2

VARIATIONAL PRINCIPLES AND LAGRANGE'S EQUATIONS

2–1 Hamilton's principle. The derivation of Lagrange's equations presented in the previous chapter has started from a consideration of the instantaneous state of the system and small virtual displacements about the instantaneous state, i.e., from a "differential principle" such as D'Alembert's principle. It is also possible to obtain Lagrange's equations from a principle which considers the entire motion of the system between times t_1 and t_2, and small virtual variations of the entire motion from the actual motion. A principle of this nature is known as an "integral principle."

Before presenting the integral principle, the meaning attached to the phrase "motion of the system between times t_1 and t_2" must first be stated in more precise language. The instantaneous configuration of a system is described by the values of the n generalized coordinates $q_1 \ldots q_n$, and corresponds to a particular point in a cartesian hyperspace where the q's form the n coordinate axes. This n-dimensional space is therefore known as configuration space. As time goes on the state of the system changes, and the system point moves in configuration space tracing out a curve, described as "the path of motion of the system." The "motion of the system," as used above, then refers to the motion of the system point along this path in *configuration space*. Time can be considered formally as a parameter of the curve; to each point on the path there is associated one or more values of the time. It must be emphasized that configuration space has no necessary connection with the physical three-dimensional space, just as the generalized coordinates are not necessarily position coordinates. The path of motion in configuration space will not have any necessary resemblance to the path in space of any actual particle; each point on the path represents the *entire* system configuration at some given instant of time.

We can now state the integral *Hamilton's Principle* for conservative systems (in the larger sense, admitting generalized potentials):

The motion of the system from time t_1 to time t_2 is such that the line integral

$$I = \int_{t_1}^{t_2} L \, dt \tag{2–1}$$

where $L = T - V$, is an extremum for the path of motion.

That is, out of all possible paths by which the system point could travel from its position at time t_1 to its position at time t_2, it will actually travel along that path for which the integral (2–1) is an extremum, whether a minimum or maximum (cf. Fig. 2–1).

We can summarize Hamilton's principle by saying the motion is such that the *variation* of the line integral I for fixed t_1 and t_2 is zero, i.e.,

$$\delta I = \delta \int_{t_1}^{t_2} L(q_1, \ldots q_n, \dot{q}_1, \ldots \dot{q}_n, t)\, dt = 0. \tag{2–2}$$

Hamilton's principle, (2–2), is both a necessary and sufficient condition for Lagrange's Eqs. (1–53). Thus one can show that Hamilton's principle necessarily follows from Lagrange's equations (cf. Whittaker's *Analytical Dynamics*, 4th Ed., p. 245). Instead, we shall prove the converse, namely, that Lagrange's equations follow from Hamilton's principle, as being the more important theorem. It enables one to construct the mechanics of conservative systems from Hamilton's principle as the basic postulate, rather than Newton's laws of motion. Such a formulation has certain advantages; for example, it is then obviously invariant to the coordinate system used to express the Lagrangian. More important, it is the route that must be followed when we try to describe apparently nonmechanical systems in the mathematical clothes of classical mechanics, as in the theory of fields.

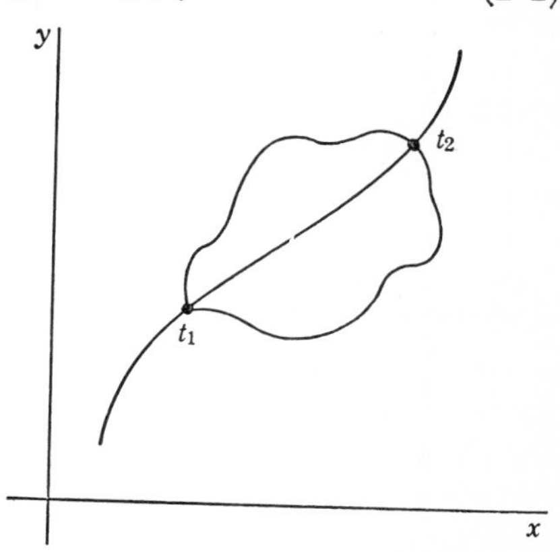

FIG. 2–1. Path of the system point in configuration space.

2–2 Some techniques of the calculus of variations. Before demonstrating that Lagrange's equations do follow from (2–2) a digression must first be made on the methods of the calculus of variations, for one of the chief problems of this calculus is to find the curve for which some given line integral is an extremum.

Consider first the problem in an essentially one dimensional form; i.e., we wish to find a path $y = y(x)$ between two values x_1 and x_2 such that the line integral of some function $f(y, \dot{y}, x)$, where $\dot{y} = \frac{dy}{dx}$, is an extremum. For the correct function y the integral

$$J = \int_{x_1}^{x_2} f(y, \dot{y}, x)\, dx \tag{2–3}$$

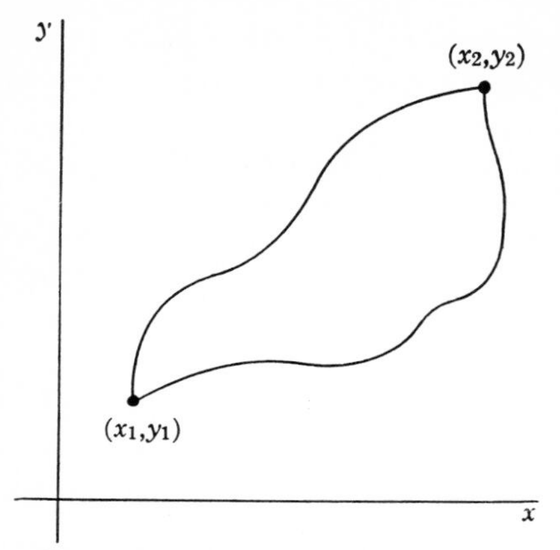

FIG. 2–2. Varied paths in the one dimensional extremum problem.

must be a maximum or minimum. The variable x here plays the role of the parameter t, and we consider only such varied paths for which $y(x_1) = y_1$, $y(x_2) = y_2$ (cf. Fig. 2–2. Note that the diagram is *not* in configuration space.).

We put the problem in a form which enables us to use the familiar apparatus of the differential calculus for obtaining an extremum value. Thus, we can label all the possible curves $y(x)$ under examination with different values of a parameter α, such that for some values of α, say $\alpha = 0$, the curve would coincide with the path or paths giving an extremum for the integral. The quantity y would then be a function of both x and the parameter α. For example, we can represent y by

$$y(x, \alpha) = y(x, 0) + \alpha\eta(x) \tag{2–4}$$

where $\eta(x)$ is any function of x which vanishes at $x = x_1$ and $x = x_2$. Eq. (2–4) is only one of the possible parametric families of curves y. Using any such parametric representation (not necessarily Eq. (2–4)), J in (2–3) is also a function of α:

$$J(\alpha) = \int_{x_1}^{x_2} f(y(x, \alpha), \dot{y}(x, \alpha), x)\, dx, \tag{2–5}$$

and the condition for obtaining an extremum is the familiar one that

$$\left(\frac{\partial J}{\partial \alpha}\right)_{\alpha=0} = 0. \tag{2–6}$$

By the usual methods of differentiating under the integral sign one finds that:

$$\frac{\partial J}{\partial \alpha} = \int_{x_1}^{x_2} \left\{ \frac{\partial f}{\partial y}\frac{\partial y}{\partial \alpha} + \frac{\partial f}{\partial \dot{y}}\frac{\partial \dot{y}}{\partial \alpha} \right\} dx. \tag{2–7}$$

Consider the second of these integrals:

$$\int_{x_1}^{x_2} \frac{\partial f}{\partial \dot{y}}\frac{\partial \dot{y}}{\partial \alpha}\, dx = \int_{x_1}^{x_2} \frac{\partial f}{\partial \dot{y}}\frac{\partial^2 y}{\partial x \partial \alpha}\, dx.$$

Integrating by parts the integral becomes

$$\int_{x_1}^{x_2} \frac{\partial f}{\partial \dot{y}} \frac{\partial^2 y}{\partial x\, \partial \alpha}\, dx = \left.\frac{\partial f}{\partial \dot{y}} \frac{\partial y}{\partial \alpha}\right|_{x_1}^{x_2} - \int_{x_1}^{x_2} \frac{d}{dx}\left(\frac{\partial f}{\partial \dot{y}}\right) \frac{\partial y}{\partial \alpha}\, dx. \tag{2-8}$$

The conditions on all the varied curves are that they pass through the points (x_1, y_1), (x_2, y_2), and hence $\partial y/\partial \alpha$ at x_1 and x_2 must vanish. Therefore the first term of (2-8) vanishes and Eq. (2-7) reduces to:

$$\frac{\partial J}{\partial \alpha} = \int_{x_1}^{x_2} \left(\frac{\partial f}{\partial y} - \frac{d}{dx} \frac{\partial f}{\partial \dot{y}}\right) \frac{\partial y}{\partial \alpha}\, dx.$$

To obtain the extremum condition we multiply through by a differential $d\alpha$, and evaluate the derivatives at $\alpha = 0$, resulting in

$$\left(\frac{\partial J}{\partial \alpha}\right)_0 d\alpha = \int_{x_1}^{x_2} \left\{\frac{\partial f}{\partial y} - \frac{d}{dx} \frac{\partial f}{\partial \dot{y}}\right\} \left(\frac{\partial y}{\partial \alpha}\right)_0 d\alpha\, dx. \tag{2-9}$$

We shall call

$$\left(\frac{\partial J}{\partial \alpha}\right)_0 d\alpha = \delta J$$

the variation of J. Similarly

$$\left(\frac{\partial y}{\partial \alpha}\right)_0 d\alpha = \delta y \tag{2-10}$$

and

$$\left(\frac{\partial \dot{y}}{\partial \alpha}\right)_0 d\alpha = \delta \dot{y},$$

although this last is not needed. Here δy represents some arbitrary variation of $y(x)$, obtained by variation of the arbitrary parameter α about its zero value. It corresponds to the previously defined virtual displacement.* Since δy is arbitrary it follows that we can only have

$$\delta J = \int_{x_1}^{x_2} \left\{\frac{\partial f}{\partial y} - \frac{d}{dx} \frac{\partial f}{\partial \dot{y}}\right\} \delta y\, dx = 0$$

if

$$\frac{\partial f}{\partial y} - \frac{d}{dx} \frac{\partial f}{\partial \dot{y}} = 0. \tag{2-11}$$

---

* These variation symbols, of course, may be used from the start, but it is well to remember always that they stand as a shorthand for the parametric procedure outlined here.

Therefore J is an extremum only for curves of $y(x)$ such that f satisfies the differential Eq. (2–11), which shows a great similarity to the Lagrange equation. Some examples of this simple type of problem may now be considered:

1. *Shortest distance between two points in a plane.* An element of arc length in a plane is

$$ds = \sqrt{dx^2 + dy^2}$$

and the total length of any curve going between points 1 and 2 is

$$I = \int_1^2 ds = \int_{x_1}^{x_2} \sqrt{1 + \left(\frac{dy}{dx}\right)^2}\, dx.$$

The condition that the curve be the shortest path is that I be a minimum. This is an example of the extremum problem as expressed by Eq. (2–3), with

$$f = \sqrt{1 + \dot{y}^2}.$$

Substituting in (2–11) with

$$\frac{\partial f}{\partial y} = 0, \qquad \frac{\partial f}{\partial \dot{y}} = \frac{\dot{y}}{\sqrt{1 + \dot{y}^2}},$$

we have

$$\frac{d}{dx}\left(\frac{\dot{y}}{\sqrt{1 + \dot{y}^2}}\right) = 0$$

or

$$\frac{\dot{y}}{\sqrt{1 + \dot{y}^2}} = c,$$

where c is some constant. This solution can be valid only if

$$\dot{y} = a$$

where a is a constant related to c by

$$a = \frac{c}{\sqrt{1 - c^2}}.$$

But this is clearly the equation of a straight line,

$$y = ax + b,$$

where b is another constant of integration. Strictly speaking, the straight line has only been proved to be an extremum path, but for this problem it is obviously also a minimum. The constants of integration, a and b, are determined by the condition that the curve pass through the two end points, (x_1, y_1), (x_2, y_2).

In a similar fashion one could obtain the shortest distance between two points on a sphere, by setting up the arc length on the surface of the sphere in terms of the angle coordinates of position on the sphere. In general, curves which give the shortest distance between two points on a given surface are called the *geodesics* of the surface.

2. *Minimum surface of revolution.* Suppose we form a surface of revolution by taking some curve passing between two fixed end points (x_1, y_1) and (x_2, y_2) and revolving it about the y axis (cf. Fig. 2-3). The problem then is to find that curve for which the surface area is a minimum. The area of a strip of the surface is $2\pi x\, ds = 2\pi x\sqrt{1+\dot{y}^2}\,dx$, and the total area is

$$2\pi\int_1^2 x\sqrt{1+\dot{y}^2}\,dx.$$

The extremum of this integral is again given by (2-11) where

$$f = x\sqrt{1+\dot{y}^2}$$

and

$$\frac{\partial f}{\partial y} = 0, \qquad \frac{\partial f}{\partial \dot{y}} = \frac{x\dot{y}}{\sqrt{1+\dot{y}^2}}.$$

Eq. (2-11) becomes in this case

$$\frac{d}{dx}\left(\frac{x\dot{y}}{\sqrt{1+\dot{y}^2}}\right) = 0$$

or

$$\frac{x\dot{y}}{\sqrt{1+\dot{y}^2}} = a,$$

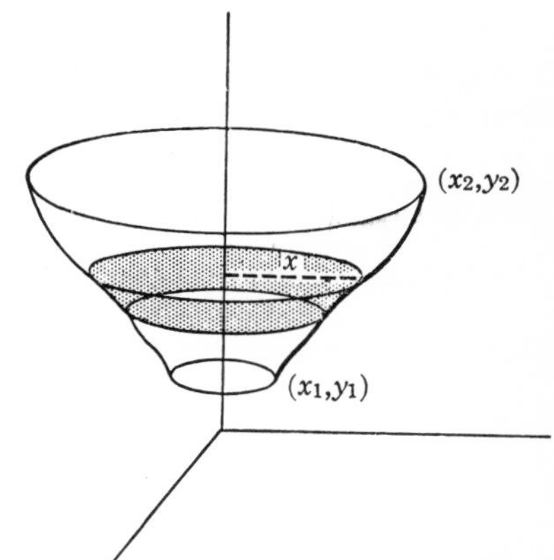

FIG. 2-3. Minimum surface of revolution.

where a is some constant of integration clearly smaller than the minimum value of x. Squaring the above equation and factoring terms we have

$$\dot{y}^2(x^2 - a^2) = a^2$$

or solving,

$$\frac{dy}{dx} = \frac{a}{\sqrt{x^2 - a^2}}.$$

The general solution of this differential equation, in light of the nature of a, is

$$y = a\int\frac{dx}{\sqrt{x^2-a^2}} + b = a \operatorname{arc\,cosh}\frac{x}{a} + b$$

or

$$x = a \cosh \frac{y - b}{a},$$

which is the equation of a catenary. Again the two constants of integration, a and b, are determined by the requirement that the curve pass through the two given end points.

3. *The brachistochrone problem.* This well-known problem is to find the curve joining two points, along which a particle falling from rest under the influence of gravity travels from the higher to the lower point in the least time.

If v is the speed along the curve, then the time required to fall an arc length ds is ds/v, and the problem is to find a minimum of the integral

$$t_{12} = \int_1^2 \frac{ds}{v}.$$

If y is measured down from the initial point of release the conservation theorem for the energy of the particle can be written as

$$\tfrac{1}{2}mv^2 = mgy$$

or

$$v = \sqrt{2gy}.$$

Then the expression for t_{12} becomes

$$t_{12} = \int_1^2 \frac{\sqrt{1 + \dot{y}^2}}{\sqrt{2gy}}\, dx,$$

and f is identified as

$$f = \sqrt{\frac{1 + \dot{y}^2}{2gy}}.$$

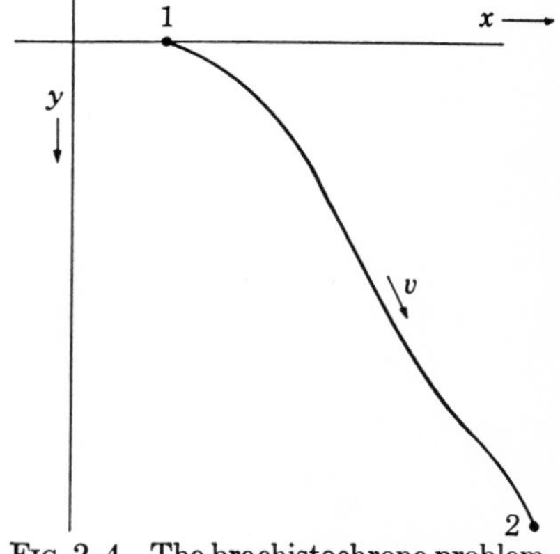

FIG. 2–4. The brachistochrone problem.

The integration of Eq. (2–11) with this form for f is straightforward and will be left as one of the exercises for this chapter. The brachistochrone problem is famous in the history of mathematics, for it was the analysis of this problem by John Bernoulli that led to the formal foundation of the calculus of variations.

2–3 Derivation of Lagrange's equations from Hamilton's principle. The fundamental problem of the calculus of variations is easily generalized to the case where f is a function of many independent variables y_i, and their derivatives $\dot{y}_i$. (Of course, all these quantities are considered as functions of the parametric variable x.) Then the variation of the integral J,

$$\delta J = \delta \int_1^2 f(y_1(x), y_2(x), \ldots \dot{y}_1(x), \dot{y}_2(x), \ldots x)\, dx, \tag{2-12}$$

is obtained, as before, by considering J as a function of a parameter α which labels all the possible curves $y_i(x,\alpha)$. Thus we may introduce α by setting

$$\begin{aligned} y_1(x, \alpha) &= y_1(x, 0) + \alpha\eta_1(x), \\ y_2(x, \alpha) &= y_2(x, 0) + \alpha\eta_2(x). \\ &\vdots \end{aligned} \tag{2-13}$$

where $y_1(x, 0)$, $y_2(x, 0)$, etc., are the solutions of the extremum problem (to be obtained) and η_1, η_2, etc., are completely arbitrary functions of x except that they vanish at the end points, since this is a variation with fixed end points. Of course, (2-13) does not represent the only way in which the curves could be labeled; it merely illustrates the procedure. The calculation proceeds as before. The variation of J is given in terms of

$$\frac{\partial J}{\partial \alpha}\, d\alpha = \int_1^2 \sum_i \left(\frac{\partial f}{\partial y_i} \frac{\partial y_i}{\partial \alpha}\, d\alpha + \frac{\partial f}{\partial \dot{y}_i} \frac{\partial \dot{y}_i}{\partial \alpha}\, d\alpha \right) dx. \tag{2-14}$$

Again we integrate by parts the integral involved in the second sum of Eq. (2-14):

$$\int_1^2 \frac{\partial f}{\partial \dot{y}_i} \frac{\partial^2 y_i}{\partial \alpha\, \partial x}\, dx = \frac{\partial f}{\partial \dot{y}_i} \frac{\partial y_i}{\partial \alpha} \bigg|_1^2 - \int_1^2 \frac{\partial y_i}{\partial \alpha} \frac{d}{dx} \left(\frac{\partial f}{\partial \dot{y}_i} \right) dx,$$

where the first term vanishes because all curves pass through the fixed end points. Substituting in (2-14), δJ becomes

$$\delta J = \int_1^2 \sum_i \left(\frac{\partial f}{\partial y_i} - \frac{d}{dx} \frac{\partial f}{\partial \dot{y}_i} \right) \delta y_i\, dx, \tag{2-15}$$

where, in analogy with (2-10), the variation δy_i is

$$\delta y_i = \left(\frac{\partial y_i}{\partial \alpha} \right)_0 d\alpha.$$

Since the y variables are independent, the variations δy_i are independent (e.g., the functions $\eta_i(x)$ will be independent of each other). Hence,

$\delta J = 0$ if and only if the coefficients of the δy_i separately vanish:

$$\frac{\partial f}{\partial y_i} - \frac{d}{dx}\frac{\partial f}{\partial \dot{y}_i} = 0, \qquad i = 1, 2, \ldots n \tag{2-16}$$

which represents the appropriate generalization of (2–11) to several variables. The set of differential equations (2–16) are known as the *Euler-Lagrange Differential Equations.* Their solutions represent those curves for which the variation of an integral of the form given in (2–12) vanishes. Further generalizations of the fundamental variational problem are easily possible. Thus one can take f as a function of higher derivatives $\ddot{y}$, $\dddot{y}$, etc., leading to equations different from (2–16). Or it can be extended to cases where there are several parameters x_j and the integral is then multiple, with f also involving as variables derivatives of y_i with respect to each of the parameters x_j. Finally, it is possible to consider variations in which the end points are *not* held fixed.

Some of these generalizations will be considered later on. For present purposes what we have here suffices, for the integral in Hamilton's principle:

$$I = \int_1^2 L(q_i, \dot{q}_i, t)\,dt \tag{2-2'}$$

has just the form stipulated in (2–12) with the transformations

$$x \rightarrow t$$
$$y_i \rightarrow q_i$$
$$f(y_i, \dot{y}_i, x) \rightarrow L(q_i, \dot{q}_i, t).$$

The Euler-Lagrange equations then become the Lagrange equations of motion:

$$\frac{d}{dt}\frac{\partial L}{\partial \dot{q}_i} - \frac{\partial L}{\partial q_i} = 0, \qquad i = 1, 2, \ldots n$$

and we have accomplished our original aim, to show that Lagrange's equations follow from Hamilton's principle — for conservative systems.

2–4 Extension of Hamilton's principle to nonconservative and nonholonomic systems. One can also generalize Hamilton's principle, formally at least, so that nonconservative forces are included and one is led to the alternative form for Lagrange's equations as given in Eq. (1–50). The extended principle appears as

$$\delta I = \delta \int_1^2 (T + W)\,dt = 0 \tag{2-17}$$

with fixed end points as before. Here W is given by

$$W = \sum_i \mathbf{F}_i \cdot \mathbf{r}_i. \tag{2–18}$$

The quantity δW has an important physical significance. It has already been remarked that the variations δq_i or $\delta \mathbf{r}_j$ are identical with virtual displacements of the coordinates since there is no variation of time involved. The varied path in configuration space can therefore be thought of as built up by a succession of virtual displacements from the actual path of motion C (cf. Fig. 2–5). Each virtual displacement occurs at some given instant of time, and at that instant the forces acting on the system have definite values. It is clear that δW represents the work done by the forces on the system during the virtual displacement from the actual to the varied path. Hamilton's principle in the form given in Eq. (2–17) can therefore be read as saying that the integral of the variation of the kinetic energy plus the virtual work involved in the variation must be zero. The variations $\delta \mathbf{r}_i$ could be evaluated in terms of δq_j by using the equations of transformation between $\mathbf{r}$ and q, each q depending on the path chosen through a parameter α:

$$\mathbf{r}_i = \mathbf{r}(q_j(\alpha, t), t).$$

We may abbreviate the process, however, by using the equivalence of $\delta \mathbf{r}_i$ with a virtual displacement; it has in fact been already shown that

$$\sum_i \mathbf{F}_i \cdot \delta \mathbf{r}_i = \sum_j Q_j \, \delta q_j.$$

Hence Eq. (2–17) can also be written as

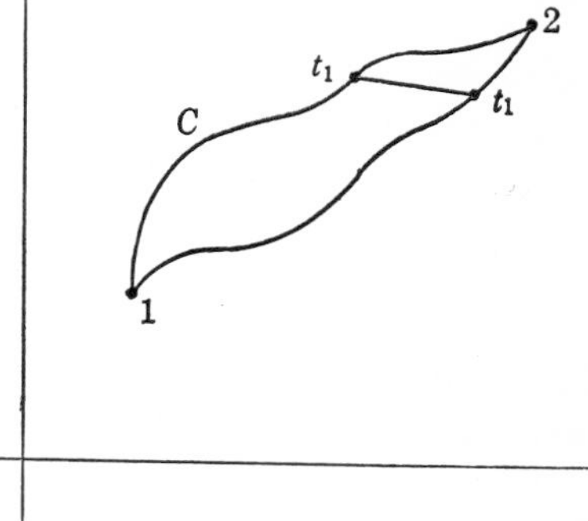

FIG. 2–5. Construction of the varied path in configuration space by virtual displacements.

$$\delta \int_1^2 T \, dt + \int_1^2 \sum_j Q_j \, \delta q_j \, dt = 0. \tag{2–19}$$

At this point it is possible to show that (2–19) reduces to the ordinary form of Hamilton's principle if the Q_j's are derivable from a generalized potential. Under these conditions the integral of the virtual work becomes:

$$\int_1^2 \sum_j Q_j \, \delta q_j \, dt = -\int_1^2 \sum_j \delta q_j \left(\frac{\partial V}{\partial q_j} - \frac{d}{dt} \frac{\partial V}{\partial \dot{q}_j} \right) dt.$$

Reversing the now familiar integration-by-parts procedure, the integral can also be written as

$$-\int_1^2 \sum_j \left(\frac{\partial V}{\partial q_j}\,\delta q_j + \frac{\partial V}{\partial \dot{q}_j}\,\delta \dot{q}_j\right) dt = -\delta \int_1^2 V\,dt.$$

In such cases (2–19) reduces to

$$\delta \int_1^2 T\,dt - \delta \int_1^2 V\,dt = \delta \int_1^2 L\,dt = 0,$$

which is Hamilton's principle as given in Eq. (2–2).

To return to the more general problem, the variation of the first integral in (2–19) can be written immediately, since T, like L for conservative systems, is a function of both q_j and $\dot{q}_j$:

$$\delta \int_1^2 T\,dt = \int_1^2 \sum_j \left(\frac{\partial T}{\partial q_j} - \frac{d}{dt}\frac{\partial T}{\partial \dot{q}_j}\right) \delta q_j\,dt.$$

Combining the two integrals, Hamilton's principle becomes

$$\int_1^2 \sum_j \left(\frac{\partial T}{\partial q_j} - \frac{d}{dt}\frac{\partial T}{\partial \dot{q}_j} + Q_j\right) \delta q_j\,dt = 0. \tag{2–20}$$

Again, since the constraints are assumed holonomic, the integral can vanish if and only if the separate coefficients vanish, or:

$$\frac{d}{dt}\frac{\partial T}{\partial \dot{q}_j} - \frac{\partial T}{\partial q_j} = Q_j, \tag{2–21}$$

so that Eq. (2–17) represents the proper extension of Hamilton's principle, Eq. (2–2), which will yield the form of Lagrange's equations when the forces are not derivable from a potential.

It is also possible to extend Hamilton's principle to cover certain types of nonholonomic systems. In deriving Lagrange's equations from either Hamilton's or D'Alembert's principle the requirement of holonomic constraints does not appear until the last step, when the variations q_j are considered as independent of each other. With nonholonomic systems the generalized coordinates are not independent of each other, and it is not possible to reduce them further by means of equations of constraint of the form $f(q_1, q_2, \ldots q_n, t) = 0$. Hence it is no longer true that the q_j's are all independent.

We can still treat nonholonomic systems providing the equations of constraint can be put in the form

$$\sum_k a_{lk}\,dq_k + a_{lt}\,dt = 0, \tag{2–22}$$

i.e., a relation connecting the *differentials* of the q's. Now, the variation process involved in Hamilton's principle is one in which the time, for each point on the path, is held constant. Hence the virtual displacements occurring in the variation must satisfy equations of constraint of the form

$$\sum_k a_{lk}\,\delta q_k = 0. \tag{2-23}$$

The index l indicates that there may be more than one such equation; it will be assumed there are m equations in all, i.e., $l = 1, 2, \ldots m$.

We can now use Eqs. (2-23) to reduce the number of virtual displacements to independent ones. The procedure for eliminating these extra virtual displacements is the method of *Lagrange undetermined multipliers.* If Eqs. (2-23) hold then it is also true that

$$\lambda_l \sum_k a_{lk}\,\delta q_k = 0, \tag{2-24}$$

where the λ_l, $l = 1, 2, \ldots m$, are some undetermined constants, functions of time in general. Let us now combine these m equations with the equation corresponding to (2-20) in the case of conservative systems:

$$\int_1^2 dt \sum_k \left(\frac{\partial L}{\partial q_k} - \frac{d}{dt}\frac{\partial L}{\partial \dot{q}_k}\right)\delta q_k = 0. \tag{2-20$'$}$$

To do this we first sum the Eqs. (2-24) over l, and then integrate the resulting equation from point 1 to point 2:

$$\int_1^2 \sum_{k,l} \lambda_l a_{lk}\,\delta q_k\,dt = 0. \tag{2-25}$$

This may be combined with Eq. (2-20′), yielding the relation

$$\int_1^2 dt \sum_{k=1}^{n}\left(\frac{\partial L}{\partial q_k} - \frac{d}{dt}\frac{\partial L}{\partial \dot{q}_k} + \sum_l \lambda_l a_{lk}\right)\delta q_k = 0. \tag{2-26}$$

The δq_k's are still not independent, of course; they are connected by the m relations (2-23). That is, while the first $n - m$ of these may be chosen independently, the last m are then fixed by the Eqs. (2-23). However, the values of the λ_l's remain at our disposal. Suppose we now choose the λ_l's to be such that

$$\frac{\partial L}{\partial q_k} - \frac{d}{dt}\frac{\partial L}{\partial \dot{q}_k} + \sum_l \lambda_l a_{lk} = 0, \qquad k = n - m + 1, \ldots n \tag{2-27}$$

which are in the nature of equations of motion for the last m of the q_k variables. With the λ_l determined by (2–27), we can write (2–26) as

$$\int_1^2 dt \sum_{k=1}^{n-m} \left(\frac{\partial L}{\partial q_k} - \frac{d}{dt} \frac{\partial L}{\partial \dot{q}_k} + \sum_l \lambda_l a_{lk} \right) \delta q_k = 0. \tag{2–28}$$

Here the only δq_k's involved are the independent ones. Hence it follows that

$$\frac{\partial L}{\partial q_k} - \frac{d}{dt} \frac{\partial L}{\partial \dot{q}_k} + \sum_l \lambda_l a_{lk} = 0, \qquad k = 1, 2, \ldots n - m. \tag{2–29}$$

Combining (2–27) and (2–29) we have finally the complete set of Lagrange's equations for nonholonomic systems:

$$\frac{d}{dt} \frac{\partial L}{\partial \dot{q}_k} - \frac{\partial L}{\partial q_k} = \sum_l \lambda_l a_{lk}, \qquad k = 1, 2, \ldots n. \tag{2–30}$$

But this is not the whole story, for now we have $n + m$ unknowns, namely the n coordinates q_k and the m λ_l's, while (2–30) gives us a total of only n equations. The additional equations needed, of course, are exactly the equations of constraint linking up the q_k's, Eqs. (2–22), except that they are now to be considered as differential equations:

$$\sum_k a_{lk} \dot{q}_k + a_{lt} = 0. \tag{2–31}$$

Eqs. (2–30) and (2–31) together constitute $n + m$ equations for $n + m$ unknowns.

In this process we have obtained more information than was originally sought. Not only do we get the q_k's we set out to find, but we also get m λ_l's. What is the physical significance of the λ_l's? Suppose one removed the constraints on the system, but instead applied external forces Q'_k in such a manner as to keep the motion of the system unchanged. The equations of motion would likewise remain the same. Clearly these extra applied forces must be equal to the forces of constraint, for they are the forces applied to the system so as to satisfy the condition of constraint. Under the influence of these forces Q'_k, the equations of motion are

$$\frac{d}{dt} \frac{\partial L}{\partial \dot{q}_k} - \frac{\partial L}{\partial q_k} = Q'_k.$$

But these must be identical with Eqs. (2–30). Hence we can identify $\Sigma \lambda_l a_{lk}$ with Q_k, the generalized forces of constraint. In this type of problem we really do not eliminate the forces of constraint from the formulation, and they are supplied as part of the answer.

Notice that Eq. (2-22) is not the most general type of nonholonomic constraint, e.g., it does not include equations of constraint in the form of inequalities. On the other hand, it does include holonomic constraints. A holonomic equation of constraint,

$$f(q_1, q_2, q_3, \ldots q_n, t) = 0,$$

is equivalent to a differential equation,

$$\sum_k \frac{\partial f}{\partial q_k} dq_k + \frac{\partial f}{\partial t} dt = 0, \tag{2-32}$$

which is identical in form with (2-22), with the coefficients

$$a_{lk} = \frac{\partial f}{\partial q_k}, \qquad a_{lt} = \frac{\partial f}{\partial t}. \tag{2-33}$$

Thus the Lagrange multiplier method can be used also for holonomic constraints when (1) it is inconvenient to reduce all the q's to independent coordinates or (2) we might wish to obtain the forces of constraint.

As an example of the method, consider the following somewhat trivial illustration — a hoop rolling, without slipping, down an inclined plane. In this example the constraint of "rolling" is actually holonomic, but this fact will be immaterial to the discussion.

The two generalized coordinates are x, θ, as in Fig. 2-6, and the equation of constraint is:

$$r d\theta = dx.$$

The kinetic energy can be resolved into kinetic energy of motion of the center of mass plus the kinetic energy of motion about the center of mass:

$$T = \tfrac{1}{2}M\dot{x}^2 + \tfrac{1}{2}Mr^2\dot{\theta}^2.$$

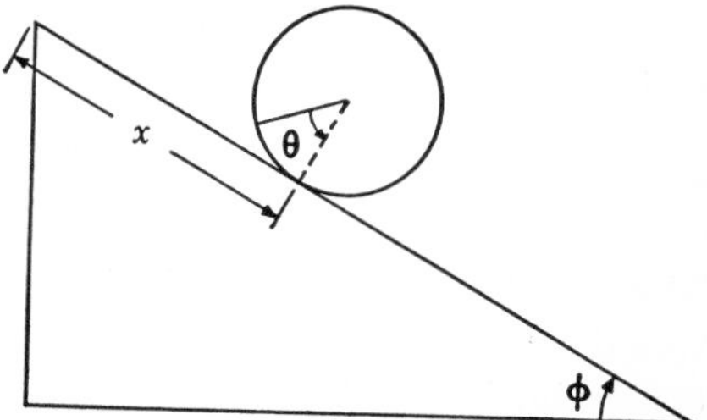

FIG. 2-6. A hoop rolling down an inclined plane.

The potential energy is

$$V = Mg\,(l - x)\sin\phi,$$

where l is the length of the inclined plane and the Lagrangian is

$$\begin{aligned} L &= T - V \\ &= \frac{M\dot{x}^2}{2} + \frac{Mr^2\dot{\theta}^2}{2} - Mg\,(l - x)\sin\phi. \end{aligned}$$

Since there is one equation of constraint, only one Lagrangian multiplier λ is needed. The coefficients appearing in the constraint equation are:

$$a_\theta = r,$$
$$a_x = -1.$$

The two Lagrange equations therefore are:

$$M\ddot{x} - Mg \sin \phi + \lambda = 0, \tag{2-34}$$
$$Mr^2\ddot{\theta} - \lambda r = 0, \tag{2-35}$$

which along with the equation of constraint,

$$r\dot{\theta} = \dot{x}, \tag{2-36}$$

constitutes three equations for three unknowns, θ, x, λ.

Differentiating (2–36) with respect to time, we have

$$r\ddot{\theta} = \ddot{x}.$$

Hence from (2–35)

$$M\ddot{x} = \lambda$$

and (2–34) becomes

$$\ddot{x} = \frac{g \sin \phi}{2}$$

along with

$$\lambda = \frac{Mg \sin \phi}{2}$$

and

$$\ddot{\theta} = \frac{g \sin \phi}{2r}.$$

Thus the hoop rolls down the incline with only one half the acceleration it would have slipping down a frictionless plane, and the friction force of constraint is $\lambda = Mg \sin \phi/2$.

From $\ddot{x} = v\dfrac{dv}{ds}$, one gets $v = \sqrt{gl \sin \phi}$ at the bottom, which can, of course, be obtained by elementary means too.

2–5 Advantages of a variational principle formulation. Although it is thus possible to extend the original formulation of Hamilton's principle (2–2) to include nonconservative systems and nonholonomic constraints, practically, this formulation of mechanics is most useful when a Lagrangian of independent coordinates can be set up for the system. The variational principle formulation has been justly described as "elegant," for in the compact Hamilton's principle is contained all of the mechanics of conservative holonomic systems. The principle has the further merit that it involves only physical quantities which can be defined without reference to a particular set of generalized coordinates, namely, the kinetic and potential energies. The formulation is therefore automatically invariant with respect to the choice of coordinates for the system.

Another advantage is that the Lagrangian formulation can be extended easily to describe systems that are not normally considered in dynamics — such as the elastic field, the electromagnetic field, field properties of elementary particles, etc. Some of these generalizations will be considered later, but as a simple example of its application outside the usual framework of mechanics, let us consider the following case.

Suppose we have a system for which there is a Lagrangian:

$$L = \frac{1}{2}\sum_j L_j\dot{I}_j^2 + \frac{1}{2}\sum_{\substack{jk \\ j\neq k}} M_{jk}\dot{I}_j\dot{I}_k - \sum_j \frac{I_j^2}{2C_j} + \sum_j \dot{E}_j(t)I_j \tag{2-37}$$

and a dissipation function

$$\mathfrak{F} = \frac{1}{2}\sum_j R_j\dot{I}_j^2, \tag{2-38}$$

where the generalized coordinates describing the system have been labeled I_j instead of q_j. The Lagrange equations are:

$$L_j\frac{d^2I_j}{dt^2} + \sum_{\substack{k \\ j\neq k}} M_{jk}\frac{d^2I_k}{dt^2} + R_j\frac{dI_j}{dt} + \frac{I_j}{C_j} = \dot{E}_j(t). \tag{2-39}$$

These equations of motion can be interpreted in at least two ways: One can say the I's are currents, the L_j's self-inductances, the M_{jk}'s mutual inductances, the R_j's resistances, the C_j's capacities, and the E_j's external emf's. Then Eqs. (2-39) are a set of equations describing a system of mutually inductively coupled networks, e.g., for $j = 1, 2, 3$ we would have three networks somewhat as in Fig. 2-7. On the other hand, it is seen that the first two terms in

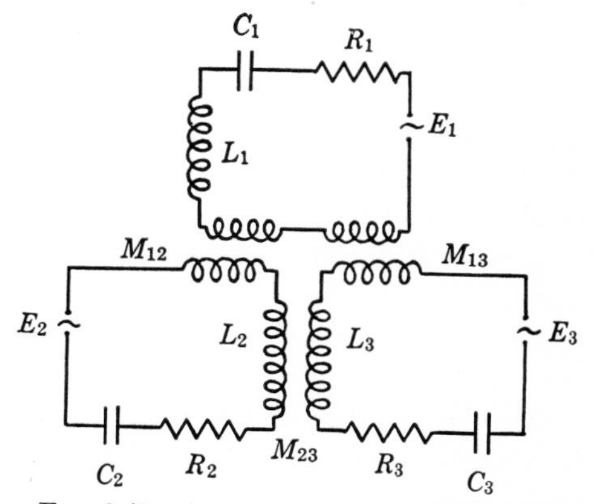

FIG. 2-7. A system of coupled circuits to which the Lagrangian formulation can be applied.

L together constitute an arbitrary homogeneous quadratic function of the generalized velocities. Whenever the system constraints (holonomic) are independent of time, the kinetic energy, T, is always of such a form. The coefficients L_j, M_{jk} then partake of the character of masses — they are *inertial terms.* The next term in the Lagrangian exactly corresponds to the potential energy of a set of springs — harmonic oscillators — where the forces obey Hooke's law,

$$F = -kx,$$

with the resulting potential

$$V = \frac{kx^2}{2},$$

so that the $1/C_j$'s represent spring constants. The last term corresponds to the potential due to driving forces $\dot{E}_j = Q_j$ which are independent of coordinates — as for example, gravitational forces — except that the E_j's may be time-varying forces. Finally, the dissipation function corresponds to the existence of dissipative or viscous forces, proportional to the generalized velocities. We are thus led to an alternative interpretation of Eqs. (2–37) and (2–38), or (2–39), which brings to mind a picture of a complicated system of masses on springs moving in some viscous fluid and driven by external forces.

This description of two different physical systems by Lagrangians of the same form means that all the results and techniques devised for investigating one of the systems can be taken over immediately and applied to the other. In this particular case the study of the behavior of electrical circuits has been pursued intensely and some special techniques have been developed; these can be directly applied to the corresponding mechanical systems. Much work has been done in formulating equivalent electrical problems for mechanical or acoustical systems, and vice versa. Terms normally reserved for electrical circuits (reactance, susceptance, etc.) are the accepted modes of expression in much of the theory of vibrations of mechanical systems.*

But, in addition, there is a type of generalization of mechanics which is due to a subtler form of equivalence. We have seen that the Lagrangian and Hamilton's principle together form a compact invariant way of implying the mechanical equations of motion. This possibility is not reserved for mechanics only; in almost every field of physics variational principles can be used to express the "equations of motion," whether they be Newton's

---

* For a detailed exposition, see H. F. Olson, *Dynamical Analogies*, D. Van Nostrand Co., Inc., New York, 1946.

equations, Maxwell's equations, or the Schrödinger equation. Consequently, when a variational principle is used as the basis of the formulation, all such fields will exhibit, at least to some degree, a *structural analogy.* When the results of experiments show the need for alteration of the physical content in the theory of one field, this degree of analogy has often indicated how similar alterations may be carried out in other fields. Thus, the experiments performed early in this century showed the need for quantization of both electromagnetic radiation and elementary particles. The methods of quantization, however, were first developed for particle mechanics, starting essentially from the Lagrangian formulation of classical mechanics. By describing the electromagnetic field by a Lagrangian and corresponding Hamilton's variational principle, it is possible to carry over the methods of particle quantization to construct a quantum electrodynamics (cf. Section 11–5).

2–6 Conservation theorems and symmetry properties. Thus far we have been concerned primarily with obtaining the equations of motion, and little has been said about how to solve them for a particular problem once they have been obtained. In general, this is a question of mathematics. A system of n degrees of freedom will have n differential equations which are second order in time. The solution of each equation will require two integrations resulting, all told, in $2n$ constants of integration. In a specific problem these constants will be determined by the initial conditions, i.e., the initial values of the n q_j's and the n $\dot{q}_j$'s. Sometimes the equations of motion will be integrable in terms of known functions, but not always. In fact, the majority of problems are not completely integrable. However, even when complete solutions cannot be obtained, it is often possible to extract a large amount of information about the physical nature of the system motion. Indeed, such information may be of greater interest to the physicist than the complete solution for the generalized coordinates as a function of time. It is important, therefore, to see how much can be stated about the motion of a given system without requiring a complete integration of the problem.*

In many problems a number of first integrals of the equations of motion can be obtained immediately; by this we mean relations of the type

$$f(q_1, q_2, \ldots \dot{q}_1, \dot{q}_2, \ldots t) = \text{constant}, \tag{2–40}$$

which are first-order differential equations. These first integrals are of interest because they tell us something physically about the system. They include, in fact, the conservation laws obtained in Chapter 1.

---

* In this and succeeding sections it will be assumed, unless otherwise specified, the system is such that its motion is completely described by a Hamilton's principle of the form (2–2).

Consider as an example a system of mass points under the influence of forces derived from potentials dependent on position only. Then

$$\frac{\partial L}{\partial \dot{x}_i} \equiv \frac{\partial T}{\partial \dot{x}_i} - \frac{\partial V}{\partial \dot{x}_i} = \frac{\partial T}{\partial \dot{x}_i} = \frac{\partial}{\partial \dot{x}_i} \sum \frac{1}{2} m_i(\dot{x}_i^2 + \dot{y}_i^2 + \dot{z}_i^2)$$
$$= m_i \dot{x}_i = p_{ix}$$

which is the x component of the linear momentum associated with the ith particle. This result suggests an obvious extension to the concept of momentum. The generalized momentum associated with the coordinate q_j shall be defined as

$$p_j = \frac{\partial L}{\partial \dot{q}_j}. \tag{2–41}$$

The terms *canonical momentum* or *conjugate momentum* are often also used for p_j. Notice that if q_j is not a cartesian coordinate p_j does not necessarily have the dimensions of a linear momentum. Further, if there is a velocity-dependent potential, then even with a cartesian coordinate q_j the associated generalized momentum will not be identical with the usual mechanical momentum. Thus in the case of a group of particles in the electromagnetic field the Lagrangian is (cf. 1–61):

$$L = \sum_i \frac{1}{2} m_i \dot{r}_i^2 - \sum_i q_i \phi(x_i) + \sum_i \frac{q_i}{c} \mathbf{A}(x_i) \cdot \dot{\mathbf{r}}_i$$

(q_i here denotes charge) and the generalized momentum conjugate to x_i is

$$p_{ix} = \frac{\partial L}{\partial \dot{x}_i} = m_i \dot{x}_i + \frac{q_i A_x}{c}, \tag{2–42}$$

i.e., mechanical momentum plus an additional term.

If the Lagrangian of a system does not contain a given coordinate q_j (although it may contain the corresponding velocity $\dot{q}_j$) then the coordinate is said to be *cyclic* or *ignorable*. This definition is not universal,* but it is the customary one and will be used here. The Lagrange equation of motion

$$\frac{d}{dt} \frac{\partial L}{\partial \dot{q}_j} - \frac{\partial L}{\partial q_j} = 0$$

* The two terms are usually taken as interchangeable and as having the meaning assigned above. However, a few authors distinguish between them, defining a cyclic coordinate as one not in the kinetic energy, T, and an ignorable coordinate as one not in the Lagrangian (cf. Webster, *The Dynamics of Particles*, and Byerly, *Generalized Coordinates*). Ames and Murnaghan (*Theoretical Mechanics*) use the two terms interchangeably but apparently confine them to meaning a coordinate not in T.

reduces, for a cyclic coordinate, to

$$\frac{d}{dt}\frac{\partial L}{\partial \dot{q}_j} = 0$$

or

$$\frac{dp_j}{dt} = 0,$$

which means that

$$p_j = \text{constant.} \tag{2–43}$$

Hence, we can state as a general conservation theorem that *the generalized momentum conjugate to a cyclic coordinate is conserved.*

Eq. (2–43) constitutes a first integral of the form (2–40) for the equations of motion. It can be used formally to eliminate the cyclic coordinate from the problem, which can then be solved entirely in terms of the remaining generalized coordinates. Briefly, the procedure, originated by Routh, consists in modifying the Lagrangian so that it is no longer a function of the generalized velocity corresponding to the cyclic coordinate, but instead involves only its conjugate momentum. The advantage in so doing is that p_j can then be considered one of the constants of integration, and the remaining integrations involve only the noncyclic coordinates. We shall defer a detailed discussion of Routh's method until the Hamiltonian formulation (to which it is quite similar) is treated.

Note that the conditions for the conservation of generalized momenta are more general than the two momentum conservation theorems previously derived. For example, they furnish a conservation theorem for a case in which the law of action and reaction is violated — namely, when electromagnetic forces are present. Suppose we have a single particle in a field in which neither ϕ nor $\mathbf{A}$ depends on x. Then x nowhere appears in L and is therefore cyclic. The corresponding canonical momentum p_x must therefore be conserved. From (1–61) this momentum now has the form

$$p_x = m\dot{x} + \frac{qA_x}{c} = \text{constant.} \tag{2–44}$$

In this case it is not the mechanical linear momentum $m\dot{x}$ which is conserved but rather its sum with $\frac{qA_x}{c}$.* Nevertheless, it should still be true that

* It can be shown from classical electrodynamics that under these conditions, i.e., neither $\mathbf{A}$ nor ϕ depending on x, that qA_x/c is exactly the x component of the electromagnetic linear momentum of the field associated with the charge q.

the conservation theorems of Chapter 1 are contained within the general rule for cyclic coordinates; with proper restrictions (2–43) should reduce to the theorems of Section 1–2.

Consider first a generalized coordinate q_j, for which a change dq_j represents a translation of the system as a whole in some given direction. An example would be one of the cartesian coordinates of the center of mass of the system. Then clearly q_j cannot appear in T, for velocities are not affected by a shift in the origin and $\frac{\partial T}{\partial q_j}$ is zero. Further we will assume conservative systems for which V is not a function of the velocities, so as to eliminate such anomalies as electromagnetic forces. The Lagrange equation of motion for a coordinate so defined then reduces to

$$\frac{d}{dt}\frac{\partial T}{\partial \dot{q}_j} \equiv \dot{p}_j = -\frac{\partial V}{\partial q_j} \equiv Q_j. \tag{2–45}$$

We will now show that (2–45) is the equation of motion for the total linear momentum, i.e., that Q_j represents the component of the total force along the direction of translation of q_j, and p_j is the component of the total linear momentum along this direction. In general, the generalized force Q_j is given by:

$$Q_j = \sum_i \mathbf{F}_i \cdot \frac{\partial \mathbf{r}_i}{\partial q_j}.$$

Since dq_j corresponds to a translation of the system along some axis, the vectors $\mathbf{r}_i(q_j)$ and $\mathbf{r}_i(q_j + dq_j)$ are related as shown in Fig. 2–8. By definition of derivative we have:

$$\frac{\partial \mathbf{r}_i}{\partial q_j} = \underset{dq_j \to 0}{\mathrm{L}} \frac{\mathbf{r}_i(q_j + dq_j) - \mathbf{r}_i(q_j)}{dq_j} = \frac{dq_j\mathbf{n}}{dq_j} = \mathbf{n}, \tag{2–46}$$

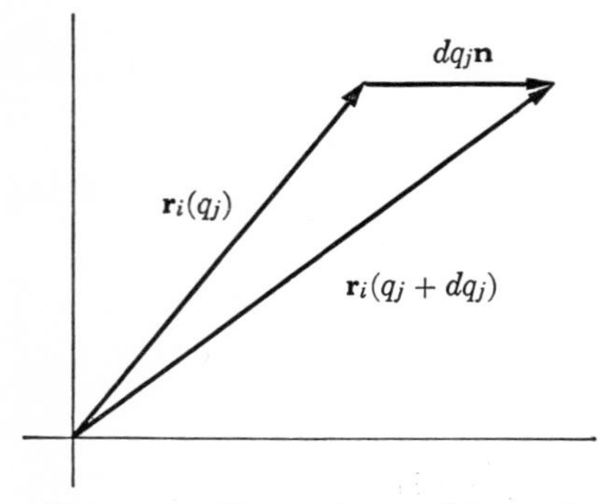

FIG. 2–8. Change in a position vector under translation of the system.

where $\mathbf{n}$ is the unit vector along the direction of translation. Hence

$$Q_j = \sum \mathbf{F}_i \cdot \mathbf{n} = \mathbf{n} \cdot \mathbf{F},$$

which, as was stated, is the component of the total force in the direction of $\mathbf{n}$. To prove the other half of the statement note that with the kinetic energy in the form

$$T = \frac{1}{2}\sum m_i \dot{\mathbf{r}}_i^2$$

the conjugate momentum is

$$p_j = \frac{\partial T}{\partial \dot{q}_j} = \sum_i m_i \dot{\mathbf{r}}_i \cdot \frac{\partial \dot{\mathbf{r}}_i}{\partial \dot{q}_j},$$
$$= \sum_i m_i \mathbf{v}_i \cdot \frac{\partial \mathbf{r}_i}{\partial q_j},$$

using Eq. (1–48). Then from Eq. (2–46)

$$p_j = \mathbf{n} \cdot \sum_i m_i \mathbf{v}_i,$$

which again, as predicted, is the component of the total system linear momentum along **n**.

Suppose now that the translation coordinate q_j which we have been discussing is cyclic. Then q_j cannot appear in V and therefore

$$-\frac{\partial V}{\partial q_j} \equiv Q_j = 0.$$

But this is just the familiar conservation theorem for linear momentum — that if a given component of the total applied force vanishes, the corresponding component of the linear momentum is conserved.

In a similar fashion it can be shown that if a cyclic coordinate q_j is such that dq_j corresponds to a rotation of the system around some axis, then the conservation of its conjugate momentum corresponds to conservation of an angular momentum. By the same argument as used above, T cannot contain q_j, for a rotation of the coordinate system cannot affect the magnitude of the velocities. Hence $\frac{\partial T}{\partial q_j}$ must be zero, and since V is independent of $\dot{q}_j$ we again get Eq. (2–45). But now we wish to show that with q_j a rotation coordinate the generalized force is the component of the total applied torque about the axis of rotation, and p_j is the component of the total angular momentum along the same axis.

The generalized force Q_j is again given by

$$Q_j = \sum_i \mathbf{F}_i \cdot \frac{\partial \mathbf{r}_i}{\partial q_j},$$

only the derivative now has a different meaning. Here the change in q_j must correspond to an infinitesimal rotation of the vector $\mathbf{r}_i$, keeping the magnitude of the vector constant. From Fig. 2–9 the magnitude of the derivative can easily be obtained:

$$|d\mathbf{r}_i| = r_i \sin\theta \, dq_j$$

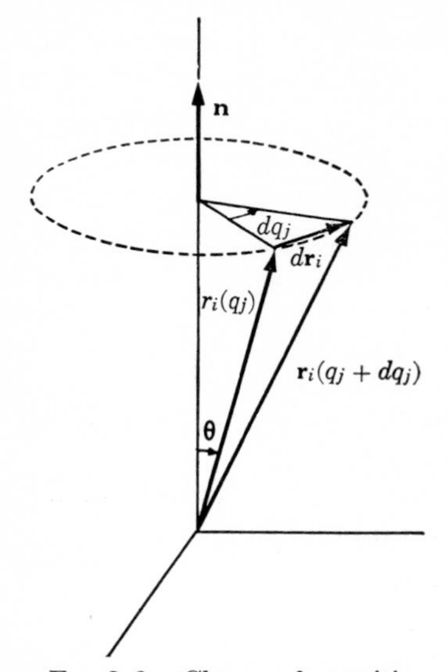

FIG. 2–9. Change of a position vector under rotation of the system.

and

$$\left|\frac{\partial \mathbf{r}_i}{\partial q_j}\right| = r_i \sin\theta;$$

and its direction is perpendicular to both $\mathbf{r}_i$ and $\mathbf{n}$. Clearly the derivative can be written in vector form as

$$\frac{\partial \mathbf{r}_i}{\partial q_j} = \mathbf{n} \times \mathbf{r}_i. \tag{2–47}$$

With this result the generalized force becomes

$$Q_j = \sum_i \mathbf{F}_i \cdot \mathbf{n} \times \mathbf{r}_i$$
$$= \sum_i \mathbf{n} \cdot \mathbf{r}_i \times \mathbf{F}_i$$

reducing to

$$Q_j = \mathbf{n} \cdot \sum_i \mathbf{N}_i = \mathbf{n} \cdot \mathbf{N}$$

which proves the first part. A similar manipulation of p_j provides proof of the second part of the statement:

$$p_j = \frac{\partial T}{\partial \dot{q}_j} = \sum_i m_i \mathbf{v}_i \cdot \frac{\partial \mathbf{r}_i}{\partial q_j} = \sum_i \mathbf{n} \cdot \mathbf{r}_i \times m_i \mathbf{v}_i$$
$$= \mathbf{n} \cdot \sum \mathbf{L}_i = \mathbf{n} \cdot \mathbf{L}.$$

Summarizing these results we see that if the rotation coordinate q_j is cyclic then Q_j, which is the component of the applied torque along $\mathbf{n}$, vanishes, and the component of $\mathbf{L}$ along $\mathbf{n}$ is constant. Here we have recovered the angular momentum conservation theorem out of the general conservation theorem relating to cyclic coordinates.

The significance of cyclic translation or rotation coordinates in relation to the properties of the system deserves some notice at this point. If a coordinate corresponding to a displacement is cyclic it means that a translation of the system, as if rigid, has no effect on the problem. In other words, if the system is *invariant* under translation along a given direction the corresponding linear momentum is conserved. Similarly, the fact that a rotation coordinate is cyclic (and therefore the conjugate angular momentum conserved) indicates that the system is invariant under rotation about the given axis. Thus the momentum conservation theorems are closely connected with the *symmetry properties* of the system. If the system is spherically symmetric we can say without further ado that all components of angular momentum are conserved. Or, if the system is symmetric only about the z-axis then only L_z will be conserved, and so on for the other axes. We will meet up with these connections between the constants of motion and the symmetry properties several times.

Another conservation theorem we should expect to obtain in the Lagrangian formulation is the conservation of total energy for conservative systems. Consider a system conservative in the sense that $\mathbf{F} = -\nabla V$, V being independent of the velocities, and with the additional restriction that the constraints are independent of time, so that L cannot be an explicit function of time. Then the total time derivative of L is

$$\frac{dL}{dt} = \sum_j \frac{\partial L}{\partial q_j}\frac{dq_j}{dt} + \sum_j \frac{\partial L}{\partial \dot{q}_j}\frac{d\dot{q}_j}{dt}. \tag{2–48}$$

From Lagrange's equations

$$\frac{\partial L}{\partial q_j} = \frac{d}{dt}\frac{\partial L}{\partial \dot{q}_j}$$

and (2–48) can be written as

$$\frac{dL}{dt} = \sum_j \frac{d}{dt}\frac{\partial L}{\partial \dot{q}_j}\dot{q}_j + \sum_j \frac{\partial L}{\partial \dot{q}_j}\frac{d\dot{q}_j}{dt}$$

or

$$\frac{dL}{dt} = \sum_j \frac{d}{dt}\left(\dot{q}_j\frac{\partial L}{\partial \dot{q}_j}\right).$$

It therefore follows that

$$\frac{d}{dt}\left(L - \sum_j \dot{q}_j\frac{\partial L}{\partial \dot{q}_j}\right) = 0,$$

indicating that the expression in brackets must be some constant, which we shall call $-H$:

$$L - \sum_j \dot{q}_j\frac{\partial L}{\partial \dot{q}_j} = -H. \tag{2–49}$$

Eq. (2–49) can be written also as

$$H = \sum_j \dot{q}_j p_j - L. \tag{2–50}$$

The right-hand side of (2–50) is thus an integral of the motion; we shall now show that it is just the total energy of the system. For conservative systems, V independent of velocities, we have that

$$p_j = \frac{\partial L}{\partial \dot{q}_j} = \frac{\partial T}{\partial \dot{q}_j}$$

and the first term on the right-hand side of (2–50) is equivalent to

$$\sum_j \dot{q}_j \frac{\partial T}{\partial \dot{q}_j}.$$

In Section 1–6 it was shown that when the constraints are independent of time or, more specifically, when the transformation Eqs. (1–36) do not contain t explicitly, T is a homogeneous quadratic function of the $\dot{q}_j$'s. It will be remembered that Euler's theorem states that if f is a homogeneous function, of order n, of a set of variables q_i, then

$$\sum_i q_i \frac{\partial f}{\partial q_i} = nf.$$

Here n is 2, so that

$$\sum_j \dot{q}_j \frac{\partial T}{\partial \dot{q}_j} = 2T \tag{2–51}$$

and we find

$$H = 2T - (T - V) = T + V, \tag{2–52}$$

which is the total energy of the system.

This form of the energy conservation theorem seems to be more stringent than that previously stated in Chapter 1, and besides requiring the forces to be conservative it seems to add the condition that the constraints be independent of time. Actually the two theorems are not talking about quite the same energy. In the previous statement, the energy change of the system included the work done by all forces, including the forces of constraint. Here, in the Lagrangian formulation, V contains only the work of the external or applied forces, excluding the forces of constraint. With time-independent constraints there is no essential difference. We have already stipulated that the forces of constraint do no work in a virtual displacement, and when the constraints are not changing with time the distinction between an actual and a virtual displacement disappears. The potential due to the forces of constraint would then be constant automatically. However, if there is a moving constraint the force of constraint need not be perpendicular to the actual displacement and the work done by such forces will not be zero. Thus if a particle is constrained to travel on a curve which is itself moving the force of constraint at a given instant will be normal to the instantaneous curve, but the displacement of the particle in time dt is not tangent to the curve, cf. Fig. (2–10). The potential of the forces of constraint will thus vary in time and it is then important whether or not the "total energy" in question includes the contribution of the forces of constraint.*

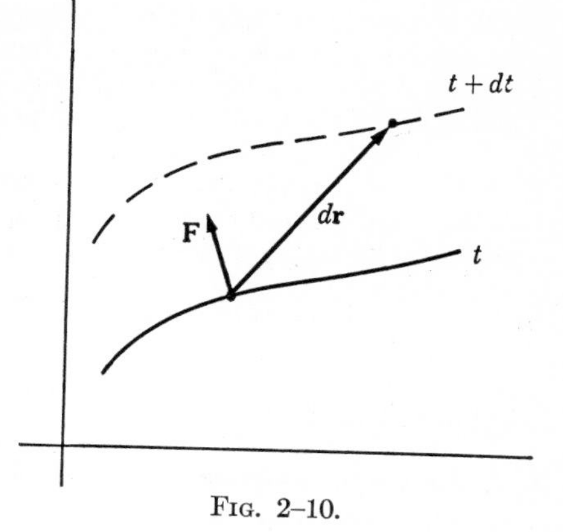

FIG. 2–10.

SUGGESTED REFERENCES

H. MARGENAU AND G. M. MURPHY, *The Mathematics of Physics and Chemistry.* The literature on the calculus of variations is naturally very extensive, but most of the references become more involved mathematically than is necessary for the simple discussion of Hamilton's principle. The short treatment of the subject in Chapter 6 of this reference is more than adequate for present purposes. A somewhat lengthier but still highly readable exposition will be found in

G. A. BLISS, *Calculus of Variations.* The extremum problems that led to the invention of the calculus of variations are discussed here in some detail.

E. T. WHITTAKER, *Analytical Mechanics.* This treatise is perhaps the best single reference for most of this chapter. Conservation theorems are discussed in Chapter III, while Hamilton's principle and its derivation from Lagrange's equations will be found in Chapter IX.

A. SOMMERFELD, *Vorlesungen über Theoretische Physik, Bd I, Mechanik.* It is to be regretted that this superb text, written in Sommerfeld's usual felicitous style, is not yet generally available here. Unfortunately, not all the subjects of interest are discussed, and one might perhaps wish for some changes in emphasis in the choice of material, but there can be little quarrel with the handling of the subjects which are included. Of special interest for this chapter are Sections 33–36.

---

* It must be mentioned here that the equations of transformation to generalized coordinates may involve the time explicitly for reasons other than moving constraints, e.g., rotating coordinate axes, and it may also be that the Lagrangian nevertheless does not contain the time explicitly. The quantity H is then conserved, but as T is not a homogeneous function of the velocities, H is no longer $T + V$. Clearly the energy of the system is also conserved, for the use of rotating coordinate axes, for example, is just a mathematical convenience which does not alter the physics of the situation. See Chapter 8 on canonical transformations.

W. E. BYERLY, *Generalized Coordinates.* This little book is of value particularly for the many detailed examples of the Lagrangian technique for setting up and solving mechanical problems. The lack of an index is a deplorable defect which makes use of the book somewhat difficult.

W. F. OSGOOD, *Mechanics.* Chapter X will be found of interest for the extension of Lagrange's equations to nonholonomic systems, with a rather involved treatment of Lagrange's undetermined multipliers.

H. F. OLSON, *Dynamical Analogies.* This book discusses in great detail the electrical circuit problems equivalent to given mechanical and acoustical systems, and illustrates the application of circuit theory to the solution of purely mechanical or acoustical problems.

J. J. THOMSON, *Applications of Dynamics to Physics and Chemistry.* This is a slender volume, representing Thomson's first researches, attempting to encompass a large variety of chemical, thermodynamical, and electrical problems within the framework of the Lagrangian formulation. He did not, however, apply the Lagrangian method to fields, which today is the most fruitful extension of Hamilton's principle. The modern reader of this book may find the exposition somewhat obscure.

EXERCISES

1. Prove that the shortest distance between two points in space is a straight line.

2. Show that the geodesics of a spherical surface are great circles, i.e., circles whose centers lie at the center of the sphere.

3. Complete the solution of the brachistochrone problem begun in Section 2–2 and show that the desired curve is a cycloid with a cusp at the initial point at which the particle is released. Show also that if the particle is projected with an initial kinetic energy $\frac{1}{2}mv_0^2$ that the brachistochrone is still a cycloid passing through the two points with a cusp at a height z above the initial point given by $v_0^2 = 2gz$.

4. In the discussion in Section 2–2 on finding the function f for which the integral $\int f\,dx$ was an extremum it was assumed that f was a function only of x, y, and the first derivative of y with respect to x, $\dot{y}$. Show that if f also involves the second derivatives of y with respect to x, $\ddot{y}$, and if y and $\dot{y}$ at the end points are not varied, then the corresponding Euler-Lagrange equation is

$$\frac{d^2}{dx^2}\frac{\partial f}{\partial \ddot{y}} - \frac{d}{dx}\frac{\partial f}{\partial \dot{y}} + \frac{\partial f}{\partial y} = 0.$$

5. Suppose that it was known experimentally that a particle fell a given distance y_0 in a time $t_0 = \sqrt{\frac{2y_0}{g}}$, but that the time of fall for distances other then y_0 were not known. Suppose further that the Lagrangian for the problem is known, but that instead of solving the equation of motion for y as a function of t, it is guessed that the functional form is

$$y = at + bt^2.$$

If the constants a and b are adjusted always so that the time to fall y_0 is correctly given by t_0, show directly that the integral

$$\int_0^{t_0} L\,dt$$

is an extremum for real values of the coefficients only when $a = 0$ and $b = g/2$.

6. When two billiard balls collide the instantaneous forces between them are very large, but act only in an infinitesimal time Δt, in such a manner that the quantity

$$\int_{\Delta t} F\,dt$$

remains finite. Such forces are described as *impulsive* forces, and the integral over Δt is known as the *impulse* of the force. Show that if impulsive forces are present Lagrange's equations may be transformed into

$$\left(\frac{\partial L}{\partial \dot{q}_j}\right)_f - \left(\frac{\partial L}{\partial \dot{q}_j}\right)_i = S_j,$$

where the subscripts i and f refer to the state of the system before and after the impulse, S_j is the impulse of the generalized impulsive force corresponding to q_j, and L is the Lagrangian including all the nonimpulsive forces.

7. A heavy particle is placed at the top of a vertical hoop. Calculate the reaction of the hoop on the particle by means of the Lagrange's undetermined multipliers and Lagrange's equations. Find the height at which the particle falls off.

8. A form of the Wheatstone impedance bridge has, in addition to the usual four resistances, an inductance in one arm and a capacitance in the opposite arm. Set up L and $\mathfrak{F}$ for the unbalanced bridge, with the currents through the elements as coordinates. Using the Kirchhoff junction conditions as constraints on the currents, obtain the Lagrange equations of motion, and show that eliminating the λ's reduces these to the usual network equations.

9. Show that if the potential in the Lagrangian contains velocity-dependent terms, the canonical momentum corresponding to a coordinate of rotation θ of the entire system is no longer the mechanical angular momentum L_θ but is given by

$$p_\theta = L_\theta - \sum_i \mathbf{n} \cdot \mathbf{r}_i \times \nabla_{v_i} U,$$

where ∇_v is the gradient operator in which the derivatives are with respect to the velocity components and $\mathbf{n}$ is a unit vector in the direction of rotation. If the forces are electromagnetic in character the canonical momentum is therefore

$$p_\theta = L_\theta + \sum_i \mathbf{n} \cdot \mathbf{r}_i \times \frac{q_i}{c} \mathbf{A}_i.$$

10. It sometimes occurs that the generalized coordinates appear separately in the kinetic energy and the potential energy in such a manner that T and V may be written in the form

$$T = \sum_i f_i(q_i)\dot{q}_i^2 \qquad \text{and} \qquad V = \sum_i V_i(q_i).$$

Show that Lagrange's equations then separate, and that the problem can always be reduced to quadratures.

CHAPTER 3

THE TWO-BODY CENTRAL FORCE PROBLEM

In this chapter we shall discuss the problem of two bodies moving under the influence of a mutual central force as an application of the Lagrangian formulation. Not all the problems of central force motion are integrable in terms of well-known functions. However, we shall attempt to explore the problem as thoroughly as is possible with the tools already developed.

3–1 Reduction to the equivalent one-body problem.

Consider a conservative system of two mass points, m_1 and m_2, where the only forces are those due to an interaction potential V. It will be assumed at first that V is any function of the vector between the two particles, $\mathbf{r}_2 - \mathbf{r}_1$, or of their relative velocity, $\dot{\mathbf{r}}_2 - \dot{\mathbf{r}}_1$, or of any higher derivatives of $\mathbf{r}_2 - \mathbf{r}_1$. Such a system has six degrees of freedom and hence six independent generalized coordinates. Let us choose these to be the three components of the radius vector to the center of mass $\mathbf{R}$, plus the three components of the difference vector $\mathbf{r} = \mathbf{r}_2 - \mathbf{r}_1$. The Lagrangian will then have the form

$$L = T(\dot{\mathbf{R}}, \dot{\mathbf{r}}) - V(\mathbf{r}, \dot{\mathbf{r}}, \ldots). \tag{3–1}$$

The kinetic energy T can be written as the sum of the kinetic energy of the motion of the center of mass, plus the kinetic energy of motion about the center of mass, T':

$$T = \tfrac{1}{2}(m_1 + m_2)\dot{\mathbf{R}}^2 + T'$$

with

$$T' = \tfrac{1}{2}m_1\dot{\mathbf{r}}_1'^2 + \tfrac{1}{2}m_2\dot{\mathbf{r}}_2'^2.$$

Here $\mathbf{r}_1'$ and $\mathbf{r}_2'$ are the radii vectors of the two particles relative to the center of mass, and are related to $\mathbf{r}$ by

$$\mathbf{r}_1' = -\frac{m_2}{m_1 + m_2}\mathbf{r}, \qquad \mathbf{r}_2' = \frac{m_1}{m_1 + m_2}\mathbf{r}. \tag{3–2}$$

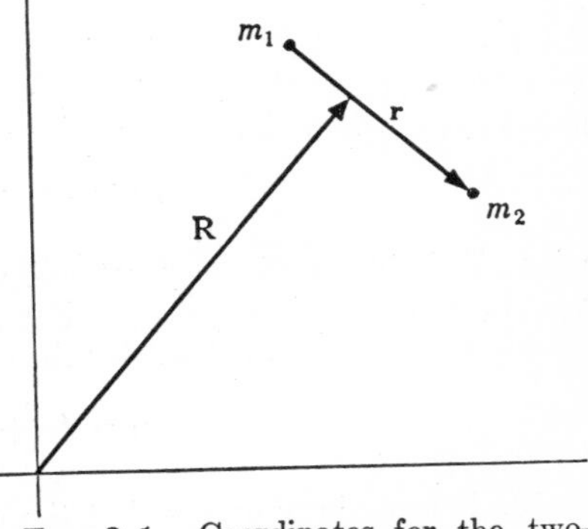

FIG. 3–1. Coordinates for the two-body problem.

Expressed in terms of $\mathbf{r}$ by means of Eq. (3–2), T' takes on the form

$$T' = \frac{1}{2}\frac{m_1 m_2}{m_1 + m_2}\dot{\mathbf{r}}^2$$

and the total Lagrangian (3–1) is:

$$L = \frac{m_1 + m_2}{2}\dot{\mathbf{R}}^2 + \frac{1}{2}\frac{m_1 m_2}{m_1 + m_2}\dot{\mathbf{r}}^2 - V(\mathbf{r}, \dot{\mathbf{r}}, \ldots). \tag{3–3}$$

It is seen that the three coordinates $\mathbf{R}$ are cyclic, so that the center of mass is either at rest or moving uniformly. None of the equations of motion for $\mathbf{r}$ will contain terms involving $\mathbf{R}$ or $\dot{\mathbf{R}}$. Consequently the process of ignoration is particularly simple here. We merely drop the first term from the Lagrangian in all subsequent discussion.

The rest of the Lagrangian is exactly what would be expected if we had a fixed center of force with a single particle at a distance $\mathbf{r}$ from it, having a mass

$$\mu = \frac{m_1 m_2}{m_1 + m_2}, \tag{3–4}$$

where μ is known as the *reduced mass*. Frequently Eq. (3–4) is written in the form

$$\frac{1}{\mu} = \frac{1}{m_1} + \frac{1}{m_2}. \tag{3–5}$$

Thus the central force motion of two bodies about their center of mass can always be reduced to an equivalent one-body problem.

3–2 The equations of motion and first integrals. We now restrict ourselves to truly central forces, where the potential V is a function of r only, so that the force is always along $\mathbf{r}$. By the results of the preceding section we need consider only the problem of a single particle of mass m moving about a fixed center of force, which will be taken as the origin of the coordinate system. Since potential energy involves only the radial distance, the problem has spherical symmetry, i.e., any rotation, about any fixed axis, can have no effect on the solution. Hence an angle coordinate representing rotation about a fixed axis must be cyclic. These symmetry properties result in a considerable simplification in the problem. Since the system is spherically symmetric, the total angular momentum vector,

$$\mathbf{L} = \mathbf{r} \times \mathbf{p},$$

is conserved. It therefore follows that $\mathbf{r}$ is always perpendicular to the fixed direction of $\mathbf{L}$ in space. This can be true only if $\mathbf{r}$ always lies in a plane whose normal is parallel to $\mathbf{L}$. While the reasoning breaks down if $\mathbf{L}$ is zero, the motion in that case must be along a straight line going through the center of force, for $\mathbf{L} = 0$ requires $\mathbf{r}$ to be parallel to $\dot{\mathbf{r}}$, which can be satisfied only in straight line motion.* Thus, central force motion is always motion in a plane. Now, the motion of a single particle in space is described by three coordinates; in spherical polar coordinates these are the azimuth angle θ, the zenith angle (or colatitude) ψ and the radial distance r. By choosing the polar axis to be in the direction of $\mathbf{L}$, the motion is always in the plane perpendicular to the polar axis. The coordinate ψ then has only the constant value $\pi/2$ and can be dropped from the subsequent discussion. The conservation of the angular momentum vector furnishes three independent constants of motion (corresponding to the three cartesian components). In effect, two of these, expressing the constant *direction* of the angular momentum, have been used to reduce the problem from three to two degrees of freedom. The third of these constants, corresponding to the conservation of the magnitude of $\mathbf{L}$, remains still at our disposal in completing the solution.

Expressed now in plane polar coordinates the Lagrangian is

$$\begin{aligned} L &= T - V \\ &= \tfrac{1}{2}m(\dot{r}^2 + r^2\dot{\theta}^2) - V(r). \end{aligned} \tag{3–6}$$

As was foreseen θ is a cyclic coordinate, whose corresponding canonical momentum is the angular momentum of the system:

$$p_\theta = \frac{\partial L}{\partial \dot{\theta}} = mr^2\dot{\theta}.$$

One of the two equations of motion is then simply

$$\dot{p}_\theta = \frac{d}{dt}(mr^2\dot{\theta}) = 0 \tag{3–7}$$

with the immediate integral

$$mr^2\dot{\theta} = l, \tag{3–8}$$

where l is the constant magnitude of the angular momentum. From (3–7) it also follows that

$$\frac{d}{dt}\left(\frac{1}{2}r^2\dot{\theta}\right) = 0. \tag{3–9}$$

The factor $\frac{1}{2}$ is inserted because $\frac{1}{2}r^2\dot{\theta}$ is just the *areal velocity* — the area swept out by the radius vector per unit time. This interpretation follows from the diagram, Fig. 3–2, the differential area swept out in time dt being

$$dA = \frac{1}{2}r(r\,d\theta),$$

* Formally: $\dot{\mathbf{r}} = \dot{r}\mathbf{n}_r + r\dot{\theta}\mathbf{n}_\theta$, hence $\mathbf{r} \times \dot{\mathbf{r}} = 0$ requires $\dot{\theta} = 0$.

and hence

$$\frac{dA}{dt} = \frac{1}{2} r^2 \frac{d\theta}{dt}.$$

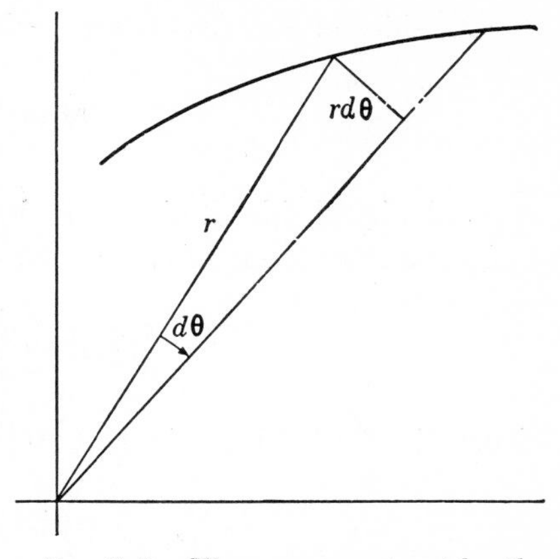

FIG. 3–2. The area swept out by the radius vector in a time dt.

The conservation of angular momentum is thus equivalent to saying the areal velocity is constant. Here we have the proof of the well-known Kepler's second law of planetary motion: the radius vector sweeps out equal areas in equal times. It should be emphasized, however, that the conservation of the areal velocity is a general property of central force motion, and is not restricted to an inverse square law of force.

The remaining Lagrange equation, for the coordinate r, is

$$\frac{d}{dt}(m\dot{r}) - mr\dot{\theta}^2 + \frac{\partial V}{\partial r} = 0. \tag{3–10}$$

Designating the force along $\mathbf{r}$, $-\frac{\partial V}{\partial r}$, by $f(r)$ the equation can be rewritten as

$$m\ddot{r} - mr\dot{\theta}^2 = f(r). \tag{3–11}$$

By making use of the first integral, Eq. (3–8), $\dot{\theta}$ can be eliminated from the equation of motion, yielding a second order differential equation involving r only:

$$m\ddot{r} - \frac{l^2}{mr^3} = f(r). \tag{3–12}$$

There is another first integral of motion available, namely the total energy, since the forces are conservative. On the basis of the general energy conservation theorem we can immediately state that a constant of the motion is

$$E = \frac{1}{2} m(\dot{r}^2 + r^2\dot{\theta}^2) + V(r), \tag{3–13}$$

where E is the energy of the system. Alternatively, this first integral could be derived again directly from the equations of motion (3–7) and (3–12). The latter can be written as

$$m\ddot{r} = -\frac{\partial}{\partial r}\left(V + \frac{1}{2}\frac{l^2}{mr^2}\right). \tag{3–14}$$

If both sides of Eq. (3–14) are multiplied by $\dot{r}$ the left-hand side becomes

$$m\ddot{r}\dot{r} = \frac{d}{dt}\left(\frac{1}{2}m\dot{r}^2\right).$$

The right-hand side similarly can be written as a total time derivative, for if $g(r)$ is any function of r, then the total time derivative of g has the form

$$\frac{d}{dt}g(r) = \frac{dg}{dr}\frac{dr}{dt}.$$

Hence Eq. (3–14) is equivalent to

$$\frac{d}{dt}\left(\frac{1}{2}m\dot{r}^2\right) = -\frac{d}{dt}\left(V + \frac{1}{2}\frac{l^2}{mr^2}\right)$$

or

$$\frac{d}{dt}\left(\frac{1}{2}m\dot{r}^2 + \frac{1}{2}\frac{l^2}{mr^2} + V\right) = 0,$$

and therefore

$$\frac{1}{2}m\dot{r}^2 + \frac{1}{2}\frac{l^2}{mr^2} + V = \text{constant.} \tag{3–15}$$

Eq. (3–15) is the statement of the conservation of total energy, for by using (3–8) for l the middle term can be written

$$\frac{1}{2}\frac{l^2}{mr^2} = \frac{1}{2mr^2}m^2r^4\dot{\theta}^2 = \frac{mr^2\dot{\theta}^2}{2},$$

and (3–15) reduces to (3–13).

These two first integrals give us in effect two of the quadratures necessary to complete the problem. As there are two variables, r and θ, a total of four integrations are needed to solve the equations of motion. The first two integrations have left the Lagrange equations as two first order equations (3–8) and (3–15); the two remaining integrations can be accomplished (formally) in a variety of ways. Perhaps the simplest procedure starts from Eq. (3–15). Solving for $\dot{r}$ we have

$$\dot{r} = \sqrt{\frac{2}{m}\left(E - V - \frac{l^2}{2mr^2}\right)}, \tag{3–16}$$

or

$$dt = \frac{dr}{\sqrt{\frac{2}{m}\left(E - V - \frac{l^2}{2mr^2}\right)}}. \tag{3–17}$$

At time $t = 0$ let r have the initial value r_0. Then the integral of both sides of the equation from the initial state to the state at time t takes the form

$$t = \int_{r_0}^{r} \frac{dr}{\sqrt{\frac{2}{m}\left(E - V - \frac{l^2}{2mr^2}\right)}}. \tag{3-18}$$

As it stands Eq. (3-18) gives t as a function of r and the constants of integration E, l, and r_0. However, it may be inverted, at least formally, to give r as a function of t and the constants. Once the solution for r is thus found, the solution θ follows immediately from Eq. (3-8), which can be written as

$$d\theta = \frac{l\,dt}{mr^2}. \tag{3-19}$$

If the initial value of θ is θ_0 then the integral of (3-19) is simply:

$$\theta = l \int_0^t \frac{dt}{mr^2(t)} + \theta_0. \tag{3-20}$$

Eqs. (3-18) and (3-20) are the two remaining integrations, and formally the problem has been reduced to quadratures, with four constants of integration E, l, r_0, θ_0. These constants are not the only ones which can be considered. We might equally as well have taken r_0, θ_0, $\dot{r}_0$, $\dot{\theta}_0$, but of course E and l can always be determined in terms of this set. For many applications, however, the set containing the energy and angular momentum is the natural one. In quantum mechanics such constants as the initial values of r and θ, or of $\dot{r}$ and $\dot{\theta}$, become meaningless, but we can still talk in terms of the system energy or of the system angular momentum. Indeed, the salient differences between classical and quantum mechanics appear in the properties of E and l in the two theories. In order to discuss the transition to quantum theories it is important therefore that the classical description of the system be in terms of its energy and angular momentum.

3-3 The equivalent one-dimensional problem, and classification of orbits. While the problem has thus been solved formally, practically speaking the integrals (3-18) and (3-20) are usually quite unmanageable and in any specific case it is often more convenient to perform the integration in some other fashion. But before obtaining the solution for specific force laws, let us see what can be learned about the motion in the general case, using only the equations of motion and the conservation theorems, without requiring explicit solutions.

For example, with a system of known energy and angular momentum, the magnitude and direction of the velocity of the particle can be immediately determined in terms of the distance r. The magnitude v follows at once from the conservation of energy in the form

$$E = \frac{1}{2}mv^2 + V(r)$$

or

$$v = \sqrt{\frac{2}{m}(E - V(r))}. \tag{3–21}$$

The radial velocity — the component of $\dot{\mathbf{r}}$ along the radius vector — has already been given in Eq. (3–16). Combined with the magnitude v this is sufficient information to furnish the direction of the velocity.* These results, and much more, can also be obtained from consideration of an equivalent one-dimensional problem.

The equation of motion in r, with $\dot{\theta}$ expressed in terms of l, Eq. (3–12), involves only r and its derivatives. It is the same equation as would be obtained for a fictitious one-dimensional problem in which a particle of mass m is subject to a force

$$f' = f + \frac{l^2}{mr^3}. \tag{3–22}$$

The significance of the additional term is clear if it is written as $mr\dot{\theta}^2 = mv_\theta^2/r$, which is the familiar centrifugal force. An equivalent statement can be obtained from the conservation theorem for energy. By Eq. (3–15) the motion of the particle in r is that of a one-dimensional problem with a fictitious potential energy:

$$V' = V + \frac{1}{2}\frac{l^2}{mr^2}. \tag{3–22′}$$

As a check we note that

$$f' = -\frac{\partial V'}{\partial r} = f(r) + \frac{l^2}{mr^3},$$

which agrees with Eq. (3–22). The energy conservation theorem (3–15) can thus also be written as

$$E = V' + \frac{1}{2}m\dot{r}^2. \tag{3–15′}$$

---

* Alternatively, the conservation of angular momentum furnishes $\dot{\theta}$, the angular velocity, and this together with $\dot{r}$ gives both the magnitude and direction of $\dot{\mathbf{r}}$.

As an illustration of this method of examining the motion, consider a plot of V' against r for the specific case of an attractive inverse square law of force:

$$f = -\frac{k}{r^2}$$

(for positive k the minus sign insures that the force is *toward* the center of force). The potential energy for this force is

$$V = -\frac{k}{r}$$

and the corresponding fictitious potential is

$$V' = -\frac{k}{r} + \frac{l^2}{2mr^2}.$$

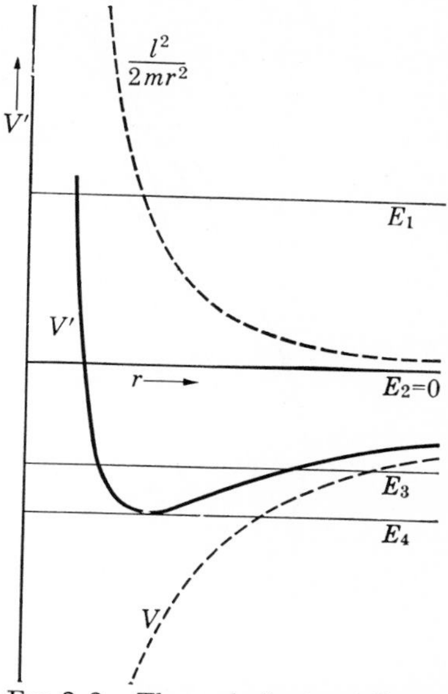

FIG. 3-3. The equivalent one-dimensional potential for attractive inverse square law of force.

Such a plot is shown in Fig. 3-3; the two dotted lines represent the separate components $-k/r$ and $+\dfrac{l^2}{2mr^2}$, and the solid line is the sum V'.

Let us consider now the motion of a particle having the energy E_1, as shown in Figs. 3-3 and 3-4. Clearly this particle can never come closer than r_1 (cf. Fig. 3-4). Otherwise with $r < r_1$, V' exceeds E_1 and by Eq. (3-15′) the kinetic energy would have to be negative, corresponding to an imaginary velocity! On the other hand, there is no upper limit to the possible value of r, so that the orbit is not closed. A particle will come in from infinity, strike the "repulsive centrifugal barrier," be repelled, and travel back out to infinity (cf. Fig. 3-5). The distance between E and V' is $\frac{1}{2}m\dot{r}^2$, i.e., proportional to the square of the radial velocity, and be-

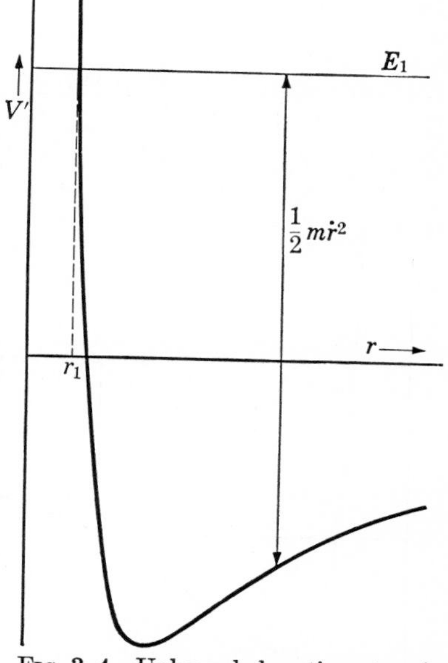

FIG. 3-4. Unbounded motion at positive energies for inverse square law of force.

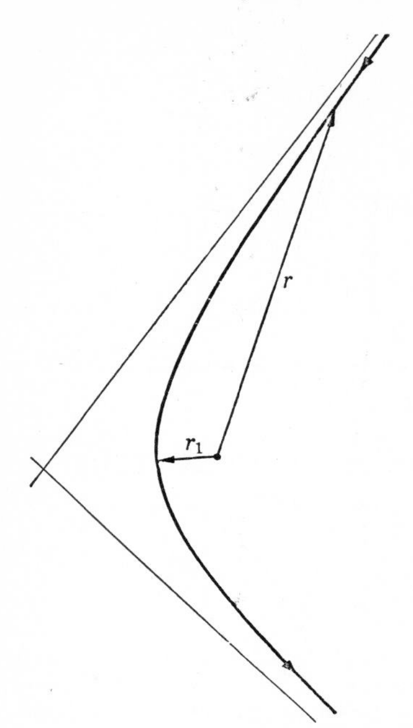

FIG. 3–5. Schematic picture of the orbit for E_1 corresponding to unbounded motion.

comes zero, naturally, at the *turning point* r_1. At the same time the distance between E and V on the plot is the kinetic energy $\frac{1}{2}mv^2$ at the given value of r. Hence the distance between the V and V' curves is $\frac{1}{2}mr^2\dot{\theta}^2$. These curves therefore supply the magnitude of the particle velocity and its components for any distance r, at the given energy and angular momentum. This information is sufficient to provide an approximate picture of the form of the orbit.

For the energy $E_2 = 0$ (cf. Fig. 3–3) a roughly similar picture of the orbit behavior is obtained. But for any lower energy, such as E_3 indicated in Fig. 3–6, we have a different story. In addition to a lower bound r_1, there is also a maximum value r_2 which cannot be exceeded by r with positive kinetic energy. The motion is then "bounded," and there are two turning points, r_1 and r_2, also known as *apsidal distances*. This does not necessarily mean that the orbits are closed. All that can be said is that they are bounded, contained between two circles of radius r_1 and r_2 with turning points always lying on the circles (cf. Fig. 3–7).

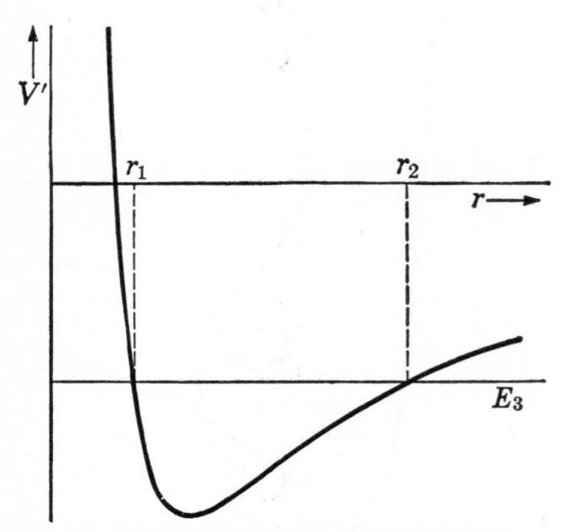

FIG. 3–6. The equivalent one-dimensional potential for inverse square law of force, illustrating bounded motion at negative energies.

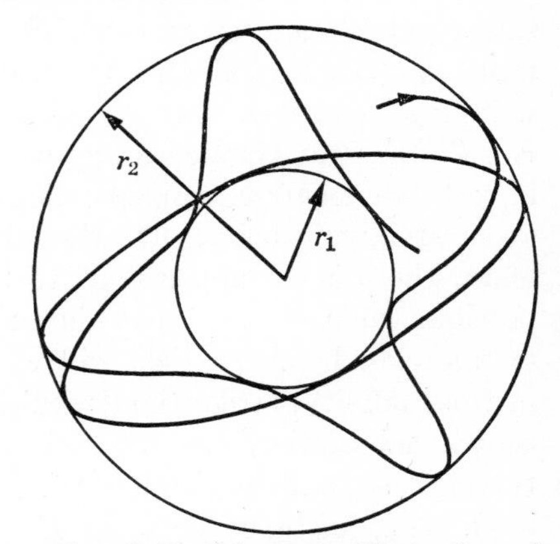

FIG. 3–7. Schematic illustration of the nature of the orbits for bounded motion.

If the energy is E_4 just at the minimum of the fictitious potential as shown in Fig. 3–8, then the two bounds coincide. In such case motion is possible at only one radius; $\dot{r} = 0$, and the orbit is a circle. Remembering that the effective "force" is the negative of the slope of the V' curve, the requirement for circular orbits is simply that f' be zero, or

$$f(r) = \frac{-l^2}{mr^3} = -mr\dot{\theta}^2.$$

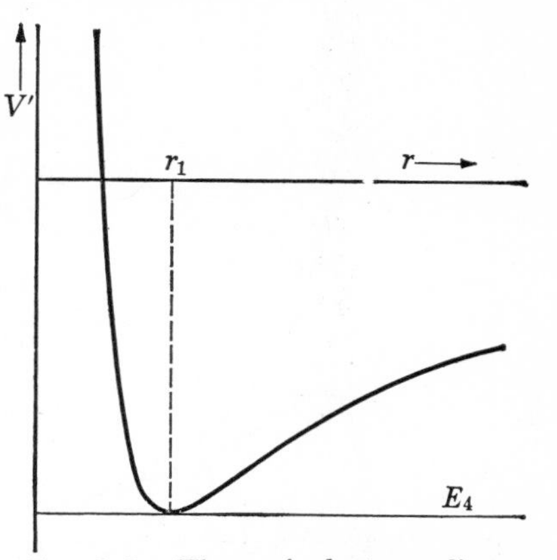

FIG. 3–8. The equivalent one-dimensional potential of inverse square law of force, illustrating the condition for circular orbits.

We have here the familiar elementary condition for a circular orbit, that the applied force just balances the "reversed effective force" of centripetal acceleration.*

It is to be emphasized that all of this discussion of the orbits for various energies has been at one value of the angular momentum. Changing l will change the quantitative details of the V' curve, but will not affect the general classification of the types of orbits.

For the attractive inverse square law of force discussed above, we shall see that the orbit for E_1 is a hyperbola, for E_2 a parabola, and for E_3 an ellipse. With other forces the orbits may not have such simple forms. However, the same general qualitative division into open, bounded, and circular orbits will be true for any attractive potential which: (1) falls off slower than $1/r^2$ as $r \to \infty$; (2) becomes infinite slower than $1/r^2$ as $r \to 0$. The first condition insures that the potential predominates over the centrifugal term for large r, while the second condition is such that for small r it is the centrifugal term which is the important one.

The qualitative nature of the motion will be altered if the potential does not satisfy these requirements, but we may still use the method of the equivalent potential to examine features of the orbits. As an example, consider the attractive potential $V(r) = -\frac{a}{r^3}$, with $f = -\frac{3a}{r^4}$. The energy diagram then is as shown in Fig. 3–9. For an energy E there are two possible types of motion, depending upon the initial value of r. If r_0 is less than r_1 the motion will be bounded, r will always remain less than r_1,

* The case $E < E_4$ does not correspond to physically possible motion for then $\dot{r}^2$ would have to be negative, or $\dot{r}$ imaginary.

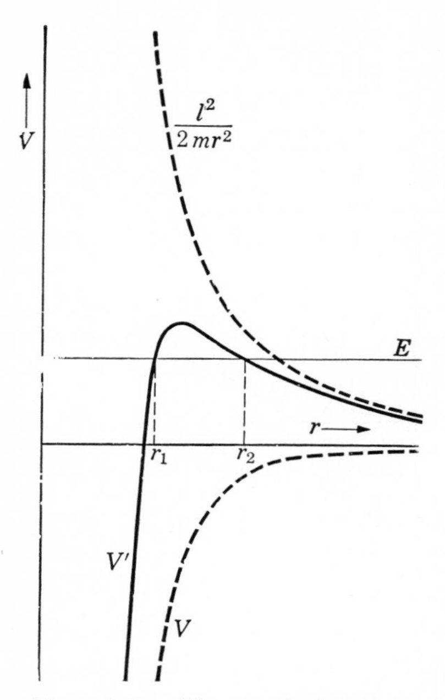

FIG. 3–9. The equivalent one-dimensional potential for an attractive inverse fourth law of force.

and the particle will eventually pass through the center of force. If r is initially greater than r_2 then it will always remain so; the motion is unbounded, and the particle can never get inside the "potential" hole. The initial condition $r_1 < r_0 < r_2$ is again not physically possible.

Another interesting example of the method occurs for a linear restoring force (harmonic oscillator):

$$f = -kr, \qquad V = \frac{1}{2}kr^2.$$

For zero angular momentum, corresponding to motion along a straight line, $V' = V$ and the situation is as shown in Fig. 3–10. For any positive energy the motion is bounded and, as we know, simple harmonic. If $l \neq 0$ we have the state of affairs shown in Fig. 3–11. The motion then is always bounded for all physically possible energies and does not pass through the center of force. In this particular case it is easily seen that the orbit is elliptic, for if $\mathbf{f} = -k\mathbf{r}$, the x and y components of the force are

$$f_x = -kx, \qquad f_y = -ky.$$

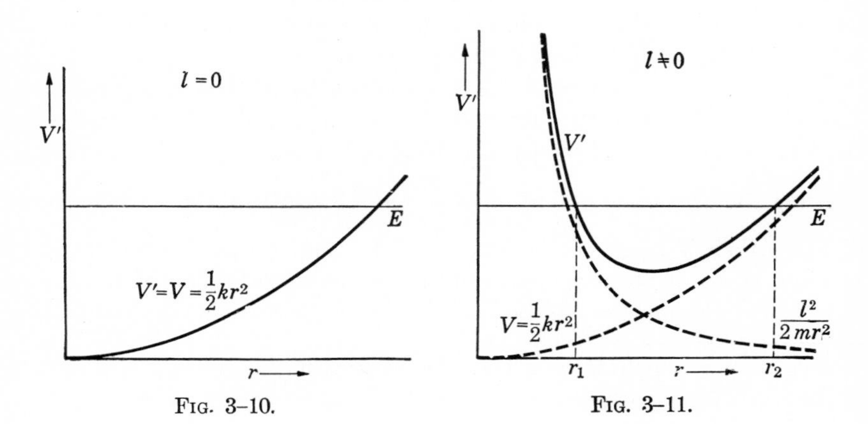

FIG. 3–10.

FIG. 3–11.

The total motion is thus the resultant of two simple harmonic oscillations at right angles, and of the same frequency, which in general leads to an elliptic orbit. A well-known example is the spherical pendulum for small amplitudes. The familiar Lissajous figures are obtained as the composition of two sinusoidal oscillations at right angles where the ratio of the frequencies is a rational number. Central force motion under a linear restoring force therefore constitutes the simplest of the Lissajous figures.

3–4 The virial theorem. Another property of central force motion can be derived as a special case of a general theorem valid for a large variety of systems — the so-called *virial theorem.* It differs in character from the theorems previously discussed in being *statistical* in nature, i.e., it is concerned with the time averages of various mechanical quantities.

Consider a general system of mass points with position vectors $\mathbf{r}_i$ and applied forces $\mathbf{F}_i$ (including any forces of constraint). The fundamental equations of motion are then

$$\dot{\mathbf{p}}_i = \mathbf{F}_i. \tag{1–1}$$

We shall be interested in the quantity

$$G = \sum_i \mathbf{p}_i \cdot \mathbf{r}_i,$$

where the summation is over all particles in the system. The total time derivative of this quantity is

$$\frac{dG}{dt} = \sum_i \dot{\mathbf{r}}_i \cdot \mathbf{p}_i + \sum_i \dot{\mathbf{p}}_i \cdot \mathbf{r}_i. \tag{3–23}$$

The first term can be transformed to

$$\sum_i \dot{\mathbf{r}}_i \cdot \mathbf{p}_i = \sum_i m_i \dot{\mathbf{r}}_i \cdot \dot{\mathbf{r}}_i = \sum_i m_i v_i^2 = 2T,$$

while the second by Eq. (1–1) is:

$$\sum_i \dot{\mathbf{p}}_i \cdot \mathbf{r}_i = \sum_i \mathbf{F}_i \cdot \mathbf{r}_i.$$

Eq. (3–23) therefore reduces to

$$\frac{d}{dt} \sum_i \mathbf{p}_i \cdot \mathbf{r}_i = 2T + \sum_i \mathbf{F}_i \cdot \mathbf{r}_i. \tag{3–24}$$

The time average of Eq. (3–24) over a time interval τ is obtained by integrating both sides with respect to t from 0 to τ, and dividing by τ:

$$\frac{1}{\tau}\int_0^\tau \frac{dG}{dt}\,dt \equiv \overline{\frac{dG}{dt}} = \overline{2T} + \overline{\sum_i \mathbf{F}_i \cdot \mathbf{r}_i}$$

or

$$\overline{2T} + \overline{\sum_i \mathbf{F}_i \cdot \mathbf{r}_i} = \frac{1}{\tau}\,[G(\tau) - G(0)]. \tag{3-25}$$

If the motion is periodic, i.e., all coordinates repeat after a certain time, and if τ is chosen to be the period, then the right-hand side of (3–25) vanishes. A similar conclusion can be reached even if the motion is not periodic, provided that the coordinates and velocities for all particles remain finite so that there is an upper bound to G. By choosing τ sufficiently long the right-hand side of Eq. (3–25) can be made as small as desired. In both cases it then follows that

$$\overline{T} = -\frac{1}{2}\overline{\sum_i \mathbf{F}_i \cdot \mathbf{r}_i}. \tag{3-26}$$

Eq. (3–26) is known as the *virial theorem*, and the right-hand side is called the *virial of Clausius*. In this form the theorem is very useful in the kinetic theory of gases. Thus, only a few further steps are required to prove Boyle's law for perfect gases (cf. Lindsay, *Physical Statistics*, p. 70). It is practically indispensable for calculating the equation of state for *imperfect gases*, where the forces $\mathbf{F}_i$ include not only the forces of constraint keeping the gas particles inside the container but also the interaction forces between molecules.

One can further show that if the forces $\mathbf{F}_i$ are the sum of nonfrictional forces $\mathbf{F}'_i$ and frictional forces $\mathbf{f}_i$ proportional to the velocity, then the virial depends only on the $\mathbf{F}'_i$; there is no contribution from the $\mathbf{f}_i$. Of course, the motion of the system must not be allowed to die down as a result of the frictional forces. Energy must constantly be pumped into the system to maintain the motion; otherwise *all* time averages would vanish as τ increases indefinitely.

If the forces are derivable from a potential then the theorem becomes

$$\overline{T} = \frac{1}{2}\overline{\sum_i \nabla V \cdot \mathbf{r}_i}, \tag{3-27}$$

and for a single particle moving under a central force it reduces to

$$\overline{T} = \frac{1}{2}\overline{\frac{\partial V}{\partial r}\,r}. \tag{3-28}$$

If V is a power-law function of r:

$$V = ar^{n+1},$$

where the exponent is chosen so that the force law goes as r^n, then

$$\frac{\partial V}{\partial r} r = (n + 1)V,$$

and Eq. (3–28) becomes

$$\overline{T} = \frac{n+1}{2} \overline{V}. \tag{3–29}$$

For the further special case of inverse square law forces n is -2 and the virial theorem takes on a well-known form:

$$\overline{T} = -\frac{1}{2} \overline{V}. \tag{3–30}$$

3–5 The differential equation for the orbit, and integrable power-law potentials. In treating specific details of actual central force problems a change in the orientation of our discussion is desirable. Hitherto solving a problem has meant finding r and θ as functions of time with E, l, and so on, as constants of integration. But most often what we really seek is the equation of the orbit, i.e., the dependence of r upon θ, eliminating the parameter t. For central force problems the elimination is particularly simple, since t occurs in the equations of motion only as a variable of differentiation. Indeed one equation of motion, (3–8), simply provides a definite relation between a differential change dt and the corresponding change $d\theta$:

$$l\, dt = mr^2\, d\theta. \tag{3–31}$$

The corresponding relation between derivatives with respect to t and θ is

$$\frac{d}{dt} = \frac{l}{mr^2} \frac{d}{d\theta}. \tag{3–32}$$

These relations may be used to convert the equation of motion (3–12) into a differential equation for the orbit. Alternatively they can be applied to the formal solution of the equations of motion, given in Eq. (3–17), to furnish the equation of the orbit directly. For the moment we shall follow the former of these possibilities.

From (3–32) a second derivative with respect to t can be written

$$\frac{d^2}{dt^2} = \frac{l}{mr^2} \frac{d}{d\theta} \left(\frac{l}{mr^2} \frac{d}{d\theta} \right),$$

and the Lagrange equation for r, (3–12), becomes

$$\frac{l}{r^2} \frac{d}{d\theta} \left(\frac{l}{mr^2} \frac{dr}{d\theta} \right) - \frac{l^2}{mr^3} = f(r). \tag{3–33}$$

Now, to simplify (3–33) we notice that

$$\frac{1}{r^2}\frac{dr}{d\theta} = -\frac{d(1/r)}{d\theta};$$

hence if the variable is changed to $u = 1/r$ we have

$$\frac{l^2u^2}{m}\left(\frac{d^2u}{d\theta^2} + u\right) = -f\left(\frac{1}{u}\right). \tag{3–34}$$

Eq. (3–34) is thus a differential equation for the orbit if the force law f is known. Conversely if the equation of the orbit is known, i.e., r is given as a function of θ, then one can work back and obtain the force law $f(r)$.

Here, however, we want to obtain some rather general results. For example, it can be shown from (3–34) that the orbit is symmetrical about the turning points. To prove this statement it will be noted that if the orbit is symmetrical it should be possible to reflect it about the direction of the turning angle without producing any change. If the coordinates are so chosen that the turning point occurs for $\theta = 0$, then the reflection can be effected mathematically by the substitution of $-\theta$ for θ. The differential equation for the orbit, (3–34), is obviously invariant under such a substitution. Further the initial conditions, here

$$u = u(0), \qquad \left(\frac{du}{d\theta}\right)_0 = 0, \qquad \text{for } \theta = 0,$$

will likewise be unaffected. Hence the orbit equation must be the same whether expressed in terms of θ or $-\theta$, which is the desired conclusion. *The orbit is therefore invariant under reflection about the apsidal vectors.* In effect this means that the complete orbit can be traced if the portion of the orbit between any two turning points is known. Reflection of the given portion about one of the apsidal vectors produces a neighboring stretch of the orbit, and this process can be repeated indefinitely until the rest of the orbit is completed, as illustrated in Fig. 3–12.

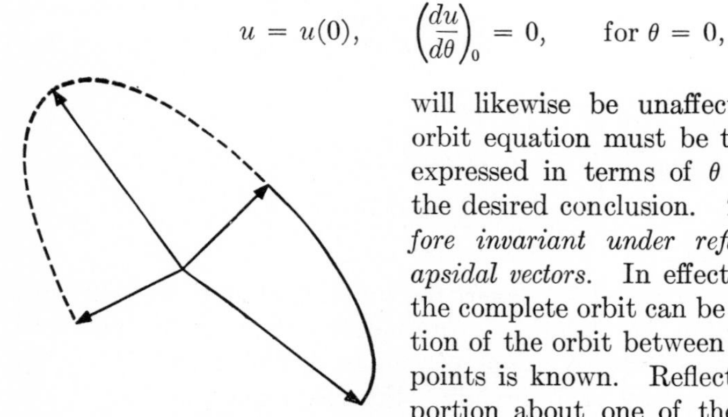

FIG. 3–12. Extension of the orbit by reflection of a portion about the apsidal vectors.

For any particular force law the actual equation of the orbit must be obtained by integrating the differential Eq. (3–34). However it is not necessary to go through all the details of the integration, as most of the work has already been done in solving the equation of motion (3–12). All that remains is to eliminate t from the solution (3-17) by means of (3-31), resulting in:

$$d\theta = \frac{l\,dr}{mr^2\sqrt{\frac{2}{m}\left(E - V(r) - \frac{l^2}{2mr^2}\right)}}. \tag{3-35}$$

With slight rearrangements the integral of (3-35) is

$$\theta = \int_{r_0}^{r} \frac{dr}{r^2\sqrt{\frac{2mE}{l^2} - \frac{2mV}{l^2} - \frac{1}{r^2}}} + \theta_0 \tag{3-36}$$

or, if the variable of integration is changed to $u = 1/r$,

$$\theta = \theta_0 - \int_{u_0}^{u} \frac{du}{\sqrt{\frac{2mE}{l^2} - \frac{2mV}{l^2} - u^2}}. \tag{3-37}$$

As in the case of the equation of motion, Eq. (3-37), while solving the problem formally, is not always a practicable solution, because the integral often cannot be expressed in terms of well-known functions. In fact, only certain types of force laws have been investigated. The most important are the power-law functions of r:

$$V = ar^{n+1} \tag{3-38}$$

so that the force varies as the nth power of r.* With this potential (3-37) becomes

$$\theta = \theta_0 - \int_{u_0}^{u} \frac{du}{\sqrt{\frac{2mE}{l^2} - \frac{2ma}{l^2}u^{-n-1} - u^2}}. \tag{3-39}$$

This again will be integrable in terms of simple functions only in certain cases. If the quantity in the radical is of no higher power in u than quadratic, the denominator has the form $\sqrt{\alpha u^2 + \beta u + \gamma}$ and the integration can be directly effected in terms of circular functions. This restriction is equivalent to requiring that

$$-n - 1 = 0, 1, 2,$$

* The case $n = -1$ is to be excluded from the following discussion. In the potential (3-38) it corresponds to a constant potential, i.e., no force at all. It is an equally anomalous case if the exponent is used in the force law directly, since a force varying as r^{-1} corresponds to a logarithmic potential, which is not a power law at all. A logarithmic potential is unusual for motion about a point; it is more characteristic of a *line* source.

or, excluding the $n = -1$ case, for

$$n = -2, -3,$$

corresponding to inverse square or inverse cube force laws. One further easily integrable case is for $n = 1$, i.e., the linear force; for then Eq. (3–39) can be written as

$$\theta = \theta_0 - \int_{u_0}^{u} \frac{du}{\sqrt{\frac{2mE}{l^2} - \frac{2ma}{l^2}\frac{1}{u^2} - u^2}}. \tag{3–39′}$$

If now we make the substitution

$$u^2 = x, \qquad du = \frac{dx}{2\sqrt{x}},$$

Eq. (3–39′) becomes

$$\theta = \theta_0 - \frac{1}{2}\int_{x_0}^{x} \frac{dx}{\sqrt{\frac{2mE}{l^2}x - \frac{2ma}{l^2} - x^2}}, \tag{3–40}$$

which again is in the desired form. Thus, a solution in terms of simple functions is obtained for the exponents

$$n = 1, -2, -3.$$

This does not mean other powers are not integrable, merely that they lead to functions not as well known. For example, there is a range of exponents for which Eq. (3–39) involves *elliptic integrals*, with the solution expressed in terms of *elliptic functions*.

By definition an elliptic integral is

$$\int R(x, \omega)\, dx,$$

where R is any rational function of x and ω, and ω is defined as

$$\omega = \sqrt{\alpha x^4 + \beta x^3 + \gamma x^2 + \delta x + \eta}.$$

Of course α and β cannot simultaneously be zero, for then the integral could be evaluated in terms of circular functions. It can be shown (Whittaker and Watson, *Modern Analysis*, 4th ed., p. 512) that any such integral can be reduced to forms involving circular functions and the Legendre elliptic integrals of the first, second, and third kind. There exist complete and detailed tables of these standard elliptic integrals, and their properties and connections with elliptic functions have been discussed exhaustively in the literature. Intrinsically they do not require any higher logical concept for their use than do circular functions, they are just not as familiar. From the definition it is seen that the integral in (3-39) can be evaluated in terms of elliptic functions if

$$n = -4, -5.$$

We can attempt to put the integral in another form also leading to elliptic integrals by multiplying numerator and denominator by u^ρ where ρ is some undetermined exponent. The integral then becomes

$$\int \frac{u^\rho \, du}{\sqrt{\frac{2mE}{l^2} u^{2\rho} - \frac{2ma}{l^2} u^{-n-1+2\rho} - u^{2(\rho+1)}}},$$

where the expression in the radical will be a polynomial of higher order than a quartic except if $\rho = 1$. The integral will therefore be no worse than elliptic only if

$$-n - 1 + 2 = 0, 1, 2, 3, 4$$

or

$$n = +1, 0, -1, -2, -3.$$

For $n = +1, -2, -3$ the solutions reduce to circular functions, the case $n = -1$ has already been eliminated, so that this procedure leads to elliptic functions only for $n = 0$.

We again can obtain integrals of the elliptic type in certain cases by changing the variable to $u^2 = x$. The integral in question then appears as

$$\frac{1}{2} \int \frac{dx}{\sqrt{\frac{2mE}{l^2} x - \frac{2ma}{l^2} x^{\frac{-n+1}{2}} - x^2}},$$

which reduces to the elliptic form for

$$(-n + 1)/2 = 3, 4$$

leading to the exponents

$$n = -5, -7.$$

Finally we again can perform the trick of multiplying numerator and denominator by x, and the condition for obtaining elliptic integrals or simpler is then

$$\frac{-n+1}{2} + 2 = 0, 1, 2, 3, 4$$

or

$$n = +5, +3, +1, -1, -3,$$

which leads to new possibilities only for $n = +5, +3$. The total number of integral exponents resulting in elliptic functions is thus

$$n = +5, +3, 0, -4, -5, -7.$$

Although this exhausts the possibilities for integral exponents, with suitable transformations some fractional exponents can also be shown to lead to elliptic integrals.

3–6 The Kepler problem: inverse square law of force. The inverse square law is the most important of all the central force laws and it deserves detailed treatment. For this case the force and potential can be written as

$$f = -\frac{k}{r^2}, \qquad V = -\frac{k}{r}. \tag{3–41}$$

There are several ways to integrate the equation for the orbit, the simplest being to substitute (3–41) in the differential equation for the orbit (3–34):

$$\frac{d^2u}{d\theta^2} + u = \frac{-mf\left(\frac{1}{u}\right)}{l^2u^2} = \frac{mk}{l^2}. \tag{3–42}$$

Changing the variable to $y = u - \frac{mk}{l^2}$, the differential equation becomes

$$\frac{d^2y}{d\theta^2} + y = 0,$$

which has the immediate solution

$$y = b \cos(\theta - \theta'),$$

b and θ' being the two constants of integration. In terms of r the solution is

$$\frac{1}{r} = \frac{mk}{l^2}(1 + \epsilon \cos(\theta - \theta')), \tag{3–43}$$

where

$$\epsilon = b\frac{l^2}{mk}.$$

It is instructive to obtain the orbit equation also from the formal solution (3–39). While this procedure is longer than the simple integration of the differential equation (3–42), it has the advantage that the significant constant of integration ϵ is automatically evaluated in terms of the energy E and the angular momentum l of the system. We write (3–39) in the form

$$\theta = \theta' - \int \frac{du}{\sqrt{\frac{2mE}{l^2} + \frac{2mku}{l^2} - u^2}}, \tag{3–44}$$

where the integral is now taken as indefinite. The quantity θ' appearing in (3–44) is a constant of integration determined by the initial conditions, and will not necessarily be the same as the initial angle θ_0 at time $t = 0$. The indefinite integral is of the standard form: *

$$\int \frac{dx}{\sqrt{a + bx + cx^2}} = \frac{1}{\sqrt{-c}} \operatorname{arc\,cos} - \frac{b + 2cx}{\sqrt{q}} \tag{3–45}$$

where

$$q = b^2 - 4ac.$$

To apply this to (3–44) we must set

$$a = \frac{2mE}{l^2}, \qquad b = \frac{2mk}{l^2}, \qquad c = -1,$$

and the discriminant q is therefore:

$$q = \left(\frac{2mk}{l^2}\right)^2 \left(1 + \frac{2El^2}{mk^2}\right).$$

With these substitutions, Eq. (3–44) becomes

$$\theta = \theta' - \operatorname{arc\,cos} \frac{\frac{l^2 u}{mk} - 1}{\sqrt{1 + \frac{2El^2}{mk^2}}}.$$

Finally, by solving for $u, \equiv \frac{1}{r}$, the equation of the orbit is found to be

$$\frac{1}{r} = \frac{mk}{l^2}\left(1 + \sqrt{1 + \frac{2El^2}{mk^2}} \cos(\theta - \theta')\right), \tag{3–46}$$

which agrees with (3–43), except that here ϵ is evaluated in terms of E and l. The constant of integration θ' can now be identified from Eq. (3–46) as one of the turning angles of the orbit. It will be noted that only three of the four constants of integration appear in the orbit equation, and this is

---

* Cf., for example, B. O. Peirce, *A Short Table of Integrals*, no. 161. A constant, $-\pi/2$, is to be added to the result as given by Peirce to obtain (3–45), which is permissible since the integral is indefinite.

always a characteristic property of the orbit. In effect, the fourth constant locates the initial position of the particle on the orbit. If we are interested solely in the orbit equation this information is clearly irrelevant, and hence does not appear in the answer. Of course, the missing constant has to be supplied if it is desired to complete the solution by finding r and θ as functions of time. Thus, if one chooses to integrate the conservation theorem for angular momentum,

$$mr^2\, d\theta = l\, dt,$$

by means of (3–46), one must specify in addition the initial angle θ_0.

Now, the general equation of a conic with one focus at the origin is

$$\frac{1}{r} = C(1 + \epsilon \cos(\theta - \theta')), \tag{3–47}$$

where ϵ is the eccentricity of the conic section. By comparison with Eq. (3–46) it follows that the orbit is always a conic section, with the eccentricity

$$\epsilon = \sqrt{1 + \frac{2El^2}{mk^2}}. \tag{3–48}$$

The nature of the orbit depends on the magnitude of ϵ, according to the following scheme:

$$\begin{aligned} &\epsilon > 1, && E > 0: && \text{hyperbola,} \\ &\epsilon = 1, && E = 0: && \text{parabola,} \\ &\epsilon < 1, && E < 0: && \text{ellipse,} \\ &\epsilon = 0, && E = -\frac{mk^2}{2l^2}: && \text{circle.} \end{aligned}$$

This classification agrees with the qualitative discussion of the orbits based on the energy diagram of the equivalent one-dimensional potential V'. The condition for circular motion appears here in a somewhat different form, but its equivalence with the previous statement is shown by writing it as

$$E = -\frac{k^2}{2mr^4\dot{\theta}^2}. \tag{3–49}$$

For a circular orbit T and V are constant in time and the virial theorem can be written

$$T = -\frac{1}{2}V,$$

so that the total energy is

$$E = \frac{1}{2}V = -\frac{k}{2r}. \tag{3–50}$$

With this value for E the condition for a circular orbit (3–49) becomes

$$\frac{k}{r^2} = mr\dot{\theta}^2,$$

which says the force of attraction is just balanced by the centrifugal force, in agreement with the condition given previously in Section 3–3.

In the case of elliptic orbits it can be shown the major axis depends solely on the energy, a theorem of considerable importance in the Bohr theory of the atom. The semimajor axis is one-half the sum of the two apsidal distances r_1 and r_2 (cf. Fig. 3–6) and in terms of Eq. (3–47) will be given by

$$a = \frac{r_1 + r_2}{2} = \frac{1}{2C(1+\epsilon)} + \frac{1}{2C(1-\epsilon)} = \frac{1}{C}\frac{1}{1-\epsilon^2}. \tag{3–51}$$

With the constants of the orbit as given in (3–46) the formula for the semimajor axis becomes

$$a = -\frac{k}{2E}, \tag{3–52}$$

which, it may be noted, agrees with Eq. (3–50) for the radius of circular orbits.

As a final item in the discussion of the inverse square law of force, we shall evaluate the period of motion in elliptic orbits. From the conservation of angular momentum the areal velocity is constant, and is given by

$$\frac{dA}{dt} = \frac{1}{2}r^2\dot{\theta} = \frac{l}{2m}. \tag{3–53}$$

The area of the orbit, A, is to be found by integrating (3–53) over a complete period τ:

$$\int_0^\tau \frac{dA}{dt}\,dt = A = \frac{l\tau}{2m}.$$

Now, the area of an ellipse is

$$A = \pi ab,$$

where the semimajor axis a is given by (3–52), and by the definition of eccentricity the semiminor axis b is related to a according to the formula

$$b = a\sqrt{1-\epsilon^2}.$$

By (3–51) it is seen that the semiminor axis can also be written as

$$b = \sqrt{\frac{a}{C}} = a^{1/2}\sqrt{\frac{l^2}{mk}},$$

and the period is therefore

$$\tau = \frac{2m}{l}\pi a^{3/2}\sqrt{\frac{l^2}{mk}} = 2\pi a^{3/2}\sqrt{\frac{m}{k}}. \tag{3-54}$$

Eq. (3–54) states that, other things being equal, the square of the period is proportional to the cube of the major axis, and this conclusion is often referred to as the third of Kepler's laws.* Actually, Kepler was concerned with the specific problem of planetary motion in the gravitational field of the sun. A more precise statement of his law would therefore be: The square of the periods of the various planets are proportional to the cube of their major axes. In this form the law is only approximately true. It must be remembered that the motion of a planet about the sun is a two-body problem and m in (3–54) must be replaced by the reduced mass:

$$\mu = \frac{m_1 m_2}{m_1 + m_2},$$

where m_1 may be taken as referring to the planet and m_2 to the sun. Further, the gravitational law of attraction is

$$f = -G\frac{m_1 m_2}{r^2},$$

so that the constant k is

$$k = Gm_1 m_2. \tag{3-55}$$

Under these conditions (3–54) becomes

$$\tau = \frac{2\pi a^{3/2}}{\sqrt{G(m_1 + m_2)}} \approx \frac{2\pi a^{3/2}}{\sqrt{Gm_2}}, \tag{3-56}$$

if we neglect the mass of the planet compared to the sun. It is the approximate version of Eq. (3–56) which is Kepler's third law, for it states that τ is proportional to $a^{3/2}$, with the same constant of proportionality for all planets. However, the planetary mass m_1 is not always completely negligible compared to the sun's; for example, Jupiter has a mass about .1% of the mass of the sun. On the other hand, Kepler's third law is rigorously true for the electron orbits in the Bohr atom, since μ and k are then the same for all orbits in a given atom.

---

* Kepler's three laws of planetary motion, published around 1610, were the result of his pioneering analysis of planetary observations, and laid the groundwork for Newton's great advances. The second law, the conservation of areal velocity, is a general theorem for central force motion, as has been noted previously. However, the first — that the planets move in elliptical orbits about the sun at one focus — and the third are restricted specifically to the inverse square law of force.

3-7 Scattering in a central force field. Historically, the interest in central forces arose out of the astronomical problems of planetary motion. There is no reason, however, why central force motion must be thought of only in terms of such problems; mention has already been made of the orbits in the Bohr atom. Another field which can be investigated in terms of classical mechanics is the *scattering* of particles by central force fields. Of course, if the particles are on the atomic scale it must be expected that the specific results of a classical treatment will often be incorrect physically, for quantum effects are usually large in such regions. Nevertheless there are many classical predictions which remain valid to a good approximation. More important, the procedures for *describing* scattering phenomena are the same whether the mechanics is classical or quantum; one can learn to speak the language equally as well on the basis of classical physics.

In its one-body formulation the scattering problem is concerned with the scattering of particles by a *center of force.* We consider a uniform beam of particles — whether electrons, or α-particles or planets is irrelevant — all of the same mass and energy incident upon a center of force. It will be assumed that the force falls off to zero for very large distances. The incident beam is characterized by specifying its *intensity* I (also called flux density), which gives the number of particles crossing unit area normal to the beam in unit time. As a particle approaches the center of force it will either be attracted or repelled, and its orbit will deviate from the incident straight line trajectory. After passing the center of force, the force acting on the particle will eventually diminish so that the orbit once again approaches a straight line. In general the final direction of motion is not the same as the incident direction, and the particle is said to be scattered. The *cross section for scattering in a given direction*, $\sigma(\Omega)$, is defined by:

$$\sigma(\Omega)\, d\Omega = \frac{\text{number of particles scattered into solid angle } d\Omega \text{ per unit time}}{\text{incident intensity}}, \tag{3-57}$$

where $d\Omega$ is an element of solid angle in the direction Ω. Often $\sigma(\Omega)$ is also designated as the *differential scattering cross section.* With central forces there must be complete symmetry around the axis of the incident beam, hence the element of solid angle can be written

$$d\Omega = 2\pi \sin\Theta\, d\Theta, \tag{3-58}$$

where Θ is the angle between the scattered and incident directions, known as the *scattering angle.* It will be noted that the name "cross section" is deserved in that $\sigma(\Omega)$ has the dimensions of an area.

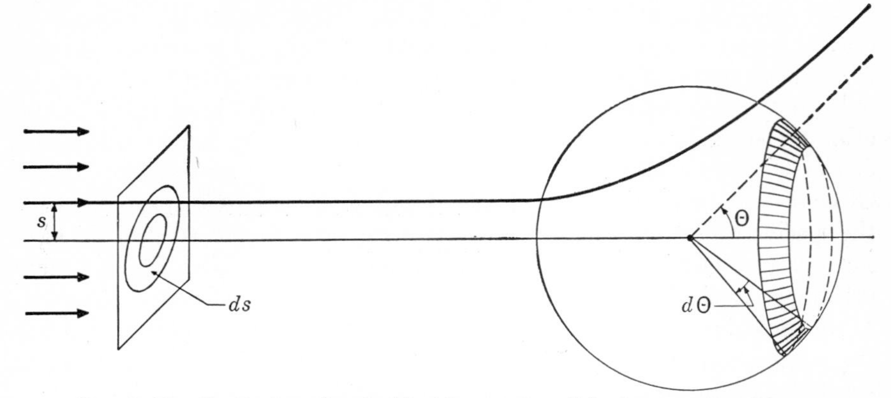

FIG. 3–13. Scattering of an incident beam of particles by a center of force.

For any given particle the constants of the orbit, and hence the amount of scattering, are determined by its energy and angular momentum. It is convenient to express the angular momentum in terms of the energy and the so-called *impact parameter*, s, defined as the perpendicular distance between the center of force and the incident velocity. If v_0 is the incident speed of the particle, then

$$l = mv_0 s = s\sqrt{2mE}. \tag{3-59}$$

Once E and s are fixed, the angle of scattering Θ is then determined uniquely.* Therefore, the number of particles scattered into a solid angle $d\Omega$ lying between Θ and $\Theta + d\Theta$ must be equal to the number of the incident particles with impact parameter lying between the corresponding s and $s + ds$:

$$2\pi Is\, ds = -2\pi\sigma(\Theta)I \sin(\Theta)\, d\Theta. \tag{3-60}$$

The minus sign is introduced in Eq. (3–60) because an increase ds in the impact parameter means less force is exerted on the particle, resulting in a decrease $d\Theta$ in the scattering angle. If s is considered as a function of the energy and the corresponding scattering angle:

$$s = s(\Theta, E), \tag{3-61}$$

then the dependence of the differential cross section on Θ is given by

$$\sigma(\Theta) = -\frac{s}{\sin\Theta}\frac{ds}{d\Theta}. \tag{3-62}$$

* It is at this point in the formulation that classical and quantum mechanics part company. Indeed it is fundamentally characteristic of quantum mechanics that one cannot unequivocally predict the trajectory of any particular particle. One can only give probabilities for scattering in various directions.

To illustrate the procedure we shall consider the historically important problem of scattering of charged particles by a coulomb field. The scattering force field is that produced by a fixed charge $-Ze$ acting on the incident particles having a charge $-Z'e$; so that the force can be written as

$$f = \frac{ZZ'e^2}{r^2},$$

i.e., a repulsive inverse square law of force. The results of the preceding section can be taken over here with no more change than writing the force constant as

$$k = -ZZ'e^2 \tag{3-63}$$

Then by (3-46) the equation for the orbit is simply

$$\frac{1}{r} = -\frac{mZZ'e^2}{l^2}(1 + \epsilon \cos \theta), \tag{3-64}$$

where the coordinates have been rotated so that $\theta' = 0$, and where ϵ is

$$\epsilon = \sqrt{1 + \frac{2El^2}{m(ZZ'e^2)^2}} = \sqrt{1 + \left(\frac{2Es}{ZZ'e^2}\right)^2}. \tag{3-65}$$

Eq. (3-64) still represents a conic section, in fact, a hyperbola, since $\epsilon > 1$. As a result of the minus sign, however, the values of θ for the orbit are restricted to angles such that

$$\cos \theta < -\frac{1}{\epsilon}, \tag{3-66}$$

cf. Fig. 3-14. While for the attractive forces the center of force is at the interior focus of the hyperbola, cf. Fig. 3-5, here the center of force lies at the exterior focus, as Fig. 3-15.

The change in θ as the particle comes in from infinity, is scattered and goes out again to infinity, is clearly the same as the angle between the asymptotes Φ, which in turn is the supplement of the scattering angle Θ. Then, by Eq. (3-66) and Fig. 3-14, Θ is given by

$$\cos \frac{\Phi}{2} = \sin \frac{\Theta}{2} = \frac{1}{\epsilon}$$

or

$$\cot^2 \frac{\Theta}{2} = \csc^2 \frac{\Theta}{2} - 1 = \epsilon^2 - 1,$$

and finally

$$\cot \frac{\Theta}{2} = \frac{2Es}{ZZ'e^2}. \tag{3-67}$$

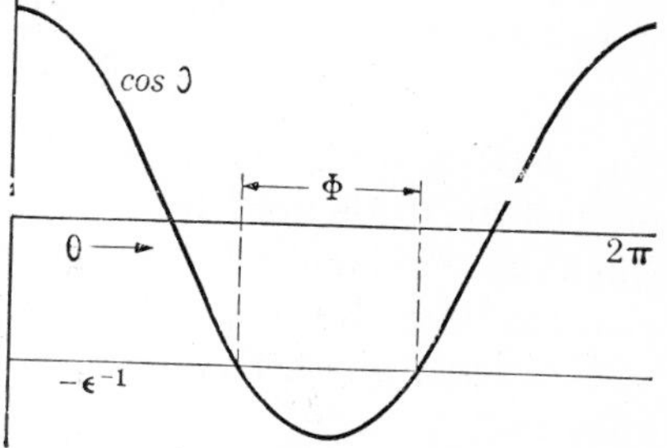

Fig. 3-14. Range of θ for repulsive coulomb scattering.

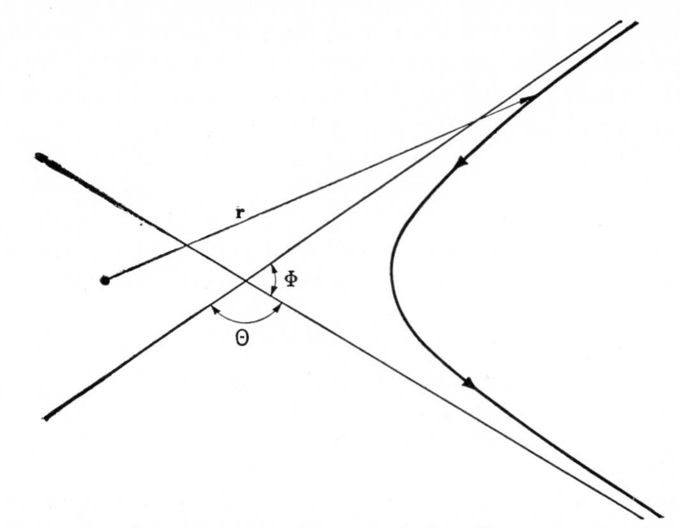

FIG. 3–15. Orbit for repulsive coulomb scattering, illustrating the connection between the angle between the asymptotes and the scattering angle.

With this result the differential cross section can easily be obtained. From Eq. (3–67) the impact parameter is given in terms of E and Θ as

$$s = \frac{ZZ'e^2}{2E} \cot \frac{\Theta}{2},$$

so that $\sigma(\Theta)$ becomes, by (3–62),

$$\sigma(\Theta) = \frac{1}{2}\left(\frac{ZZ'e^2}{2E}\right)^2 \frac{\cot \frac{\Theta}{2}}{\sin \Theta} \frac{1}{\sin^2 \frac{\Theta}{2}}$$

or

$$\sigma(\Theta) = \frac{1}{4}\left(\frac{ZZ'e^2}{2E}\right)^2 \frac{1}{\sin^4 \frac{\Theta}{2}}. \tag{3–68}$$

Eq. (3–68) gives the famous Rutherford scattering cross section, originally derived by Rutherford for the scattering of α particles by atomic nuclei. Quantum mechanics in the nonrelativistic limit yields a cross section identical with this classical result.

In atomic physics the concept of a *total scattering cross section* σ_t, defined as

$$\sigma_t = \int_{4\pi} \sigma(\Omega)\, d\Omega = 2\pi \int_0^{\pi} \sigma(\Theta) \sin \Theta \, d\Theta,$$

is of considerable importance. However, if we attempt to calculate the total cross section for coulomb scattering by substituting Eq. (3–68) in this definition we obtain an infinite result! The physical reason behind this behavior is not far to seek. From its definition the total cross section is the number of particles scattered in all directions per unit time for unit incident intensity. Now, the coulomb field is an example of a "long range" force; its effects extend to infinity. The very small deflections occur only for particles with very large impact parameters. Hence all particles in an incident beam of infinite cross section will be scattered to some extent and must be included in the total scattering cross section. It is clear, therefore, that the infinite value for σ_t is not peculiar to the coulomb field; it occurs in classical mechanics whenever the scattering field is different from zero at all distances no matter how large.* Only if the force field "cuts off," i.e., is zero beyond a certain distance, will the scattering cross section be finite. Physically, such a cut-off occurs for the coulomb field of a nucleus as a result of the presence of the atomic electrons, which "screen" the nucleus and effectively cancel its charge at large distances.

3–8 Transformation of the scattering problem to laboratory coordinates. The previous section has been concerned with the one-body problem of the scattering of a particle by a fixed center of force. In practice the scattering always involves two bodies, e.g., in Rutherford scattering we have the α particle and the atomic nucleus. The second particle is not fixed, but recoils from its initial position as a result of the scattering. Since it has been shown that any two-body central force problem can be reduced to a one-body problem it might be thought that the only change is to replace m by the reduced mass μ. However, the matter is not quite that simple. The scattering angle actually measured in the laboratory, which we shall denote by ϑ, is the angle between the final and incident directions of the scattered particle.† On the other hand, the angle Θ calculated from the equivalent one-body problem is the angle between the final and initial directions of the relative vector between the two particles. These two angles would be the same only if the second particle remains stationary

---

* σ_t is also infinite for the coulomb field in quantum mechanics, since it has been stated that Eq. (3–68) remains valid there. However, not all "long range" forces give rise to infinite total cross sections in quantum mechanics. It turns out that all potentials which fall off faster at larger distances than $1/r^2$ produce a finite quantum-mechanical total scattering cross section.

† The scattering angle ϑ must not be confused with the angle coordinate θ of the relative vector, $\mathbf{r}$, between the two particles.

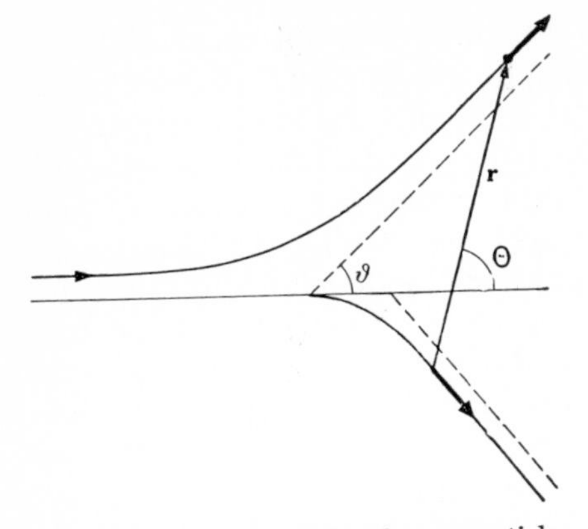

FIG. 3–16. Scattering of two particles as viewed in the laboratory system.

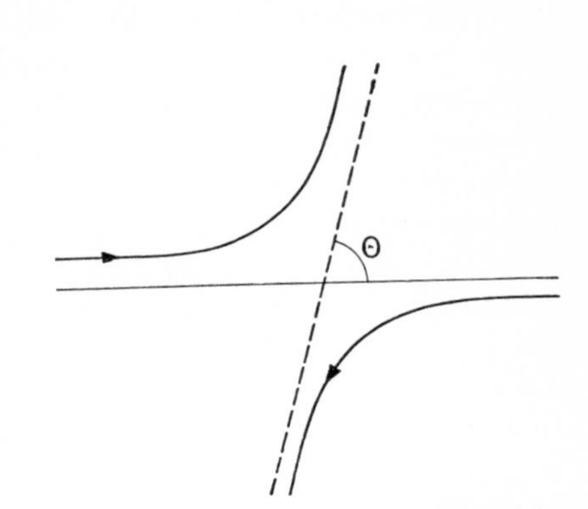

FIG. 3–17. Scattering of two particles as viewed in the center of mass system.

through the scattering process. In general, however, the second particle, though initially at rest, is itself set in motion by the mutual force between the two particles, and, as is indicated in Fig. 3–16, the two angles then have different values. The equivalent one-body problem thus does not directly furnish the scattering angle as measured in the laboratory coordinate system.

However, the situation is quite different when viewed in a coordinate system moving with the center of mass of both particles. In such a system the total linear momentum is zero, of course, and the two particles always move with equal and opposite momenta. Fig. 3–17 illustrates the appearance of the scattering process to an observer in the center of mass system. Before the scattering the particles are moving directly towards each other; after, they are moving directly away from each other. The angle between the initial and final directions of the relative vector, Θ, must therefore be the same as the scattering angle of either particle in the center of mass system. The connection between the two scattering angles Θ and ϑ can thus be obtained by considering the transformation between the center of mass system and the laboratory system. Following the terminology of Section 3–1:

- $\mathbf{r}_1$ and $\mathbf{v}_1$ are the position and velocity of the incident particle 1 in the laboratory system,
- $\mathbf{r}'_1$ and $\mathbf{v}'_1$ are the position and velocity of particle 1 in the center of mass system, and
- $\mathbf{R}$ and $\dot{\mathbf{R}}$ are the position and (constant) velocity of the center of mass in the laboratory system.

At any instant, by definition

$$\mathbf{r}_1 = \mathbf{R} + \mathbf{r}_1',$$

and consequently

$$\mathbf{v}_1 = \dot{\mathbf{R}} + \mathbf{v}_1'. \tag{3–68}$$

Fig. 3–18 graphically portrays this vector relation evaluated *after* the scattering has taken place; at which time $\mathbf{v}_1$ and $\mathbf{v}_1'$ make the angles ϑ and Θ respectively with the vector $\dot{\mathbf{R}}$ lying along the initial direction. From the diagram we readily obtain the equation

$$\tan\vartheta = \frac{v_1' \sin\Theta}{v_1' \cos\Theta + \dot{R}}. \tag{3–69}$$

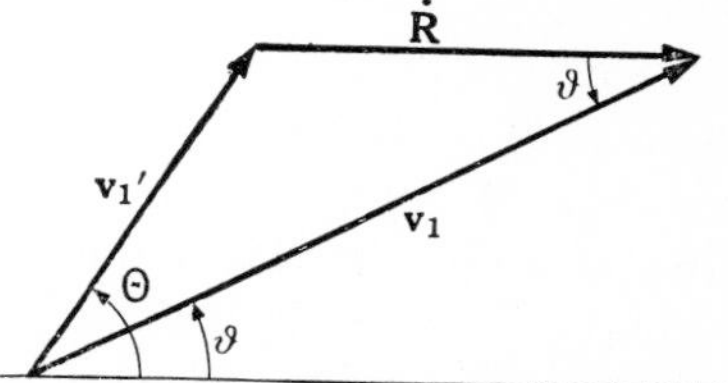

Fig. 3–18. The relations between the velocities in the center of mass and laboratory coordinates.

Now, by Eq. (3–2) $\mathbf{v}_1'$ is related to the relative velocity by

$$\mathbf{v}_1' = \frac{\mu}{m_1}\dot{\mathbf{r}}.$$

Since the system is conservative the relative velocity after scattering, i.e., when the two particles are no longer in each other's potential, must have the same magnitude as the initial velocity $\mathbf{v}_0$. Hence, after scattering,

$$v_1' = \frac{\mu}{m_1} v_0. \tag{3–70}$$

The constant velocity of the center of mass can be determined by the conservation theorem for total linear momentum:

$$(m_1 + m_2)\dot{\mathbf{R}} = m_1\mathbf{v}_0$$

or

$$\dot{R} = \frac{\mu}{m_2} v_0. \tag{3–71}$$

Substituting (3–70) and (3–71) in (3–69) the relation between the two scattering angles becomes

$$\tan\vartheta = \frac{\sin\Theta}{\cos\Theta + \dfrac{m_1}{m_2}}. \tag{3–72}$$

When m_1 is much less than m_2 the two angles are approximately equal; the massive scatterer m_2 suffers little recoil and acts practically as a fixed center of mass.

Because of the transformation of scattering angles, the scattering cross sections will be different in the center of mass and laboratory coordinates. The connection between the two is obtained from the fact that the number of particles scattered into a given element of solid angle must be the same in both systems:

$$2\pi I \sigma(\Theta) \sin \Theta \, d\Theta = 2\pi I \sigma'(\vartheta) \sin \vartheta \, d\vartheta$$

or

$$\sigma'(\vartheta) = \sigma(\Theta) \frac{\sin \Theta}{\sin \vartheta} \frac{d\Theta}{d\vartheta} = \sigma(\Theta) \frac{d \cos \Theta}{d \cos \vartheta}, \tag{3-73}$$

where $\sigma'(\vartheta)$ is the cross section in the laboratory system. In principle at least, Eq. (3-72) can be solved for Θ in terms of ϑ, and $\sigma'(\vartheta)$ evaluated in terms of $\sigma(\Theta)$; but this is quite a job for arbitrary m_1/m_2 ratio. For the Rutherford α-particle scattering it is clear the corrections are small, since m_1 is 4 atomic units and m_2 is usually 100 atomic units or larger. But if the two masses are equal, as in neutron-proton scattering, then the effects are at a maximum and quite large. In this case Eq. (3-72) becomes

$$\tan \vartheta = \frac{\sin \Theta}{\cos \Theta + 1} = \tan \frac{\Theta}{2},$$

so that

$$\vartheta = \frac{\Theta}{2}. \qquad (m_1/m_2 = 1). \tag{3-74}$$

With equal masses the maximum scattering angle in the laboratory system is 90° — there can be no scattering back of that! The scattering cross section in the laboratory system is then:

$$\sigma'(\vartheta) = 4 \cos \vartheta \, \sigma(2\vartheta). \qquad (m_1/m_2 = 1). \tag{3-75}$$

The scattering described here is *elastic* in the sense that the total kinetic energy before and after scattering remains the same. Nevertheless the speeds of the particles in the laboratory system will not remain unchanged. Thus the *scattering* particle is initially at rest, but as a result of the scattering undergoes a recoil and in general ends up with a finite velocity and kinetic energy. To maintain the total kinetic energy constant the *scattered* particle must suffer a decrease of velocity and kinetic energy. The scattering process thus results in a transfer of kinetic energy from the incident particle to the scattering particle. Mathematically, the decrease can be calculated from Fig. 3-18 by the cosine law:

$$v_1'^2 = v_1^2 + \dot{R}^2 - 2v_1\dot{R} \cos \vartheta,$$

or, using (3-70) and (3-71):

$$\left(\frac{v_1}{v_0}\right)^2 - \frac{2\mu}{m_2}\left(\frac{v_1}{v_0}\right)\cos\vartheta - \frac{m_2 - m_1}{m_2 + m_1} = 0, \tag{3-76}$$

which is a quadratic equation for v_1/v_0. In the special case $m_1 = m_2$ the solution is extremely simple:

$$\frac{v_1}{v_0} = \cos\vartheta, \qquad (m_1/m_2 = 1).$$

Thus for $\vartheta = 90°$, corresponding to back scattering ($\Theta = \pi$) in the center of mass system, there is maximum transfer, and the recoil particle takes all the incident energy.

This transfer of kinetic energy by scattering is, of course, the principle behind the "moderator" in a slow neutron pile. Fast neutrons produced by fission make successive elastic scatterings until their kinetic energy is reduced to the level at which the neutron is more liable to cause fission than to be captured. Clearly the best moderators will be the light elements, ideally hydrogen. For a pile hydrogen is not practical (because it captures neutrons) but deuterium, of mass 2, and carbon, of mass 12, serve as approximations. In the laboratory, however, hydrogen in the form of paraffin hydrocarbons is constantly used to slow down neutrons.

Despite their very up-to-date applications these calculations of the transformation from laboratory to center of mass coordinates, and of the transfer of kinetic energy, are not particularly "modern" or "quantum" in nature. Nor is the classical mechanics involved particularly advanced or difficult. All that has been used, essentially, is the conservation of momentum and energy. Indeed, similar calculations may be found in freshman textbooks, usually in terms of elastic collisions between, say, billiard balls. But it is their very elementary nature which results in the widespread validity of these calculations. So long as momentum is conserved (and this will be true in quantum mechanics) and the scattering is elastic, the details of the scattering process are irrelevant. In effect the vicinity of the scattering particle is a "black box," and we are concerned only with what goes in and what comes out. It matters not at all whether the phenomena occurring inside the box are "classical" or "quantum." Consequently the formulae of this section may be used without fear in the experimental analysis of phenomena essentially quantum in nature, as for example, neutron-proton scattering.

CHAPTER 3

SUGGESTED REFERENCES

E. T. WHITTAKER, *Analytical Dynamics.* Almost every text on mechanics devotes considerable time to central force motion and only a few of the many references can be listed here. Sections 47–49 of Whittaker's treatise form a concise introduction to the subject on an advanced level.

W. D. MACMILLAN, *Statics and the Dynamics of a Particle.* Chapter XII of this reference provides a most elaborate discussion of central force motion, including detailed consideration of the orbits for some force laws other than the customary inverse square law. The treatment is elementary and does not use the Lagrangian formulation.

J. C. SLATER AND N. H. FRANK, *Mechanics.* The qualitative discussion of central force motion by means of an equivalent potential including the centrifugal barrier is quite common in modern physics, but it is rarely discussed in connection with classical mechanics. A notable exception is this text of Slater and Frank, and the reader will find many interesting applications of the method in Chapters III and IV.

A. SOMMERFELD, *Vorlesungen über Theoretische Physik, Bd. 1, Mechanik.* The kinematics of scattering phenomena is common knowledge amongst most physicists, but it is difficult to give any comprehensive reference to the subject. Many texts on atomic or nuclear physics discuss Rutherford scattering and a few consider the scattering of particles of equal mass. The rest of the material appears dispersed in individual papers, chiefly on nuclear research. In Section 3, part 5 of his book, Sommerfeld presents an interesting discussion of collisions based solely on the conservation of momentum and energy, and devotes a brief notice to inelastic collisions. The exercises to this section are of especial value.

R. B. LINDSAY, *Introduction to Physical Statistics.* A brief discussion of the virial theorem and its use in deriving the equation of state for ideal and real gases is presented in Chapter V. Reference may also be made to the monumental treatise on statistical mechanics by R. H. Fowler.

EXERCISES

1. Two particles move about each other in circular orbits under the influence of gravitational forces, with a period τ. Their motion is suddenly stopped at a given instant of time, and they are then released and allowed to fall into each other. Prove that they collide after a time $\tau/4\sqrt{2}$.

2. A particle moves in a central force field given by the potential

$$V = -k\frac{e^{-ar}}{r},$$

where k and a are positive constants. Discuss qualitatively the nature of the motion, using the method of the equivalent one-dimensional potential.

3. Consider a system in which the total forces acting on the particles consist of conservative forces $\mathbf{F}_i'$ and frictional forces $\mathbf{f}_i$ proportional to the velocity. Show that for such a system the virial theorem holds in the form

$$\overline{T} = -\frac{1}{2}\overline{\sum_i \mathbf{F}_i' \cdot \mathbf{r}_i},$$

providing the motion reaches a steady state and is not allowed to die down as a result of the frictional forces.

4. Show that if a particle describes a circular orbit under the influence of an attractive central force directed toward a point on the circle, then the force varies as the inverse fifth power of the distance.

5. Show that the central force problem is soluble in terms of elliptic functions when the force is a power law function of the distance with the following fractional exponents:

$$n = -\frac{3}{2}, \quad -\frac{5}{2}, \quad -\frac{1}{3}, \quad -\frac{5}{3}, \quad -\frac{7}{3}.$$

6. Evaluate approximately the ratio of the mass of the sun to that of the earth, using only the lengths of the year and of the lunar month (27.3 days), and the mean radii of the earth's orbit (1.49×10^8 km) and of the moon's orbit (3.8×10^5 km).

7. Discuss the motion of a particle in a central force field

$$f = -\frac{k}{r^2} + \frac{C}{r^3}.$$

In particular show that the equation of the orbit can be put in the form

$$r = \frac{a(1 - \epsilon^2)}{1 + \epsilon \cos \alpha\theta},$$

which is an ellipse for $\alpha = 1$, but is a *precessing* ellipse if $\alpha \neq 1$. The precessing motion may be described in terms of the *rate of precession of the perihelion,* where the term perihelion is used loosely to denote any one of the turning points of the orbit. Derive an approximate expression for the rate of precession when α is close to unity, in terms of the dimensionless quantity

$$\eta = \frac{C}{ka}.$$

The ratio η is a measure of the strength of the perturbing inverse cube term relative to the main inverse square term. The perihelion of Mercury is observed to precess at the rate of 40″ of arc per century. Show that this precession could be accounted for classically if the value of η were as small as 1.42×10^{-7}. (The eccentricity of Mercury's orbit is 0.206, and its period is 0.24 year.)

8. What changes, if any, would there be in Rutherford scattering if the coulomb force were attractive, instead of repulsive?

9. Examine the scattering produced by a repulsive central force $f = kr^{-3}$. Show that the differential cross section is given by

$$\sigma(\Theta)\, d\Theta = \frac{k}{2E} \frac{(1 - x)\, dx}{x^2(2 - x)^2 \sin \pi x},$$

where x is the ratio Θ/π and E is the energy.

10. A central force potential frequently encountered in nuclear physics is the so-called *rectangular well*, defined by the potential:

$$\begin{aligned} V &= 0, & r &> a \\ &= -V_0 & r &\leq a. \end{aligned}$$

Show that the scattering produced by such a potential in classical mechanics is identical with the refraction of light rays by a sphere of radius a and relative index of refraction

$$n = \sqrt{\frac{E + V_0}{E}}.$$

(This equivalence demonstrates why it was possible to explain refraction phenomena both by Huygens' waves and by Newton's mechanical corpuscles.) Show also that the differential cross section is

$$\sigma(\Theta) = \frac{n^2a^2}{4\cos\frac{\Theta}{2}} \frac{\left(n\cos\frac{\Theta}{2} - 1\right)\left(n - \cos\frac{\Theta}{2}\right)}{\left(1 + n^2 - 2n\cos\frac{\Theta}{2}\right)^2}.$$

What is the total cross section?

11. Show that for any repulsive central force a formal solution for the angle of scattering is given by

$$\Theta = \pi + 2\int_0^{u_0} \frac{s\,du}{\sqrt{1 - \frac{V(u)}{E} - s^2u^2}},$$

where V is the potential energy, $u = 1/r$, and u_0 corresponds to the turning point of the orbit. What is the corresponding expression for attractive potentials?

12. a. Show that the angle of recoil of the scattering particle relative to the incident direction of the scattered particle is simply $\phi = \frac{1}{2}(\pi - \Theta)$.

b. Consider a scattering system of two particles of equal mass. It is observed that the energy distribution of the recoil particles is constant up to a certain energy and zero above. Show that the scattering must be isotropic in the center of mass system.

13. For elliptical central force motion in an inverse-square-law attractive force, show by use of the virial theorem and Eq. (3–54) that

$$\oint p_r\,dr + \oint p_\theta\,d\theta = \pi k\sqrt{\frac{2m}{-E}}.$$

In the Bohr quantum model of the hydrogen atom $k = e^2$, and the circuital integrals of the conjugate momenta are quantized with the values

$$\oint p_i\,dq_i = n_ih, \quad n_i \text{ integral},$$

where h is Planck's constant. The energy levels on this model are therefore given by

$$E = -\frac{2\pi^2me^4}{(n_r + n_\theta)^2h^2}.$$

CHAPTER 4

THE KINEMATICS OF RIGID BODY MOTION

A rigid body was defined previously as a system of mass points subject to the holonomic constraints that the distances between all pairs of points remain constant throughout the motion. Although something of an idealization, the concept is quite useful, and the mechanics of rigid body motion deserves a full exposition. In this chapter we shall discuss principally the *kinematics* of rigid bodies, i.e., the nature and characteristics of their motions. We shall devote some time to developing the mathematical techniques involved, which are of considerable interest in themselves, and have many important applications to other fields of physics. Having learned how to describe the motion of rigid bodies, the next chapter will then discuss, within the framework of the Lagrangian formulation, how such motion is generated by applied forces and torques.

4–1 The independent coordinates of a rigid body. Before discussing the motion of a rigid body we must first establish how many independent coordinates are necessary to specify its configuration. A rigid body with N particles can at most have $3N$ degrees of freedom, but these are greatly reduced by the constraints, which can be expressed as equations of the form

$$r_{ij} = c_{ij}. \tag{4–1}$$

Here r_{ij} is the distance between the ith and jth particles and the c's are constants. The actual number of degrees of freedom cannot be obtained simply by subtracting the number of constraint equations from $3N$, for there are $\frac{1}{2}N(N-1)$ possible equations of the form of Eq. (4–1), which exceeds $3N$ for large N. In truth, the Eqs. (4–1) are not all independent. To fix a point in the rigid body it is not necessary to specify its distances to *all* other points in the body; one need only state the distances to any three other noncollinear points, cf. Fig. 4–1.

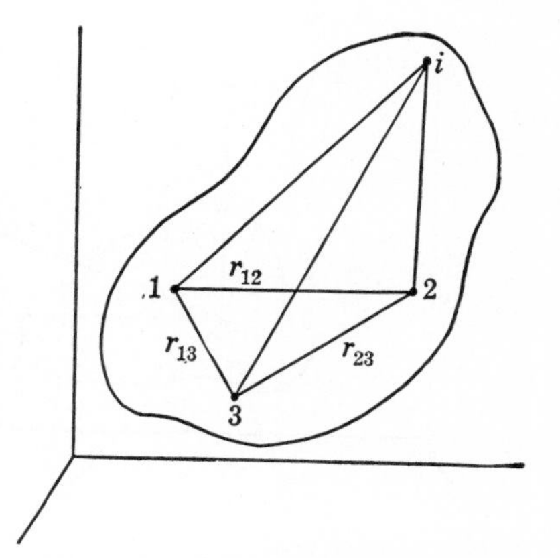

FIG. 4–1. Diagram illustrating the location of a point in a rigid body by its distances from three reference points.

Thus, once the positions of three of the particles of the rigid body are determined the constraints fix the positions of all remaining particles. The number of degrees of freedom therefore cannot be more than nine. But the three reference points are themselves not independent, there are in fact three equations of rigid constraint imposed on them:

$$r_{12} = c_{12} \qquad r_{23} = c_{23} \qquad r_{13} = c_{13},$$

which reduce the number of degrees of freedom to *six*. That only six coordinates are needed can also be seen from the following considerations. To establish the position of one of the reference points three coordinates must be supplied. But once point 1 is fixed, point 2 can be specified by only two coordinates, since it is constrained to move on the surface of a sphere centered at point 1. With these two points determined point 3 has only one degree of freedom, for it can only rotate about the axis joining the other two points. Hence a total of six coordinates is sufficient.

A rigid body in space thus needs six independent generalized coordinates to specify its configuration, no matter how many particles it may contain — even in the limit of a continuous body. Of course, there may be additional constraints on the body besides the constraint of rigidity. For example, the body may be constrained to move on a surface, or with one point fixed. In such case the additional constraints will further reduce the number of degrees of freedom, and hence the number of independent coordinates.

How shall these coordinates be assigned? It will be noticed that the configuration of a rigid body is completely specified by locating a cartesian set of coordinates fixed in the rigid body (the primed axes shown in Fig. 4–2)

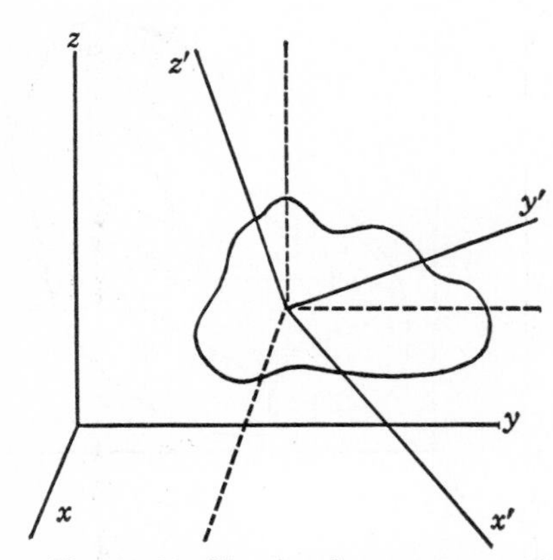

FIG. 4–2. Unprimed axes represent an external reference set of axes; the primed axes are fixed in the rigid body.

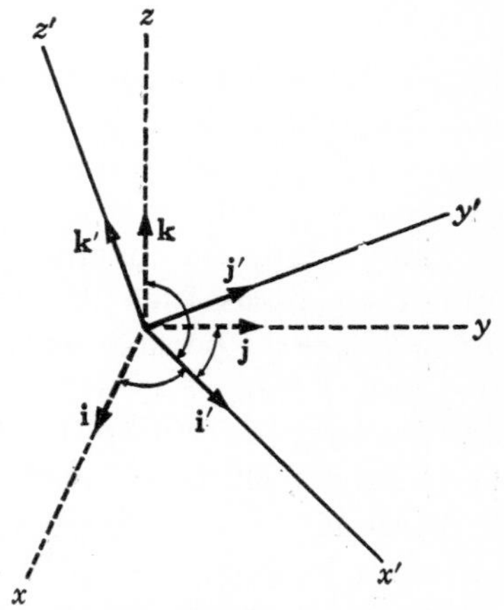

FIG. 4–3. Direction cosines of the body set of axes relative to an external set of axes.

relative to the coordinate axes of the external space. Clearly three of the coordinates are needed to specify the coordinates of the origin of this "body" set of axes. The remaining three coordinates must then specify the orientation of the primed axes relative to a coordinate system parallel to the external axes, but with the same origin as the primed axes.

There are many ways of specifying the orientation of a cartesian set of axes relative to another set with common origin. A most fruitful procedure is to state the direction cosines of the primed axes relative to the unprimed. Thus the x' axis could be specified by its three direction cosines α_1, α_2, α_3 with respect to the x, y, z axes. If, as customary, $\mathbf{i}$, $\mathbf{j}$, $\mathbf{k}$ are three unit vectors along x, y, z, and $\mathbf{i}'$, $\mathbf{j}'$, $\mathbf{k}'$ perform the same function in the primed system, then these direction cosines are defined as

$$\begin{aligned} \alpha_1 &= \cos(\mathbf{i}', \mathbf{i}) = \mathbf{i}' \cdot \mathbf{i} \\ \alpha_2 &= \cos(\mathbf{i}', \mathbf{j}) = \mathbf{i}' \cdot \mathbf{j} \\ \alpha_3 &= \cos(\mathbf{i}', \mathbf{k}) = \mathbf{i}' \cdot \mathbf{k}. \end{aligned} \tag{4–2}$$

The vector $\mathbf{i}'$ can be expressed in terms of $\mathbf{i}$, $\mathbf{j}$, $\mathbf{k}$ by the relation

$$\mathbf{i}' = (\mathbf{i}' \cdot \mathbf{i})\mathbf{i} + (\mathbf{i}' \cdot \mathbf{j})\mathbf{j} + (\mathbf{i}' \cdot \mathbf{k})\mathbf{k}$$

or

$$\mathbf{i}' = \alpha_1\mathbf{i} + \alpha_2\mathbf{j} + \alpha_3\mathbf{k}. \tag{4–3}$$

Similarly the direction cosines of the y' axis with x, y, z may be designated by β_1, β_2 and β_3, and these will be the components of $\mathbf{j}'$ in the unprimed reference frame:

$$\mathbf{j}' = \beta_1\mathbf{i} + \beta_2\mathbf{j} + \beta_3\mathbf{k}. \tag{4–4}$$

An equation analogous to (4–4) can be written for $\mathbf{k}'$, with the direction cosines of the z' axis designated by γ's. These sets of nine direction cosines then completely specify the orientation of the x', y', z' axes relative to the x, y, z set. One can equally well invert the process, and use the direction cosines to express the $\mathbf{i}$, $\mathbf{j}$, $\mathbf{k}$ unit vectors in terms of their components along the primed axes. Thus we can write

$$\mathbf{i} = (\mathbf{i} \cdot \mathbf{i}')\mathbf{i}' + (\mathbf{i} \cdot \mathbf{j}')\mathbf{j}' + (\mathbf{i} \cdot \mathbf{k}')\mathbf{k}'$$

or

$$\mathbf{i} = \alpha_1\mathbf{i}' + \beta_1\mathbf{j}' + \gamma_1\mathbf{k}', \tag{4–5}$$

with analogous equations for $\mathbf{j}$ and $\mathbf{k}$.

The direction cosines also furnish directly the relations between the coordinates of a given point in one system and the coordinates in the other system. Thus, the coordinates of a point in a given reference frame are the components of the position vector, $\mathbf{r}$, along the axes of the system. The x' coordinate is then given in terms of x, y, z by

$$x' = (\mathbf{r} \cdot \mathbf{i}') = \alpha_1 x + \alpha_2 y + \alpha_3 z,$$

while for the other coordinates we obtain

$$\begin{aligned} y' &= \beta_1 x + \beta_2 y + \beta_3 z \\ z' &= \gamma_1 x + \gamma_2 y + \gamma_3 z. \end{aligned} \tag{4–6}$$

What has been done here for the components of the **r** vector can obviously be done for any arbitrary vector. If **G** is some vector, then the component of **G** along the x' axis will be related to its x, y, z components by

$$G_{x'} = (\mathbf{G} \cdot \mathbf{i}') = \alpha_1 G_x + \alpha_2 G_y + \alpha_3 G_z,$$

and so on. The set of nine direction cosines thus completely spells out the transformation between the two coordinate systems.

If the primed axes are taken as fixed in the body, then the nine direction cosines will be functions of time as the body changes its orientation in the course of the motion. In this sense the α's, β's and γ's can be considered as coordinates describing the instantaneous orientation of the body. But, clearly, they are not independent coordinates, for there are nine of them and it has been shown that only three coordinates are needed to specify an orientation.

The connections between the direction cosines arise from the fact that the basis vectors in both coordinate systems are orthogonal to each other and have unit magnitude; in symbols:

$$\mathbf{i} \cdot \mathbf{j} = \mathbf{j} \cdot \mathbf{k} = \mathbf{k} \cdot \mathbf{i} = 0,$$

and

$$\mathbf{i} \cdot \mathbf{i} = \mathbf{j} \cdot \mathbf{j} = \mathbf{k} \cdot \mathbf{k} = 1, \tag{4–7}$$

with similar relations for $\mathbf{i}'$, $\mathbf{j}'$, and $\mathbf{k}'$. We can obtain the conditions satisfied by the nine coefficients by forming all possible dot products among the three equations for **i**, **j**, and **k** in terms of $\mathbf{i}'$, $\mathbf{j}'$, and $\mathbf{k}'$ (as in Eq. 4–5), making use of the Eqs. (4–7):

$$\begin{aligned} \alpha_l \alpha_m + \beta_l \beta_m + \gamma_l \gamma_m &= 0, \qquad l, m = 1, 2, 3; l \neq m, \\ \alpha_l^2 + \beta_l^2 + \gamma_l^2 &= 1, \qquad l = 1, 2, 3. \end{aligned} \tag{4–8}$$

These two sets of three equations each are exactly sufficient to reduce the number of independent quantities from nine to three. Formally, the six equations can be combined into one by using the Kronecker δ-symbol δ_{lm}, defined by

$$\begin{aligned} \delta_{lm} &= 1 \qquad l = m \\ &= 0 \qquad l \neq m. \end{aligned}$$

Eqs. (4–8) can then be written as:

$$\alpha_l \alpha_m + \beta_l \beta_m + \gamma_l \gamma_m = \delta_{lm}. \tag{4–9}$$

It is not possible, therefore, to set up a Lagrangian and subsequent equations of motion with the nine direction cosines as generalized coordinates. For this purpose we must use some set of three independent functions of the direction cosines. A number of such sets of independent variables will be described later, the most important being the Euler angles. The use of direction cosines to describe the connections between two cartesian coordinate systems nevertheless has a number of important advantages. With their aid many of the theorems about the motion of rigid bodies can be expressed with great elegance and generality, and in a form naturally leading to the procedures necessarily used in special relativity and quantum mechanics. Such a mode of description therefore merits an extended discussion here.

4-2 Orthogonal transformations. To better study the properties of the nine direction cosines it is convenient to change the notation and denote all coordinates by x, distinguishing the axes by subscripts:

$$\begin{aligned} x &\rightarrow x_1 \\ y &\rightarrow x_2 \\ z &\rightarrow x_3. \end{aligned} \tag{4-10}$$

Thus the Eqs. (4-6) become

$$\begin{aligned} x_1' &= \alpha_1 x_1 + \alpha_2 x_2 + \alpha_3 x_3 \\ x_2' &= \beta_1 x_1 + \beta_2 x_2 + \beta_3 x_3 \\ x_3' &= \gamma_1 x_1 + \gamma_2 x_2 + \gamma_3 x_3. \end{aligned} \tag{4-11}$$

Eqs. (4-11) constitute a group of transformation equations from a set of coordinates x_1, x_2, x_3 to a new set x_1', x_2', x_3'. In particular they form an example of a *linear* or *vector* transformation, defined by transformation equations of the form

$$\begin{aligned} x_1' &= a_{11}x_1 + a_{12}x_2 + a_{13}x_3 \\ x_2' &= a_{21}x_1 + a_{22}x_2 + a_{23}x_3 \\ x_3' &= a_{31}x_1 + a_{32}x_2 + a_{33}x_3, \end{aligned} \tag{4-12}$$

where the $a_{11}, a_{12}, \ldots$ are any set of constant (independent of x, x') coefficients.* The Eqs. (4-11) are only a special case of (4-12) since the direction cosines are not all independent.

The connections between the coefficients, Eqs. (4-8) may be rederived here in terms of the newer notation. Since both coordinate systems are cartesian, the magnitude of a vector is given in terms of the sum of squares

* Eqs. (4-12), of course, are not the most general set of transformation equations, cf., for example, those from the **r**'s to the q's (1-36).

of the components. Further, since the actual vector remains unchanged no matter which coordinate system is used, the magnitude of the vector must be the same in both systems. In symbols we can state the invariance of the magnitude as:

$$\sum_{i=1}^{3} x_i'^2 = \sum_{i=1}^{3} x_i^2. \tag{4-13}$$

Now, Eqs. (4-12) can be written in brief form as

$$x_i' = \sum_{j=1}^{3} a_{ij}x_j \qquad i = 1, 2, 3. \tag{4-14}$$

The left-hand side of Eq. (4-13) is therefore

$$\sum_{i=1}^{3}\left(\sum_{j=1}^{3} a_{ij}x_j\right)\left(\sum_{k=1}^{3} a_{ik}x_k\right) = \sum_{i=1}^{3}\sum_{j,k=1}^{3} a_{ij}a_{ik}x_jx_k.$$

By rearranging the summations this expression can be written as

$$\sum_{j,k}\left(\sum_{i} a_{ij}a_{ik}\right) x_jx_k,$$

and it will reduce to the right-hand side of Eq. (4-13) if, and only if

$$\begin{aligned}\sum_i a_{ij}a_{ik} &= 1 \qquad j = k\\ &= 0 \qquad j \neq k,\end{aligned}$$

or, in a more compact form, if

$$\sum_i a_{ij}a_{ik} = \delta_{jk} \qquad j, k = 1, 2, 3. \tag{4-15}$$

When the a_{ij} coefficients are expressed in terms of the α, β, γ's the six equations contained in Eq. (4-15) become identical with the Eqs. (4-9).

Any linear transformation, (4-12), which has the properties required by Eq. (4-15) is called an *orthogonal* transformation, and Eq. (4-15) itself is known as the *orthogonality condition.* Thus, the transition from coordinates fixed in space to coordinates fixed in the rigid body is accomplished by means of an orthogonal transformation. The array of transformation quantities (the direction cosines), written as

$$\begin{pmatrix} a_{11} & a_{12} & a_{13}\\ a_{21} & a_{22} & a_{23}\\ a_{31} & a_{32} & a_{33}\end{pmatrix}, \tag{4-16}$$

is called the *matrix of transformation*, and will be denoted by a capital letter A. The quantities a_{ij} are correspondingly known as the *matrix elements* of the transformation.

To make these formal considerations more meaningful consider the simple example of motion in a plane, so that we are restricted to two dimensional coordinate systems. Then the indices in the above relations can only take on the values 1, 2, and the transformation matrix reduces to the form

$$\begin{pmatrix} a_{11} & a_{12} \\ a_{21} & a_{22} \end{pmatrix}.$$

The four matrix elements are connected by three orthogonality conditions:

$$\sum_{i=1}^{2} a_{ij}a_{ik} = \delta_{jk} \qquad j, k = 1, 2,$$

so that only one independent parameter is needed to specify the transformation. But this conclusion is not surprising. A two-dimensional transformation from one cartesian coordinate system to another corresponds to a rotation of the axes in the plane, cf. Fig. 4-4, and such a rotation can be specified completely by only one quantity, the rotation angle ϕ. Expressed in terms of this single parameter, the transformation equations become

$$\begin{aligned} x_1' &= x_1 \cos\phi + x_2 \sin\phi \\ x_2' &= -x_1 \sin\phi + x_2 \cos\phi. \end{aligned}$$

The matrix elements are therefore:

$$\begin{aligned} a_{11} &= \cos\phi, & a_{12} &= \sin\phi, \\ a_{21} &= -\sin\phi, & a_{22} &= \cos\phi, \end{aligned} \qquad (4\text{-}17)$$

so that the matrix A can be written:

$$\mathsf{A} = \begin{pmatrix} \cos\phi & \sin\phi \\ -\sin\phi & \cos\phi \end{pmatrix}. \qquad (4\text{-}17')$$

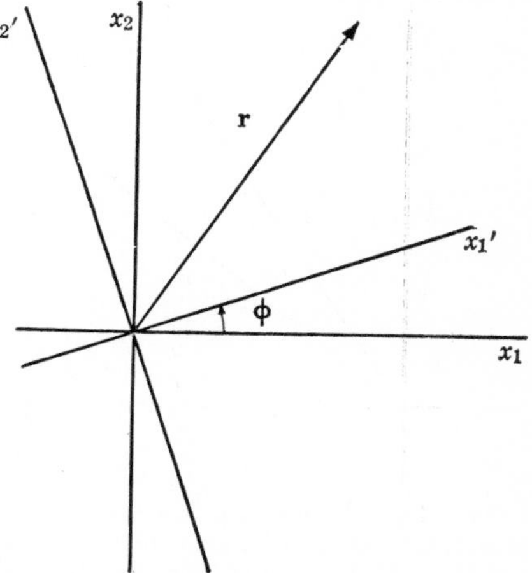

FIG. 4-4. Rotation of the coordinate axes, as equivalent to a two-dimensional orthogonal transformation.

The three orthogonality conditions expand into the equations:

$$\begin{aligned} a_{11}a_{11} + a_{21}a_{21} &= 1 \\ a_{12}a_{12} + a_{22}a_{22} &= 1 \\ a_{11}a_{12} + a_{21}a_{22} &= 0. \end{aligned}$$

These conditions are obviously satisfied by the matrix (4-17′), for in terms of the matrix elements (4-17) they reduce to the identities

$$\begin{aligned} \cos^2\phi + \sin^2\phi &= 1 \\ \sin^2\phi + \cos^2\phi &= 1 \\ \cos\phi \sin\phi - \sin\phi\cos\phi &= 0. \end{aligned}$$

The transformation matrix A can be thought of as an *operator* which acting on the unprimed system transforms it into the primed system. Symbolically the process might be written

$$(\mathbf{r})' = \mathsf{A}\mathbf{r}, \tag{4–18}$$

which is to be read: The matrix A operating on the components of a vector in the unprimed system yields the components of the vector in the primed system. It is to be emphasized that in the development of the subject so far, A acts on the coordinate system only, the vector is unchanged, and we ask merely for its components in two different coordinate frames. A parenthesis has therefore been placed around $\mathbf{r}$ on the left in Eq. (4–18) to make clear that the same vector is involved on both sides of the equation. Only the components have changed. In two dimensions the transformation of coordinates, it has been seen, is simply a rotation, and A is then identical with the *rotation* operator in a plane.

Despite this, it must be pointed out that without changing the formal mathematics, A can also be thought of as an operator acting on the *vector* $\mathbf{r}$, changing it to a different vector $\mathbf{r}'$:

$$\mathbf{r}' = \mathsf{A}\mathbf{r}, \tag{4–19}$$

with both vectors expressed in the same coordinate system. Thus, in two dimensions, instead of rotating the coordinate system one can rotate the vector $\mathbf{r}$ clockwise by an angle ϕ to a new vector $\mathbf{r}'$. The components of the new vector will then be related to the components of the old by the same Eqs. (4–12) which describe the transformation of coordinates. From a formal standpoint it is therefore not necessary to use the parenthesis in Eq. (4–18); rather, it can be written as in Eq. (4–19) and interpreted equally as an operation on the coordinate system or on the vector. The algebra remains the same no matter which of these two points of view is followed. The interpretation as an operator acting on the coordinates is the more pertinent one when using the orthogonal transformation to specify the orientation of a rigid body. On the other hand, the notion of an operator changing one vector into another has the more widespread application. In the mathematical discussion either interpretation will be freely used, as

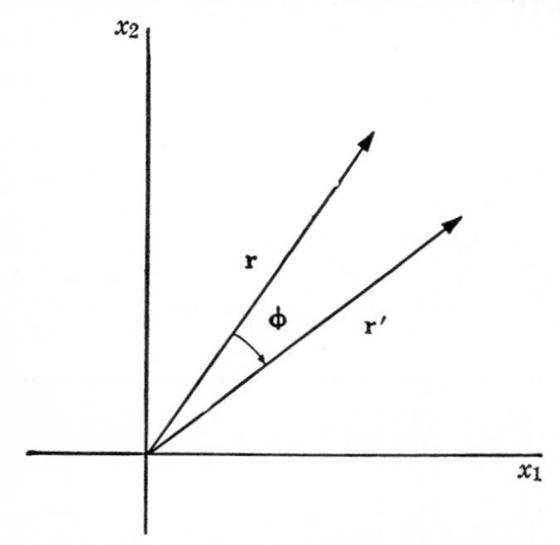

FIG. 4–5. Illustrating the interpretation of an orthogonal transformation as a rotation of the vector, leaving the coordinate system unchanged.

suits the convenience of the situation. Of course, it should be kept in mind that the nature of the operation represented by A will change according to which interpretation is selected. Thus if A corresponds to a *counterclockwise* rotation by an angle ϕ when applied to the coordinate system, it will correspond to a *clockwise* rotation when applied to the vector.

4-3 Formal properties of the transformation matrix. Consider what happens when two successive transformations are made — corresponding to two successive displacements of the rigid body. Let the first transformation from r to r' be denoted by B:

$$x'_k = \sum_j b_{kj} x_j, \tag{4-20}$$

and the succeeding transformation from r' to a third coordinate set r'' by A:

$$x''_i = \sum_k a_{ik} x'_k. \tag{4-21}$$

The relation between x''_i and x_j can then be obtained by combining the two Eqs. (4-20) and (4-21)

$$\begin{aligned} x''_i &= \sum_k a_{ik} \sum_j b_{kj} x_j \\ &= \sum_j \left(\sum_k a_{ik} b_{kj} \right) x_j, \end{aligned}$$

which may also be written as

$$x''_i = \sum_j c_{ij} x_j, \tag{4-22}$$

where

$$c_{ij} = \sum_k a_{ik} b_{kj}. \tag{4-23}$$

The successive application of two orthogonal transformations A, B is thus equivalent to a third linear transformation C. It can be shown that C is also an orthogonal transformation in consequence of the orthogonality of A and B. The detailed proof will be left for the exercises. Symbolically the resultant operator C can be considered as the product of the two operators A and B:

$$\mathsf{C} = \mathsf{AB},$$

and the matrix elements c_{ij} are by definition the elements of the matrix obtained by multiplying the two matrices A and B.

Note that this "matrix" or operator multiplication is not commutative,

$$\mathsf{BA} \neq \mathsf{AB},$$

for, by definition, the elements of the transformation $\mathsf{D} = \mathsf{BA}$ are:

$$d_{ij} = \sum_k b_{ik}a_{kj}, \tag{4–24}$$

which generally do not agree with the matrix elements of C, Eq. (4–23). Thus, the final coordinate system depends on the order of application of the operators A and B, i.e., whether first A then B, or first B and then A. However, matrix multiplication is associative; in a product of three or more matrices the order of the *multiplications* is unimportant:

$$(\mathsf{AB})\mathsf{C} = \mathsf{A}(\mathsf{BC}). \tag{4–25}$$

In Eq. (4–19) the juxtaposition of A and $\mathbf{r}$, to indicate the operation of A on the coordinate system (or on the vector), was said to be merely symbolic. But, by extending our concept of matrices, it may also be taken as indicating an actual matrix multiplication. Up to the present the matrices used have been square, i.e., with equal number of rows and columns. However, we may also have one-column matrices, such as x and x' defined by

$$\mathsf{x} = \begin{pmatrix} x_1 \\ x_2 \\ x_3 \end{pmatrix}, \qquad \mathsf{x}' = \begin{pmatrix} x_1' \\ x_2' \\ x_3' \end{pmatrix}. \tag{4–26}$$

The product Ax, by definition of matrix multiplication, is a one-column matrix, with the elements

$$(\mathsf{Ax})_i = \sum_j a_{ij}x_j = x_i'.$$

Hence Eq. (4–19) can also be written as the matrix equation

$$\mathsf{x}' = \mathsf{Ax}. \tag{4–27}$$

The *addition* of two matrices, while not as important a concept as multiplication, is a frequently used operation. The sum $\mathsf{A} + \mathsf{B}$ is a matrix C whose elements are the sum of the corresponding elements of A and B:

$$c_{ij} = a_{ij} + b_{ij}.$$

Of greater importance is the transformation inverse to A, the operation which changes r' back to r. This transformation will be called A^{-1} and its matrix elements designated by a_{ij}'. We then have the set of equations:

$$x_i = \sum_j a_{ij}'x_j', \tag{4–28}$$

which must be consistent with:

$$x_k' = \sum_i a_{ki}x_i. \tag{4–29}$$

Substituting x_i from (4–28), Eq. (4–29) becomes:

$$\begin{aligned} x'_k &= \sum_i a_{ki} \sum_j a'_{ij} x'_j \\ &= \sum_j \Big(\sum_i a_{ki} a'_{ij} \Big) x'_j. \end{aligned} \tag{4–29′}$$

Since the components of $\mathbf{r}'$ are independent, Eq. (4–29′) is correct only if the summation over j reduces identically to x'_k. The coefficient of x'_j must therefore be 1 for $j = k$ and zero for $j \neq k$; in symbols:

$$\sum_i a_{ki} a'_{ij} = \delta_{kj}. \tag{4–30}$$

The left-hand side of Eq. (4–30) is easily recognized as the matrix element for the product AA^{-1}, while the right-hand side is the element of the matrix known as the unit matrix $\mathsf{1}$:

$$\mathsf{1} = \begin{pmatrix} 1 & 0 & 0 \\ 0 & 1 & 0 \\ 0 & 0 & 1 \end{pmatrix}. \tag{4–31}$$

Eq. (4–30) can therefore be written as

$$\mathsf{AA}^{-1} = \mathsf{1}, \tag{4–32}$$

which indicates the reason for the designation of the inverse matrix by A^{-1}. The transformation corresponding to $\mathsf{1}$ is known as the *identity transformation*, producing no change in the coordinate system:

$$\mathsf{x} = \mathsf{1x}.$$

Similarly multiplying any matrix A by $\mathsf{1}$, in any order, leaves A unaffected:

$$\mathsf{1A} = \mathsf{A1} = \mathsf{A}.$$

By slightly changing the order of the proof of Eq. (4–32) it can be shown that A and A^{-1} commute. Instead of substituting x_i in Eq. (4–29) in terms of x', one could equally as well demand consistency by eliminating x' from the two equations, leading in analogous fashion to

$$\sum_j a'_{ij} a_{jk} = \delta_{ik}.$$

In matrix notation this reads

$$\mathsf{A}^{-1}\mathsf{A} = \mathsf{1}, \tag{4–33}$$

which proves the statement.

Consider now the double sum

$$\sum_{k,i} a_{kl} a_{ki} a'_{ij}$$

which can be evaluated either by summing over k first,

$$\sum_i \left(\sum_k a_{kl} a_{ki} \right) a'_{ij},$$

or over i first:

$$\sum_k \left(\sum_i a_{ki} a'_{ij} \right) a_{kl}.$$

Applying the orthogonality conditions, Eq. (4–15), the sum from the first point of view reduces to

$$\sum_i \delta_{il} a'_{ij} = a'_{lj}.$$

On the other hand, the same sum from the second point of view, and with the help of Eq. (4–30), can be written

$$\sum_k \delta_{kj} a_{kl} = a_{jl}.$$

Thus the elements of the direct matrix A and the reciprocal A^{-1} are related by

$$a'_{ij} = a_{ji}. \tag{4–34}$$

In general, the matrix obtained from A by interchanging rows and columns is known as the *transposed matrix*, indicated by the tilde thus: $\tilde{\mathsf{A}}$. Eq. (4–34) therefore states that for *orthogonal matrices* the reciprocal matrix is to be identified as the transposed matrix; symbolically:

$$\mathsf{A}^{-1} = \tilde{\mathsf{A}}. \tag{4–35}$$

If this result is substituted in Eq. (4–33) we obtain

$$\tilde{\mathsf{A}}\mathsf{A} = 1, \tag{4–36}$$

which is identical with the set of orthogonality conditions, Eq. (4–15), written in abbreviated form, as can be verified by direct expansion.* Similarly, an alternative form of the orthogonality conditions can be obtained from Eq. (4–30) by substituting (4–34):

$$\sum_i a_{ki} a_{ji} = \delta_{kj}. \tag{4–37}$$

---

* Indeed one may obtain (4–35) directly from the orthogonality conditions in the form (4–36) and the brevity of the proof is indicative of the power of the symbolic procedures. Multiply (4–36) by A^{-1} from the right

$$\tilde{\mathsf{A}}\mathsf{A}\mathsf{A}^{-1} = \mathsf{A}^{-1},$$

and by (4–32) there results

$$\tilde{\mathsf{A}} = \mathsf{A}^{-1}.$$

In symbolic form (4–37) can be written

$$\mathsf{A}\tilde{\mathsf{A}} = 1$$

and may be derived directly from (4–36) by multiplying it from the left by A and from the right by A^{-1}.

Associated with the notion of the transposed matrix is its complex conjugate known to physicists as the *adjoint matrix*, and indicated by a dagger,†:

$$\mathsf{A}^\dagger = (\tilde{\mathsf{A}})^*. \tag{4–38}$$

Analogous to the definition (4–36) for an orthogonal matrix, a *unitary matrix* A satisfies the condition

$$\mathsf{A}^\dagger\mathsf{A} = 1. \tag{4–39}$$

In the problem of specifying the orientation of a rigid body the transformation matrix must be real, for both x and x' are real. There is then no distinction between the orthogonality and the unitary property or between transposed and adjoint matrices. In short, a real orthogonal matrix is unitary. But we shall soon have occasion in this chapter, and later in connection with relativity, to introduce complex matrices. There the difference is significant.

The two interpretations of an operator as transforming the vector, or alternatively the coordinate system, are both involved if we seek to find the transformation of an operator under a change of coordinates. Let A be considered an operator acting upon a vector $\mathbf{F}$ (or a single-column matrix F) to produce a vector $\mathbf{G}$:

$$\mathsf{G} = \mathsf{AF}.$$

If the coordinate system is transformed by a matrix B the components of the vector $\mathbf{G}$ in the new system will be given by

$$\mathsf{BG} = \mathsf{BAF},$$

which can also be written

$$\mathsf{BG} = \mathsf{BAB}^{-1}\mathsf{BF}. \tag{4–40}$$

Eq. (4–40) can be interpreted as stating that the operator BAB^{-1} acting upon the vector $\mathbf{F}$, expressed in the new system, produces the vector $\mathbf{G}$, likewise expressed in the new coordinates. We may therefore consider BAB^{-1} to be the form taken by the operator A when transformed to a new set of axes:

$$\mathsf{A}' = \mathsf{BAB}^{-1}. \tag{4–41}$$

Any transformation of a matrix having the form of Eq. (4–41) is known as a *similarity transformation*.

It is appropriate at this point to consider the properties of the determinant formed from the elements of a square matrix. As is customary, we shall denote such a determinant by vertical bars, thus: $|\mathsf{A}|$. It will be noticed that the definition of matrix multiplication is identical with that for the multiplication of determinants (cf. Bôcher, *Introduction to Higher Algebra*, p. 26). Hence

$$|\mathsf{AB}| = |\mathsf{A}| \cdot |\mathsf{B}|.$$

Since the determinant of the unit matrix is 1, the determinantal form of the orthogonality conditions, Eq. (4–36), can be written

$$|\tilde{\mathsf{A}}| \cdot |\mathsf{A}| = 1.$$

Further, as the value of a determinant is unaffected by interchanging rows and columns we can write

$$|\mathsf{A}|^2 = 1, \tag{4–42}$$

which implies that the determinant of an orthogonal matrix can only be $+1$ or -1. We shall have occasion later to dwell at length on the geometrical significance of these two values.

When the matrix is not orthogonal the determinant does not have these simple values, of course. It can be shown however that the value of the determinant is invariant under a similarity transformation. Multiplying the equation (4–41) for the transformed matrix from the right by B, we obtain the relation

$$\mathsf{A'B} = \mathsf{BA},$$

or in determinantal form

$$|\mathsf{A'}| \cdot |\mathsf{B}| = |\mathsf{B}| \cdot |\mathsf{A}|.$$

Since the determinant of B is merely a number, and not zero,* we can divide by $|\mathsf{B}|$ on both sides to obtain the desired result:

$$|\mathsf{A'}| = |\mathsf{A}|.$$

In discussing rigid body motion later, all these properties of matrix transformations, especially of orthogonal matrices, will be employed. In addition, other properties are needed, and they will be derived as the occasion requires.

---

* If it were zero there could be no inverse operator B^{-1} (by Cramer's rule), which is required in order that Eq. (4–41) make sense.

4–4 The Eulerian angles. It has already been noted that since the elements a_{ij} are not independent they are not suitable as generalized coordinates. Before setting up the motion of rigid bodies in the Lagrangian formulation of mechanics it will therefore be necessary to seek three independent parameters specifying the orientation of a rigid body. Only when such generalized coordinates have been found, can one write a Lagrangian for the system and obtain the Lagrangian equations of motion. A number of such sets of parameters have been described in the literature, but the most common and useful are the *Eulerian angles*. We shall therefore define these angles at this point, and show how the elements of the orthogonal transformation matrix can be expressed in terms of them.

One can carry out the transformation from a given cartesian coordinate system to another by means of three successive rotations performed in a specific sequence. The Eulerian angles are then defined as the three successive angles of rotation. The sequence will be started by rotating the initial system of axes, xyz, by an angle ϕ counterclockwise about the z axis, and the resultant coordinate system will be labelled the $\xi\eta\zeta$ axes. In the second stage the intermediate axes, $\xi\eta\zeta$, are rotated about the ξ axis counterclockwise by an angle θ to produce another intermediate set, the

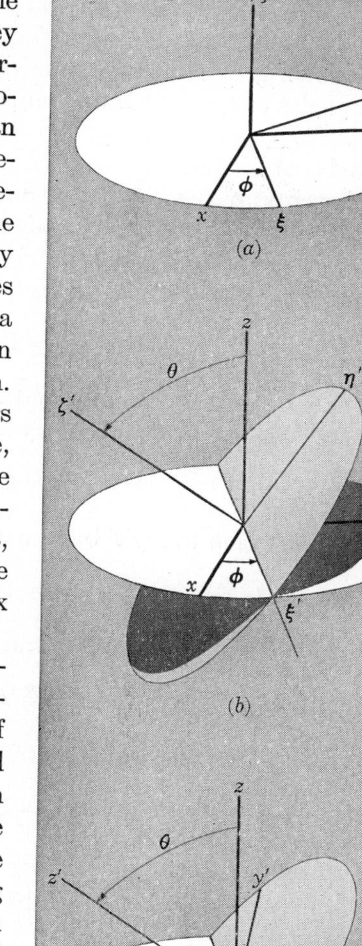

Fig. 4–6. The rotations defining the Eulerian angles.

$\xi'\eta'\zeta'$ axes. The ξ' axis is at the intersection of the xy and $\xi'\eta'$ planes and is known as the line of nodes. Finally the $\xi'\eta'\zeta'$ axes are rotated counterclockwise by an angle ψ about the ζ' axis to produce the desired $x'y'z'$ system of axes. Fig. 4–6 illustrates the various stages of the sequence. The Eulerian angles θ, ϕ, and ψ thus completely specify the orientation of the $x'y'z'$ system relative to the xyz and can therefore act as the three needed generalized coordinates.*

The elements of the complete transformation A can be obtained by writing the matrix as the triple product of the separate rotations, each of which has a relatively simple matrix form. Thus, the initial rotation about z can be described by a matrix D:

$$\boldsymbol{\xi} = \mathsf{D}\mathsf{x},$$

where $\boldsymbol{\xi}$ and x stand for column matrices. Similarly the transformation from $\xi\eta\zeta$ to $\xi'\eta'\zeta'$ can be described by a matrix C:

$$\boldsymbol{\xi}' = \mathsf{C}\boldsymbol{\xi},$$

and the last rotation to $x'y'z'$ by a matrix B

$$\mathsf{x}' = \mathsf{B}\boldsymbol{\xi}'.$$

Hence the matrix of the complete transformation:

$$\mathsf{x}' = \mathsf{A}\mathsf{x}$$

is the product of the successive matrices:

$$\mathsf{A} = \mathsf{BCD}.$$

* Unfortunately there is no unanimity in the literature about the definition of the Eulerian angles. The differences are not very great, but often are sufficient to frustrate any easy comparison of the end formulae such as the matrix elements. Greatest confusion, perhaps, arises from the occasional use of left-handed coordinate systems (as by Osgood and Margenau & Murphy). More frequent is the custom of measuring the angle of the line of nodes from the y axis and not the x axis. This practice, which seems characteristic of the British school (cf. Whittaker, Newboult, Ames and Murnaghan) is due to taking the second rotation about η instead of ξ. Our angles ϕ and ψ are then equal to the angles $\phi + \frac{\pi}{2}$ and $\frac{\pi}{2} - \psi$ respectively in such a notation. Continental authors usually follow the definitions given here, except that the meanings of ϕ and ψ are often interchanged. This does not end the confusion, however; many authors of quantum-mechanical discussions appear to use *clockwise* rotations, rather than the counterclockwise convention observed here.

Now the D transformation is a rotation about z, and hence has a matrix of the form (cf. Eq. (4–17)):

$$\mathsf{D} = \begin{pmatrix} \cos\phi & \sin\phi & 0 \\ -\sin\phi & \cos\phi & 0 \\ 0 & 0 & 1 \end{pmatrix}. \tag{4–43}$$

The C transformation corresponds to a rotation about ξ, with the matrix:

$$\mathsf{C} = \begin{pmatrix} 1 & 0 & 0 \\ 0 & \cos\theta & \sin\theta \\ 0 & -\sin\theta & \cos\theta \end{pmatrix}, \tag{4–44}$$

and finally B is a rotation about ζ' and therefore has the same form as D:

$$\mathsf{B} = \begin{pmatrix} \cos\psi & \sin\psi & 0 \\ -\sin\psi & \cos\psi & 0 \\ 0 & 0 & 1 \end{pmatrix}. \tag{4–45}$$

The product matrix $\mathsf{A} = \mathsf{BCD}$ then follows as

$$\mathsf{A} = \begin{pmatrix} \cos\psi\cos\phi - \cos\theta\sin\phi\sin\psi & \cos\psi\sin\phi + \cos\theta\cos\phi\sin\psi & \sin\psi\sin\theta \\ -\sin\psi\cos\phi - \cos\theta\sin\phi\cos\psi & -\sin\psi\sin\phi + \cos\theta\cos\phi\cos\psi & \cos\psi\sin\theta \\ \sin\theta\sin\phi & -\sin\theta\cos\phi & \cos\theta \end{pmatrix}. \tag{4–46}$$

The inverse transformation from body coordinates to space axes

$$\mathsf{x} = \mathsf{A}^{-1}\mathsf{x}'$$

is then given immediately by the transposed matrix $\tilde{\mathsf{A}}$:

$$\mathsf{A}^{-1} = \tilde{\mathsf{A}} = \begin{pmatrix} \cos\psi\cos\phi - \cos\theta\sin\phi\sin\psi & -\sin\psi\cos\phi - \cos\theta\sin\phi\cos\psi & \sin\theta\sin\phi \\ \cos\psi\sin\phi + \cos\theta\cos\phi\sin\psi & -\sin\psi\sin\phi + \cos\theta\cos\phi\cos\psi & -\sin\theta\cos\phi \\ \sin\theta\sin\psi & \sin\theta\cos\psi & \cos\theta \end{pmatrix}. \tag{4–47}$$

Verification of the multiplications and of the orthogonality of the A matrix will be left as an exercise.

4–5 The Cayley-Klein parameters. Various other groups of variables have been used to describe the orientation of a rigid body, usually for convenience in some specialized calculation. One such is of sufficient interest to deserve further mention — the so-called *Cayley-Klein parameters.* As there are four quantities involved, these parameters are not all independent, and therefore are not suitable as generalized coordinates. Originally, they were introduced into classical mechanics by Felix Klein mainly to facilitate the integration of complicated gyroscopic problems. Today our principal interest in them lies in their intimate connection with the treatment of spatial rotation in advanced quantum mechanics.

In the previous sections we employed on occasion a two-dimensional real space with axes x_1 and x_2 to illustrate the properties of orthogonal transformations. We shall now consider a different two-dimensional space, this time having complex axes denoted by u and v. A general linear transformation in such a space appears as

$$\begin{aligned} u' &= \alpha u + \beta v, \\ v' &= \gamma u + \delta v, \end{aligned} \tag{4–48}$$

with the corresponding transformation matrix

$$\mathsf{Q} = \begin{pmatrix} \alpha & \beta \\ \gamma & \delta \end{pmatrix}. \tag{4–49}$$

For the remainder of the discussion we will restrict Q to transformations which are unitary and where the determinant is $+1$. It must be emphasized that these are two separate conditions. The unitary property, Eq. (4–39):

$$\mathsf{Q}^\dagger\mathsf{Q} = \mathsf{1} = \mathsf{Q}\mathsf{Q}^\dagger \tag{4–50}$$

only requires that the determinant shall have unit magnitude:

$$|\mathsf{Q}|^*|\mathsf{Q}| = 1,$$

but does not restrict its phase angle. The condition that the determinant be $+1$,

$$\alpha\delta - \beta\gamma = +1, \tag{4–51}$$

is therefore an additional requirement not contained in the unitary property.

The matrix of a general two-dimensional linear transformation has *eight* quantities to be specified, since each of the four elements is complex. However, the conditions imposed reduce the number of *independent* quantities. Expanded, the unitary condition, Eq. (4–50), becomes the equations

$$\begin{aligned} \alpha^*\alpha + \beta^*\beta &= 1 \\ \gamma^*\gamma + \delta^*\delta &= 1 \\ \alpha^*\gamma + \beta^*\delta &= 0. \end{aligned} \tag{4–52}$$

The first two of Eqs. (4–52) are real, while the last is complex, so that together they comprise four conditions. A fifth condition is provided by the requirement imposed on the determinant, Eq. (4-51). Therefore Q contains only three independent quantities, the same number required to specify the orientation of a rigid body.

Some of the reduction to independent parameters can be performed without much difficulty. Thus, from the last of Eqs. (4-52) one may write

$$\frac{\delta}{\gamma} = -\frac{\alpha^*}{\beta^*}, \tag{4-53}$$

which when substituted in the determinant condition yields

$$-(\alpha\alpha^* + \beta\beta^*)\frac{\gamma}{\beta^*} = 1.$$

The first of Eqs. (4-52) states that the quantity in parentheses is unity, and hence

$$\gamma = -\beta^*. \tag{4-54}$$

It follows then from (4-53) that

$$\delta = \alpha^*. \tag{4-55}$$

As a result of these four conditions (Eqs. 4-54 and 4-55), the matrix Q could be written also as

$$\mathsf{Q} = \begin{pmatrix} \alpha & \beta \\ -\beta^* & \alpha^* \end{pmatrix} \tag{4-56}$$

with the one remaining condition

$$\alpha\alpha^* + \beta\beta^* = 1. \tag{4-57}$$

However, we often prefer to leave the matrix in the form (4-49).

Let P be a matrix operator in this space with the specific form

$$\mathsf{P} = \begin{pmatrix} z & x - iy \\ x + iy & -z \end{pmatrix}. \tag{4-58}$$

Mathematically x, y, z can be considered simply as three real quantities; physically they will be interpreted as coordinates of a point in space. Suppose the P matrix is transformed by means of the Q matrix in the following manner:

$$\mathsf{P}' = \mathsf{QPQ}^\dagger. \tag{4-59}$$

From the unitary property of Q, the adjoint $\mathsf{Q}^\dagger$ is the same as the inverse Q^{-1} and Eq. (4-59) merely represents the similarity transformation of P when the uv space is subjected to the unitary transformation Q. It will be noted that the adjoint of P is identical with P itself; the matrix is called *self-adjoint* or *hermitean*. Further, the sum of the diagonal elements of $\mathbf{P}$, known as the *spur* or *trace* of the matrix, is here zero. Now, it can be shown that both the hermitean property and the spur of a matrix are invariant under similarity transformation (cf. the exercises at the end of the chapter). Hence $\mathbf{P}'$ must likewise be self-adjoint and have a vanishing spur, which can be true only if it has the form

$$\mathbf{P}' = \begin{pmatrix} z' & x' - iy' \\ x' + iy' & -z' \end{pmatrix}, \tag{4–60}$$

where x', y', and z' are real quantities. The determinant of $\mathbf{P}$ is also invariant under the similarity transformation (4–59), so that we can write the following equality:

$$|\mathbf{P}| = -(x^2 + y^2 + z^2) = -(x'^2 + y'^2 + z'^2) = |\mathbf{P}'|.$$

This statement will be recognized as the orthogonality condition; it requires that the length of the vector $\mathbf{r} = x\mathbf{i} + y\mathbf{j} + z\mathbf{k}$ shall be unchanged by the transformation. To each unitary matrix $\mathbf{Q}$ in the complex two-dimensional space there is therefore associated some real orthogonal transformation in ordinary three-dimensional space.

Some insight into the nature of this association is provided by the following considerations. Let the real orthogonal matrix transforming from coordinates $\mathbf{x}$ to $\mathbf{x}'$ be designated by $\mathbf{B}$:

$$\mathbf{x}' = \mathbf{B}\mathbf{x}$$

and denote the associated unitary matrix by $\mathbf{Q}_1$:

$$\mathbf{P}' = \mathbf{Q}_1\mathbf{P}\mathbf{Q}_1^{\dagger}.$$

A second orthogonal transformation from $\mathbf{x}'$ to $\mathbf{x}''$ may be accomplished by the matrix $\mathbf{A}$:

$$\mathbf{x}'' = \mathbf{A}\mathbf{x}',$$

with the associated matrix $\mathbf{Q}_2$:

$$\mathbf{P}'' = \mathbf{Q}_2\mathbf{P}'\mathbf{Q}_2^{\dagger}.$$

Now, the direct transformation from $\mathbf{x}$ to $\mathbf{x}''$ is produced by the matrix $\mathbf{C}$ defined by

$$\mathbf{C} = \mathbf{A}\mathbf{B}.$$

Correspondingly the direct transformation from $\mathbf{P}$ to $\mathbf{P}''$ will be effected by a similarity transformation with some matrix $\mathbf{Q}_3$ which must therefore be associated with C. However we can also obtain the transformation from P to P'' from the equation

$$\mathsf{P}'' = \mathsf{Q}_2\mathsf{Q}_1\mathsf{P}\mathsf{Q}_1^\dagger\mathsf{Q}_2^\dagger.$$

It can easily be shown that

$$\mathsf{Q}_1^\dagger\mathsf{Q}_2^\dagger = (\mathsf{Q}_2\mathsf{Q}_1)^\dagger.$$

Since the product of two unitary matrices is also unitary it follows that Q_3 must be identified with the product $\mathsf{Q}_2\mathsf{Q}_1$:

$$\mathsf{Q}_3 = \mathsf{Q}_2\mathsf{Q}_1.$$

Thus the correspondence between the 2×2 complex unitary matrices and the 3×3 real orthogonal matrices is such that any relation among the matrices of one set is satisfied also by the corresponding matrices of the other set. The two sets of matrices are said to be *isomorphic.*

One can write the elements of an orthogonal matrix in terms of the elements of the isomorphic Q matrix. From (4–54) and (4–55) the adjoint to Q is

$$\mathsf{Q}^\dagger = \begin{pmatrix} \alpha^* & \gamma^* \\ \beta^* & \delta^* \end{pmatrix} = \begin{pmatrix} \delta & -\beta \\ -\gamma & \alpha \end{pmatrix}.$$

To simplify the calculation we shall introduce the notational abbreviations x_+ and x_- defined as

$$\begin{aligned} x_+ &= x + iy \\ x_- &= x - iy. \end{aligned}$$

The transformed matrix P' is then written as

$$\mathsf{P}' = \begin{pmatrix} z' & x'_- \\ x'_+ & -z' \end{pmatrix} = \begin{pmatrix} \alpha & \beta \\ \gamma & \delta \end{pmatrix}\begin{pmatrix} z & x_- \\ x_+ & -z \end{pmatrix}\begin{pmatrix} \delta & -\beta \\ -\gamma & \alpha \end{pmatrix},$$

or upon performing the indicated multiplications:

$$\mathsf{P}' = \begin{pmatrix} (\alpha\delta + \beta\gamma)z - \alpha\gamma x_- + \beta\delta x_+ & -2\alpha\beta z + \alpha^2 x_- - \beta^2 x_+ \\ 2\gamma\delta z - \gamma^2 x_- + \delta^2 x_+ & -(\alpha\delta + \beta\gamma)z + \alpha\gamma x_- - \beta\delta x_+ \end{pmatrix}. \tag{4–61}$$

By equating matrix elements the transformation equations between the primed and unprimed coordinate systems can be written in the form

$$\begin{aligned} x'_+ &= 2\gamma\delta z - \gamma^2 x_- + \delta^2 x_+ \\ x'_- &= -2\alpha\beta z + \alpha^2 x_- - \beta^2 x_+ \\ z' &= (\alpha\delta + \beta\gamma)z - \alpha\gamma x_- + \beta\delta x_+. \end{aligned} \tag{4–62}$$

Finally, the matrix elements a_{ij} can be obtained in terms of α, β, γ and δ by comparing Eqs. (4–62) with the customary transformation equations (4–14). Thus the last of Eqs. (4–62) may be written as

$$z' = (\beta\delta - \alpha\gamma)x + i(\alpha\gamma + \beta\delta)y + (\alpha\delta + \beta\gamma)z,$$

from which it follows immediately that

$$a_{31} = (\beta\delta - \alpha\gamma) \quad a_{32} = i(\alpha\gamma + \beta\delta) \quad a_{33} = \alpha\delta + \beta\gamma.$$

By this process the complete transformation matrix is easily found to be

$$\mathsf{A} = \begin{pmatrix} \frac{1}{2}(\alpha^2 - \gamma^2 + \delta^2 - \beta^2) & \frac{i}{2}(\gamma^2 - \alpha^2 + \delta^2 - \beta^2) & \gamma\delta - \alpha\beta \\ \frac{i}{2}(\alpha^2 + \gamma^2 - \beta^2 - \delta^2) & \frac{1}{2}(\alpha^2 + \gamma^2 + \beta^2 + \delta^2) & -i(\alpha\beta + \gamma\delta) \\ \beta\delta - \alpha\gamma & i(\alpha\gamma + \beta\delta) & \alpha\delta + \beta\gamma \end{pmatrix}. \tag{4–63}$$

Eq. (4–63) provides a matrix which specifies the orientation of a rigid body, and which is expressed entirely in terms of the quantities α, β, γ, and δ. Like the Eulerian angles these four thus furnish a way of establishing the body's orientation; they are customarily known as the *Cayley-Klein parameters*.* The reality of the matrix elements follows as a consequence of the hermitean property of P, or may be shown directly by examining the matrix elements with the help of Eqs. (4–54, 4–55).

One can express the Cayley-Klein parameters in terms of the corresponding Euler angles; if need be, by direct comparison of the elements of Eq. (4–63) with the elements expressed in terms of ϕ, θ, and ψ. However it is a simpler procedure, and more instructive, to first construct the Q matrices corresponding to the separate successive rotations that define the Euler angles and then combine them to form the complete matrix. Thus the angle ϕ has been defined in terms of a rotation about the z axis, where the transformation in terms of x_+, x_- and z appears as:

$$\begin{aligned} x'_+ &= e^{-i\phi}x_+ \\ x'_- &= e^{i\phi}x_- \\ z' &= z. \end{aligned}$$

---

* The matrix A, Eq. (4–63), does not agree with the corresponding form as given, say, in Whittaker, p. 12. Essentially this is because of a different initial choice of the matrix P. Clearly there are many ways of setting up a matrix whose determinant will be $-r^2$, and the specific choice is a matter of convention. The form used here, (4–58), was chosen to agree with customary usage in quantum mechanics.

Comparing these equations with (4-62) it is clear that for this simple rotation the elements of the matrix $\mathbf{Q}$ must have the form

$$\gamma = \beta = 0 \qquad \alpha^2 = e^{i\phi} \qquad \delta^2 = e^{-i\phi}$$

or

$$\mathbf{Q}_\phi = \begin{pmatrix} e^{i\phi/2} & 0 \\ 0 & e^{-i\phi/2} \end{pmatrix}. \tag{4-64}$$

It will be noted that these matrix elements automatically satisfy the conditions (4-54), (4-55), and (4-57).

The next rotation is about the new x axis *counterclockwise* by an angle θ, and the identification of the corresponding matrix elements proceeds in a similar fashion, but the calculations become rather tedious. It will simply be stated that the corresponding $\mathbf{Q}$ matrix is

$$\mathbf{Q}_\theta = \begin{pmatrix} \cos\frac{\theta}{2} & i\sin\frac{\theta}{2} \\ i\sin\frac{\theta}{2} & \cos\frac{\theta}{2} \end{pmatrix}. \tag{4-65}$$

To check, one can directly verify that

$$\begin{pmatrix} \cos\frac{\theta}{2} & i\sin\frac{\theta}{2} \\ i\sin\frac{\theta}{2} & \cos\frac{\theta}{2} \end{pmatrix} \begin{pmatrix} z & x_- \\ x_+ & -z \end{pmatrix} \begin{pmatrix} \cos\frac{\theta}{2} & -i\sin\frac{\theta}{2} \\ -i\sin\frac{\theta}{2} & \cos\frac{\theta}{2} \end{pmatrix}$$

$$= \begin{pmatrix} z\cos\theta - y\sin\theta & x - i(y\cos\theta + z\sin\theta) \\ x + i(y\cos\theta + z\sin\theta) & -z\cos\theta + y\sin\theta \end{pmatrix},$$

which leads to the desired transformation:

$$\begin{aligned} x' &= x \\ y' &= y\cos\theta + z\sin\theta \\ z' &= -y\sin\theta + z\cos\theta. \end{aligned}$$

The final rotation, defining ψ, is again about a z axis so that

$$\mathbf{Q}_\psi = \begin{pmatrix} e^{i\psi/2} & 0 \\ 0 & e^{-i\psi/2} \end{pmatrix}. \tag{4-66}$$

In Section 4-4 the orthogonal matrix for the complete transformation was obtained as the product of the separate matrices for each of the three rotations. It follows from the isomorphism of the 3×3 real orthogonal matrices with the **Q** matrices that **Q** for the complete transformation is likewise given by the product of the three rotation matrices $\mathbf{Q}_\psi$, $\mathbf{Q}_\theta$, $\mathbf{Q}_\phi$:

$$\mathbf{Q} = \mathbf{Q}_\psi \mathbf{Q}_\theta \mathbf{Q}_\phi = \begin{pmatrix} e^{i\psi/2} & 0 \\ 0 & e^{-i\psi/2} \end{pmatrix} \begin{pmatrix} \cos\frac{\theta}{2} & i\sin\frac{\theta}{2} \\ i\sin\frac{\theta}{2} & \cos\frac{\theta}{2} \end{pmatrix} \begin{pmatrix} e^{i\phi/2} & 0 \\ 0 & e^{-i\phi/2} \end{pmatrix},$$

or

$$\mathbf{Q} = \begin{pmatrix} e^{i(\psi+\phi)/2}\cos\frac{\theta}{2} & ie^{i(\psi-\phi)/2}\sin\frac{\theta}{2} \\ ie^{-i(\psi-\phi)/2}\sin\frac{\theta}{2} & e^{-i(\psi+\phi)/2}\cos\frac{\theta}{2} \end{pmatrix}. \tag{4-67}$$

The Cayley-Klein parameters in terms of the Euler angles are then:

$$\begin{aligned} \alpha &= e^{i(\psi+\phi)/2}\cos\frac{\theta}{2} & \beta &= ie^{i(\psi-\phi)/2}\sin\frac{\theta}{2} \\ \gamma &= ie^{-i(\psi-\phi)/2}\sin\frac{\theta}{2} & \delta &= e^{-i(\psi+\phi)/2}\cos\frac{\theta}{2}, \end{aligned} \tag{4-68}$$

completing the desired identification.

It will be noted that the **P** matrix can be written as the sum of three matrices:

$$\mathbf{P} = x\boldsymbol{\sigma}_x + y\boldsymbol{\sigma}_y + z\boldsymbol{\sigma}_z, \tag{4-69}$$

where $\boldsymbol{\sigma}_x$, $\boldsymbol{\sigma}_y$, and $\boldsymbol{\sigma}_z$ are the three so-called *Pauli spin matrices:*

$$\boldsymbol{\sigma}_x = \begin{pmatrix} 0 & 1 \\ 1 & 0 \end{pmatrix}, \quad \boldsymbol{\sigma}_y = \begin{pmatrix} 0 & -i \\ i & 0 \end{pmatrix}, \quad \boldsymbol{\sigma}_z = \begin{pmatrix} 1 & 0 \\ 0 & -1 \end{pmatrix}. \tag{4-70}$$

These three, together with the unit matrix

$$\mathbf{1} = \begin{pmatrix} 1 & 0 \\ 0 & 1 \end{pmatrix},$$

form a set of four independent matrices. Consequently, any 2×2 matrix involving four independent quantities can be expressed as a linear function of them. The **Q** matrices for rotation about a coordinate axis can be expressed in terms of the σ's in a particularly simple form. For example, $\mathbf{Q}_\theta$ for rotation about the x axis, Eq. (4-65), may be written as

$$\mathbf{Q}_\theta = \mathbf{1}\cos\frac{\theta}{2} + i\boldsymbol{\sigma}_x \sin\frac{\theta}{2}. \tag{4-71}$$

Similarly the $\mathbf{Q}_\phi$ matrix for rotation about the z axis has the form

$$\mathbf{Q}_\phi = \begin{pmatrix} \cos\frac{\phi}{2} + i \sin\frac{\phi}{2} & 0 \\ 0 & \cos\frac{\phi}{2} - i \sin\frac{\phi}{2} \end{pmatrix} = \mathbf{1} \cos\frac{\phi}{2} + i\boldsymbol{\sigma}_z \sin\frac{\phi}{2}, \quad (4\text{–}72)$$

and it can be directly verified that a rotation about the y axis has the same matrix form as in (4–72) with $\boldsymbol{\sigma}_z$ replaced by $\boldsymbol{\sigma}_y$. All the elementary rotation matrices are thus given by similar expressions, involving only the unit matrix and the corresponding $\boldsymbol{\sigma}$-matrix. Each of the Pauli spin matrices is therefore associated with rotation about one particular axis and may be thought of as the *unit rotator* for that axis.

Characteristic of the Cayley-Klein parameters, and of the matrices containing them, is the ubiquitous presence of half angles, and this feature leads to some peculiar properties for the uv space. For example, a rotation in ordinary space about the z axis through the angle 2π merely reproduces the original coordinate system. Thus, if in the $\mathbf{D}$ matrix of the preceding section, ϕ is set equal to 2π, then $\cos\phi = 1$, $\sin\phi = 0$, and $\mathbf{D}$ properly reduces to the unit matrix $\mathbf{1}$ corresponding to the identity transformation. On the other hand if the same substitution is made in $\mathbf{Q}_\phi$, Eq. (4–64), we obtain

$$\mathbf{Q}_{2\pi} = \begin{pmatrix} e^{i\pi} & 0 \\ 0 & e^{-i\pi} \end{pmatrix} = \begin{pmatrix} -1 & 0 \\ 0 & -1 \end{pmatrix},$$

which is $-\mathbf{1}$ and not $\mathbf{1}$. At the same time the 2×2 $\mathbf{1}$ matrix must also correspond to the three-dimensional identity transformation. Hence there are two $\mathbf{Q}$-matrices, $\mathbf{1}$ and $-\mathbf{1}$, corresponding to the 3×3 unit matrix. In general, if a matrix $\mathbf{Q}$ corresponds to some real orthogonal matrix then $-\mathbf{Q}$ also corresponds to the same matrix. The isomorphism between the two sets thus involves, in this case, a one-to-one correspondence between the single 3×3 matrix and the *pair* of matrices $(\mathbf{Q}, -\mathbf{Q})$, and not between the individual matrices. In this sense one may say that the $\mathbf{Q}$ matrix is a *double-valued* function of the corresponding three-dimensional orthogonal matrix.

Such a paradoxical situation plays no havoc with our common sense. As here presented the uv space is entirely a mathematical construct, devised solely to establish a correspondence between 3×3 and 2×2 matrices of a certain type. One would not require nor expect such a space to have the same properties as physical three-dimensional space. Mathematicians have paid considerable attention to the properties of the uv space and have designated the two-dimensional complex vector in the space by the term *spinor*. It turns out that in quantum mechanics the spinor space comes a bit closer to physical reality, for to include the effects of the "spin" of the electron, the wave function, or parts of it, must be made a spinor. Indeed, the half angles and resultant double-valued property are intimately connected with the fact that the spin is half integral.* To pursue the subject further would clearly take us outside the scope of classical mechanics.

4–6 Euler's theorem on the motion of a rigid body. The discussions of the previous sections provide a complete mathematical technique for describing the motions of a rigid body. At any instant the orientation of the body can be specified by an orthogonal transformation, the elements of which may be expressed in terms of some suitable set of parameters. As time progresses the orientation will change and hence the matrix of transformation will be a function of time and may be written $\mathsf{A}(t)$. If the body axes are chosen coincident with the space axes at the time $t = 0$ then the transformation is initially simply the identity transformation:

$$\mathsf{A}(0) = 1$$

At any later time $\mathsf{A}(t)$ will in general differ from the identity transformation, but since the physical motion must be continuous $\mathsf{A}(t)$ must be a continuous function of time. The transformation may thus be said to evolve *continuously from the identity transformation.*

With this method of describing the motion, and using only the mathematical apparatus already introduced, we are now in a position to obtain the important characteristics of rigid body motion. Of basic importance is the so-called *Euler's theorem: the general displacement of a rigid body with one point fixed is a rotation about some axis.*

If the fixed point is taken as the origin of the body set of axes then the displacement of the rigid body involves no translation of the body axes, the only change is in orientation. The theorem then states that the body set of axes at any time t can always be obtained by a single rotation of the initial set of axes (taken as coincident with the space set). In other words, the *operation* implied in the matrix A describing the physical motion of the rigid body is a *rotation.* Now it is characteristic of a rotation that one direction, namely the axis of rotation, is left unaffected by the operation. Thus any vector lying along the axis of rotation must have the same components in both the initial and final axes. The other necessary condition

---

* Although the wave function may be double valued under rotation, all physically observable properties remain single valued, of course.

for a rotation, that the magnitude of the vectors be unaffected, is automatically provided by the orthogonality conditions. Hence Euler's theorem will be proven if it can be shown that there exists a vector **R** having the same components in both systems. Using matrix notation for the vector,

$$\mathsf{R}' = \mathsf{AR} = \mathsf{R}. \tag{4-73}$$

Eq. (4–73) constitutes a special case of the more general equation:

$$\mathsf{R}' = \mathsf{AR} = \lambda\mathsf{R}, \tag{4-74}$$

where λ is some constant, which may be complex. The values of λ for which Eq. (4–74) is soluble are known as the characteristic values, or *eigenvalues*,* of the matrix. The problem of finding the vectors satisfying Eq. (4–74) is therefore called the *eigenvalue problem* for the given matrix, and Eq. (4–74) itself is referred to as the *eigenvalue equation.* Correspondingly, the vector solutions are the *eigenvectors* of A. Euler's theorem can now be restated in the following language:

The real orthogonal matrix specifying the physical motion of a rigid body with one point fixed always has the eigenvalue $+1$.

The eigenvalue Eqs. (4–74) may be written

$$(\mathsf{A} - \lambda \mathsf{1})\mathsf{R} = 0, \tag{4-75}$$

or, in expanded form:

$$\begin{aligned}(a_{11} - \lambda)X + a_{12}Y + a_{13}Z &= 0,\\ a_{21}X + (a_{22} - \lambda)Y + a_{23}Z &= 0,\\ a_{31}X + a_{32}Y + (a_{33} - \lambda)Z &= 0.\end{aligned} \tag{4-76}$$

Eqs. (4–76) comprise a set of 3 homogeneous simultaneous equations for the components X, Y, Z of the eigenvector **R**. As such they can never furnish definite values for the three components, but only ratios of components. Physically, this corresponds to the circumstance that only the *direction* of the eigenvector can be fixed, the magnitude remains undetermined. The product of a constant with an eigenvector is also an eigenvector. In any case, being homogeneous, the Eqs. (4–76) can have a solution only when the determinant of the coefficients vanishes:

$$|\mathsf{A} - \lambda\mathsf{1}| = \begin{vmatrix} a_{11} - \lambda & a_{12} & a_{13} \\ a_{21} & a_{22} - \lambda & a_{23} \\ a_{31} & a_{32} & a_{33} - \lambda \end{vmatrix} = 0. \tag{4-77}$$

* This term is derived from the German *eigenwerte,* literally "proper values."

Eq. (4–77) is known as the *characteristic* or *secular* equation of the matrix, and the values of λ for which the equation is satisfied are the desired eigenvalues. Euler's theorem reduces to the statement that, for the real orthogonal matrices under consideration, the secular equation must have the root $\lambda = +1$.

In general the secular equation will have three roots with three corresponding eigenvectors. For convenience in discussion, the notation X_1, X_2, X_3 will often be used instead of X, Y, Z. In such a notation the components of the eigenvectors might be labelled as X_{ik}; the first subscript indicating the particular component, the second denoting which of the three eigenvectors is involved. A typical member of the group of equations (4–76) would then be written as

$$\sum_j a_{ij}X_{jk} = \lambda_k X_{ik}$$

or, alternatively as:

$$\sum_j a_{ij}X_{jk} = \sum_j X_{ij}\delta_{jk}\lambda_k. \tag{4–78}$$

Both sides of Eq. (4–78) then have the form of a matrix product element; the left side as the product of A with a matrix X having the elements X_{jk}, the right side as the product of X with a matrix whose jkth element is $\delta_{jk}\lambda_k$. The last matrix is diagonal, and its diagonal elements are the eigenvalues of A. We shall therefore designate the matrix by $\boldsymbol{\lambda}$:

$$\boldsymbol{\lambda} = \begin{pmatrix} \lambda_1 & 0 & 0 \\ 0 & \lambda_2 & 0 \\ 0 & 0 & \lambda_3 \end{pmatrix}. \tag{4–79}$$

Eq. (4–78) thus implies the matrix equation:

$$\mathsf{AX} = \mathsf{X}\boldsymbol{\lambda},$$

or, multiplying from the left by X^{-1}:

$$\mathsf{X}^{-1}\mathsf{AX} = \boldsymbol{\lambda}. \tag{4–80}$$

Now, the left side is in the form of a similarity transformation operating on A. (One has only to denote X^{-1} by the symbol Y to reduce it to the form Eq. (4–41).) Thus Eq. (4–80) provides the following alternate approach to the eigenvalue problem: we seek a matrix which transforms A into a diagonal matrix, the elements of which are then the desired eigenvalues.

In proving Euler's theorem it is convenient first to derive some simple lemmas about the nature of the eigenvalues:

1. *The eigenvalues all have unit magnitude.* This follows from the orthogonality of **A**. Although all the elements of **A** are real one must include the possibility that the secular equation has complex roots. In such case the corresponding eigenvectors relate to a complex space, and do not exist in the real, physical space. The magnitude of a complex vector is determined by the sum of the squares, not of the components, but of the *magnitudes* of the components:

$$|X|^2 + |Y|^2 + |Z|^2 = \mathbf{R}^* \cdot \mathbf{R} = |\mathbf{R}|^2. \tag{4–81}$$

The orthogonality condition then requires that the transformation leave the magnitude of the vector **R** unchanged:

$$\mathbf{R}^{*\prime} \cdot \mathbf{R}' = \mathbf{R}^* \cdot \mathbf{R}.$$

However, if **R** is an eigenvector it is also true that:

$$\mathbf{R}^{*\prime} \cdot \mathbf{R}' = \lambda^* \lambda \mathbf{R}^* \cdot \mathbf{R}$$

and hence we have the desired result

$$\lambda^* \lambda = 1. \tag{4–82}$$

Naturally, this lemma provides no information about the phase angle of λ.

2. *A real orthogonal* 3×3 *matrix has at least one real eigenvalue.* The secular equation, (4–77), is a cubic equation for λ of the conventional form:

$$\lambda^3 + b\lambda^2 + c\lambda + d = 0. \tag{4–83}$$

Since the matrix **A** is real the coefficients will all be real. For large negative λ the left side of Eq. (4–83) will be large and negative, while for large positive λ it will be large and positive. Hence a graph of the cubic polynomial must cross the axis at least once between $\lambda = -\infty$ and $\lambda = +\infty$, proving the lemma. By lemma 1 this real root can only be ± 1.

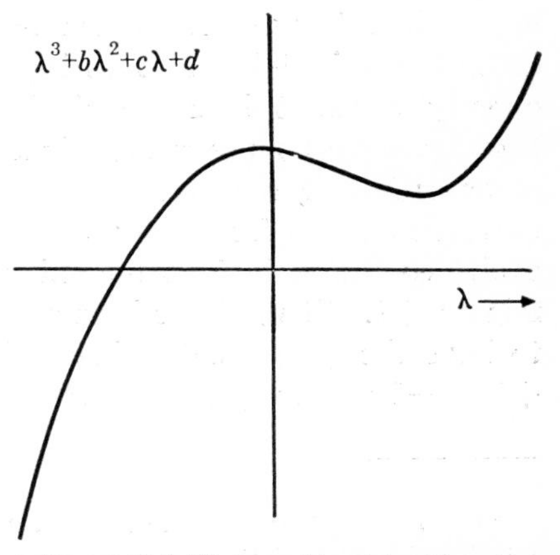

FIG. 4–7. The secular determinant as a function of λ.

The determinant of the matrix $\boldsymbol{\lambda}$ is the product of the three eigenvalues $\lambda_1\lambda_2\lambda_3$. Since the determinant of a matrix is invariant under similarity transformations, it follows that the product is also the determinant of A:

$$\lambda_1\lambda_2\lambda_3 = |\mathsf{A}|.$$

Hence the product of the roots can be at most $+1$ or -1. Lemma 3 now states that the value -1 must be excluded.

3. *The product of the roots of the secular determinant must be* $+1$ *for the possible displacements of a rigid body.* Consider a simple matrix with the determinant -1:

$$\mathsf{S} = \begin{pmatrix} -1 & 0 & 0 \\ 0 & -1 & 0 \\ 0 & 0 & -1 \end{pmatrix} = -1.$$

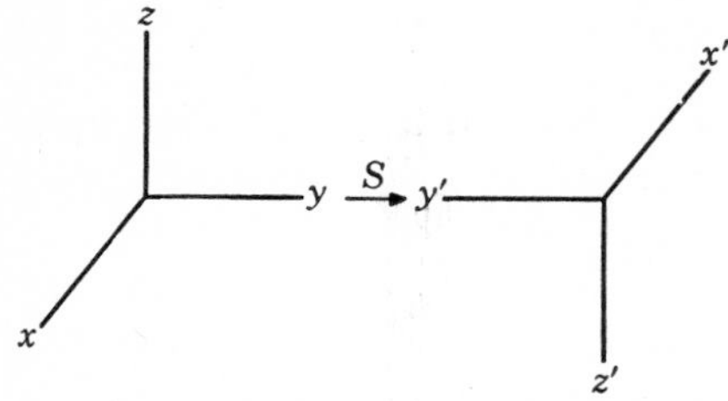

FIG. 4–8. Inversion of the coordinate axes.

The transformation S has the effect of changing the sign of each of the components or coordinate axes (cf. Fig. 4–8). Such an operation transforms a right-handed coordinate system into a left-handed one, and is known as an *inversion* or *reflection* of the coordinate axes.

From the nature of this operation it is clear that an inversion of a right-handed system into a left-handed one cannot be accomplished by any *rigid* change in the orientation of the coordinate axes. An inversion therefore never corresponds to a physical displacement of a rigid body. What is true for S is equally valid for any matrix whose determinant is -1, for any such matrix can be written as the product of S with a matrix whose determinant is $+1$, and thus includes the inversion operation. Consequently it cannot describe a rigid change in orientation. Therefore, the transformations representing rigid body motion must be restricted to matrices having the determinant $+1$. Another method of reaching this conclusion starts from the fact that the matrix of transformation must evolve continuously from the unit matrix, which of course has the determinant $+1$. It would be incompatible with the continuity of the motion to have the matrix determinant change suddenly from its initial value $+1$ to -1 at some given time.*

---

* Orthogonal transformations with a determinant -1 are said to be *improper* rotations in contradistinction to matrices of determinant $+1$ which, it will be shown by Euler's theorem, are proper rotations.

One further lemma will be needed:

4. *The complex conjugate of an eigenvalue is also an eigenvalue.* This conclusion follows directly from the reality of the coefficients in the secular Eq. (4–83). If λ is some solution, then the complex conjugate of Eq. (4–83) shows λ^* is a solution of the same equation. Thus, complex eigenvalues always occur in *pairs*, which are complex conjugates of each other.

With these lemmas established, here now is the proof of Euler's theorem. Consider the possible eigenvalues of the real orthogonal matrix with determinant $+1$. All three of the roots cannot be real and unequal, for real roots can only be $+1$ or -1. If all of the roots are real, and two of them are equal, then the single root must be $+1$, else the determinant would not be $+1$. Except for the trivial case of three equal roots $+1$ (which is the identity transformation), the only remaining possibility is one real and two complex roots. But the two complex roots are complex conjugate and their product is $+1$, so that the single real root must be $+1$ to give the correct determinant. In any nontrivial physical transformation there is thus one, and only one, eigenvalue $+1$, which is the statement of Euler's theorem.

The direction cosines of the axis of rotation can then be obtained by setting $\lambda = 1$ in the eigenvalue Eqs. (4–76) and solving for X, Y, and Z.* The angle of rotation Φ can likewise be obtained without difficulty. By means of some similarity transformation it is always possible to transform the matrix A to a system of coordinates where the z axis lies along the axis of rotation. In such a system of coordinates A' represents a rotation about the z axis through the angle Φ, and therefore has the form

$$\mathsf{A}' = \begin{pmatrix} \cos\Phi & \sin\Phi & 0 \\ -\sin\Phi & \cos\Phi & 0 \\ 0 & 0 & 1 \end{pmatrix}.$$

The trace (cf. Section 4–5) of A' is simply

$$1 + 2\cos\Phi.$$

---

* If there are multiple roots to the secular equation then the corresponding eigenvectors cannot be found as simply, cf. Sections 5–4 and 10–2. Indeed, it is not always possible to completely diagonalize a general matrix if the eigenvalues are not all distinct. These exceptions are of no importance for the present considerations, as Euler's theorem shows that for all nontrivial orthogonal matrices $+1$ is a single root.

Since a trace is always invariant under similarity transformations, this must also be the value of the trace of A in its original form:

$$\sum_i a_{ii} = 1 + 2\cos\Phi, \tag{4-84}$$

which furnishes the rotation angle in terms of the matrix elements. For example, by writing a_{ii} in terms of the Euler angles the rotation angle Φ can be expressed as a function of the successive rotation angles ϕ, θ, ψ.

An immediate corollary of Euler's theorem is

Chasles' theorem: the most general displacement of a rigid body is a translation plus a rotation. Detailed proof is hardly necessary. Simply stated, removing the constraint of motion with one point fixed introduces three translatory degrees of freedom for the origin of the body system of axes.

4-7 Infinitesimal rotations. It is tempting to try to associate a vector with the finite rotation represented by an orthogonal transformation. Certainly a direction can be assigned — that of the axis of rotation — and also a magnitude, e.g., the angle of rotation. But it soon becomes evident that such a correspondence cannot be made successfully. Suppose **A** and **B** are two such "vectors" associated with transformations A and B. Then to qualify as vectors they must be commutative in addition:

$$\mathbf{A} + \mathbf{B} = \mathbf{B} + \mathbf{A}.$$

But the addition of two rotations, i.e., one rotation performed after another, it has been seen, corresponds to the product AB of the two matrices. However, matrix multiplication is not commutative, $\mathsf{AB} \neq \mathsf{BA}$, and hence **A**, **B** are not commutative in addition and cannot be accepted as vectors. This conclusion, that the sum of finite rotations depends on the order of the rotations, is strikingly demonstrated by a simple experiment. Thus Fig. 4-9 illustrates the sequence of events in rotating a block first through

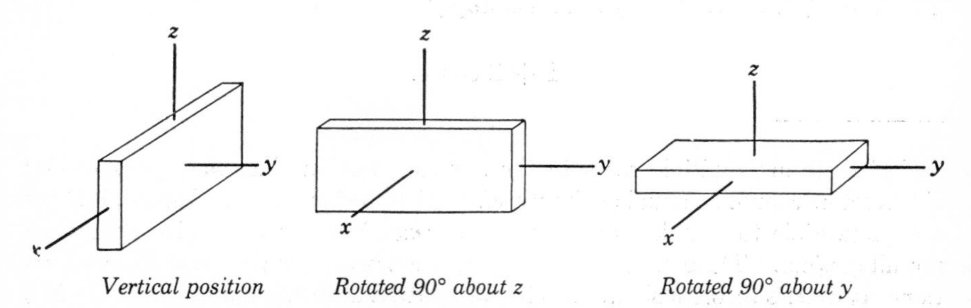

FIG. 4-9. Illustrating the effect of two rotations performed in a given order.

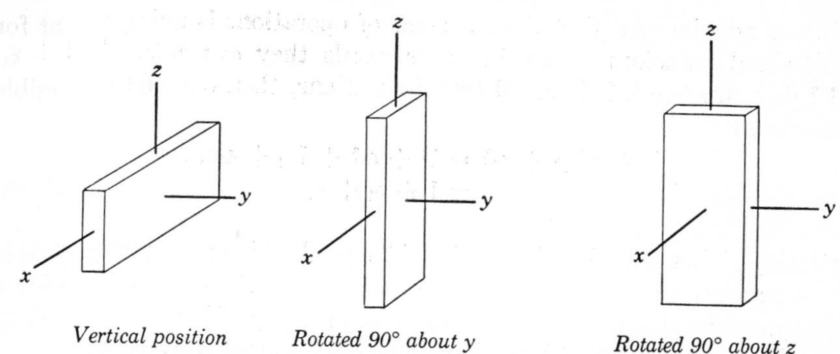

Vertical position *Rotated 90° about y* *Rotated 90° about z*

FIG. 4–10. The two rotations shown in Fig. 4–9, but performed in reverse order.

90° about the z axis and then 90° about the y axis, while Fig. 4–10 presents the same rotations in reverse order. The final position is markedly different in the two sequences.

While a finite rotation thus cannot be represented by a single vector, the same objections do not hold if only *infinitesimal rotations* are considered. An infinitesimal rotation is an orthogonal transformation of coordinate axes in which the components of a vector are almost the same in both sets of axes — the change is infinitesimal. Thus the x_1' component of some vector $\mathbf{r}$ would be practically the same as x_1, the difference being extremely small:

$$x_1' = x_1 + \epsilon_{11}x_1 + \epsilon_{12}x_2 + \epsilon_{13}x_3. \tag{4-85}$$

The matrix elements ϵ_{11}, ϵ_{12} and so on are to be considered as infinitesimals, so that in subsequent calculations only the first nonvanishing order in ϵ_{ij} need be retained. For any general component x_i' the equations of infinitesimal transformation can be written as

$$x_i' = x_i + \sum_j \epsilon_{ij}x_j$$

or

$$x_i' = \sum_j (\delta_{ij} + \epsilon_{ij})x_j. \tag{4-86}$$

The quantity δ_{ij} will be recognized as the element of the unit matrix, and Eq. (4–86) appears in matrix notation as

$$\mathsf{x}' = (\mathsf{1} + \boldsymbol{\epsilon})\mathsf{x}. \tag{4-87}$$

Eq. (4–87) states that the typical form for the matrix of an infinitesimal transformation is $\mathsf{1} + \boldsymbol{\epsilon}$, i.e., it is almost the identity transformation, differing at most by an infinitesimal operator.

It can now be seen that the sequence of operations is unimportant for infinitesimal transformations, in other words they *commute*. If $\mathsf{1} + \boldsymbol{\epsilon}_1$, and $\mathsf{1} + \boldsymbol{\epsilon}_2$ are two infinitesimal transformations, then one of the possible products is:

$$\begin{aligned}(\mathsf{1} + \boldsymbol{\epsilon}_1)(\mathsf{1} + \boldsymbol{\epsilon}_2) &= \mathsf{1}^2 + \boldsymbol{\epsilon}_1\mathsf{1} + \mathsf{1}\boldsymbol{\epsilon}_2 + \boldsymbol{\epsilon}_1\boldsymbol{\epsilon}_2 \\ &= \mathsf{1} + \boldsymbol{\epsilon}_1 + \boldsymbol{\epsilon}_2,\end{aligned} \tag{4–88}$$

neglecting higher order infinitesimals. The product in reverse order merely interchanges $\boldsymbol{\epsilon}_1$ and $\boldsymbol{\epsilon}_2$ which has no effect on the result, as matrix addition is always commutative. The commutative property of infinitesimal transformations removes the objection to their representation by vectors.

The inverse matrix for an infinitesimal transformation is readily obtained. If $\mathsf{A} = \mathsf{1} + \boldsymbol{\epsilon}$ is the matrix of the transformation, then the inverse is

$$\mathsf{A}^{-1} = \mathsf{1} - \boldsymbol{\epsilon}. \tag{4–89}$$

As proof note that the product AA^{-1} reduces to the unit matrix,

$$\mathsf{AA}^{-1} = (\mathsf{1} + \boldsymbol{\epsilon})(\mathsf{1} - \boldsymbol{\epsilon}) = \mathsf{1},$$

in agreement with the definition for the inverse matrix, Eq. (4–32).

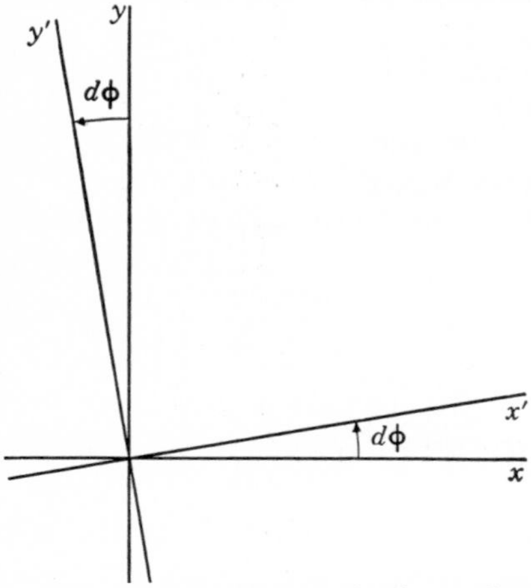

The concept of an infinitesimal transformation may become more explicit if we consider a specific infinitesimal rotation about the z-axis. For a *finite* rotation the matrix has the form (cf. Eq. (4–43)):

$$\mathsf{A} = \begin{pmatrix} \cos\phi & \sin\phi & 0 \\ -\sin\phi & \cos\phi & 0 \\ 0 & 0 & 1 \end{pmatrix}.$$

FIG. 4–11. An infinitesimal rotation about the z-axis.

The matrix of the infinitesimal rotation can be obtained by replacing ϕ in the finite rotation by the infinitesimal $d\phi$ and retaining only first order infinitesimals:

$$\mathsf{1} + \boldsymbol{\epsilon} = \begin{pmatrix} 1 & d\phi & 0 \\ -d\phi & 1 & 0 \\ 0 & 0 & 1 \end{pmatrix}.$$

Correspondingly, the infinitesimal matrix $\boldsymbol{\epsilon}$ is

$$\boldsymbol{\epsilon} = \begin{pmatrix} 0 & d\phi & 0 \\ -d\phi & 0 & 0 \\ 0 & 0 & 0 \end{pmatrix} = d\phi \begin{pmatrix} 0 & 1 & 0 \\ -1 & 0 & 0 \\ 0 & 0 & 0 \end{pmatrix}. \tag{4–90}$$

Notice that the diagonal elements of $\boldsymbol{\epsilon}$ are zero, and the nonvanishing off-diagonal elements are the negative of elements situated symmetrically across the diagonal. A matrix with these characteristics is said to be *antisymmetric* or *skew-symmetric.* This property is not peculiar to the particular matrix considered, for the $\boldsymbol{\epsilon}$ matrix of every infinitesimal rotation is antisymmetric. By (4–89), the inverse matrix A^{-1} is equal to $\mathbf{1} - \boldsymbol{\epsilon}$. But, for an orthogonal transformation the inverse matrix is identical with the transpose $\tilde{\mathsf{A}}$, $= \mathbf{1} + \tilde{\boldsymbol{\epsilon}}$. Consequently $\boldsymbol{\epsilon}$ is the same as the negative transposed matrix:

$$\boldsymbol{\epsilon} = -\tilde{\boldsymbol{\epsilon}}$$

or

$$\epsilon_{ij} = -\epsilon_{ji},$$

which is the definition of an antisymmetric matrix.*

Since the diagonal elements of an antisymmetric matrix are necessarily zero there can only be three distinct elements in any 3×3 antisymmetric matrix. Hence there is no loss of generality in writing $\boldsymbol{\epsilon}$ in the form

$$\boldsymbol{\epsilon} = \begin{pmatrix} 0 & d\Omega_3 & -d\Omega_2 \\ -d\Omega_3 & 0 & d\Omega_1 \\ d\Omega_2 & -d\Omega_1 & 0 \end{pmatrix}. \tag{4–91}$$

The three quantities $d\Omega_1$, $d\Omega_2$, $d\Omega_3$ are clearly to be identified with the three independent parameters specifying the rotation. It will now be shown that these three quantities also form the components of a vector. The *change* in the components of a vector under the infinitesimal transformation will be expressed by the matrix equation

$$\mathsf{x}' - \mathsf{x} = d\mathsf{x} = \boldsymbol{\epsilon}\mathsf{x}, \tag{4–92}$$

---

* It has been assumed implicitly in this section that an infinitesimal orthogonal transformation corresponds to a rotation. In a sense this assumption is obvious; an "infinitesimal inversion" is a contradiction in terms. Formally, the statement follows from the antisymmetry of $\boldsymbol{\epsilon}$. All the diagonal elements of $\mathbf{1} + \boldsymbol{\epsilon}$ are then unity and to first order in small quantities the determinant of the transformation is always $+1$, which is the mark of a rotation.

which in expanded form, with $\boldsymbol{\epsilon}$ given by (4–91), becomes

$$\begin{aligned} dx_1 &= x_2\, d\Omega_3 - x_3\, d\Omega_2 \\ dx_2 &= x_3\, d\Omega_1 - x_1\, d\Omega_3 \\ dx_3 &= x_1\, d\Omega_2 - x_2\, d\Omega_1. \end{aligned} \tag{4–93}$$

The right-hand side of each of Eqs. (4–93) is in the form of a component of the cross product of two vectors, namely, the cross product of $\mathbf{r}$ with a vector $d\boldsymbol{\Omega}$ having components $d\Omega_1$, $d\Omega_2$, $d\Omega_3$. We can therefore write Eq. (4–93) as

$$d\mathbf{r} = \mathbf{r} \times d\boldsymbol{\Omega}. \tag{4–94}$$

However, the mere statement of Eqs. (4–93) in a vector form is not sufficient to prove the vector nature of $d\boldsymbol{\Omega}$. The essential test of a vector is that it have the correct transformation properties under an orthogonal transformation. If $d\boldsymbol{\Omega}$ is truly a vector, its components must transform under an orthogonal matrix B according to the equations

$$d\Omega_i' = \sum_l b_{il}\, d\Omega_l. \tag{4–95}$$

Now, the quantities $d\Omega_l$ have been introduced as the elements of an antisymmetric matrix, and it is not at all obvious that the matrix elements will transform according to Eqs. (4–95). As will be seen, the formal derivation of the actual transformation equations for $d\Omega_l$ is quite complicated. But some simple considerations are enough to show that $d\boldsymbol{\Omega}$ passes most of this test for a vector, although in one respect it fails to make the grade.

Under an orthogonal change of coordinates B, Eq. (4–92), becomes

$$d\mathsf{x}' = \boldsymbol{\epsilon}'\mathsf{x}', \tag{4–92$'$}$$

where $d\mathsf{x}'$ and x' are the transformed column matrices, and $\boldsymbol{\epsilon}'$ is the result of transforming $\boldsymbol{\epsilon}$ by a similarity transformation with the matrix B:

$$\boldsymbol{\epsilon}' = \mathsf{B}\boldsymbol{\epsilon}\mathsf{B}^{-1}.$$

It is proved in one of the exercises that the antisymmetry property is preserved under similarity transformation by an orthogonal matrix. The matrix $\boldsymbol{\epsilon}'$ is therefore antisymmetric with the three distinct elements $d\Omega_1'$, $d\Omega_2'$, $d\Omega_3'$. Consequently Eq. (4–92$'$) can be written in the same form as (4–93), and we are led to a vector relation corresponding to (4–94):

$$d\mathbf{r}' = \mathbf{r}' \times d\boldsymbol{\Omega}'. \tag{4–94$'$}$$

Thus, the elements of an antisymmetric matrix form a vector in *all* cartesian coordinate systems and must therefore transform like the components of a vector.

However, consider the behavior of Eqs. (4-93) under an inversion S (cf. Section 4-6). The components of $\mathbf{r}$ and $d\mathbf{r}$ change sign under such an inversion of the coordinate axes, as must the components of $d\mathbf{\Omega}$, if it is truly to be a vector. But the Eqs. (4-93) remain the same in form in all coordinate systems, and this can be true only if the components of $d\mathbf{\Omega}$ do *not* change sign. Thus $d\mathbf{\Omega}$ has the properties of a vector in all respects except with regard to its behavior under improper rotations.

These conclusions are verified by the actual transformation equations for $d\mathbf{\Omega}$, which we shall now derive. Formally, the quantities $d\Omega_i$ are related to the elements of $\boldsymbol{\epsilon}$ by the equation

$$d\Omega_i = \frac{1}{2}\sum_{j,k=1}^{3} \delta_{ijk}\epsilon_{jk}. \tag{4-96}$$

The quantity δ_{ijk} is known as the *Levi-Civita density*, defined to be zero if any two of the indices ijk are equal, and either $+1$ or -1 otherwise, according as ijk is an even or odd permutation of 123. For example, if $i = 1$, the terms in the summation vanish except for j, k being 2 or 3. Eq. (4-96) then has only two terms:

$$d\Omega_1 = \frac{1}{2}(\delta_{123}\epsilon_{23} + \delta_{132}\epsilon_{32}).$$

By definition $\delta_{123} = 1$, but $\delta_{132} = -1$. However $\epsilon_{32} = -\epsilon_{23}$ so that (4-96) reduces to

$$d\Omega_1 = \frac{1}{2}(\epsilon_{23} + \epsilon_{23}) = \epsilon_{23},$$

in agreement with (4-91).

Similarly $d\Omega_i'$ in the new coordinate system can be written

$$d\Omega_i' = \frac{1}{2}\sum_{j,k} \delta_{ijk}\epsilon_{jk}'.$$

Since $\mathsf{B}^{-1} = \tilde{\mathsf{B}}$, the transformed matrix element, ϵ_{jk}' is related to ϵ_{mn} by

$$\epsilon_{jk}' = \sum_{m,n} b_{jm}\epsilon_{mn}b_{kn},$$

so that $d\Omega_i'$ can be put in the form

$$d\Omega_i' = \frac{1}{2}\sum_{\substack{j,k\\m,n}} \delta_{ijk}b_{jm}b_{kn}\epsilon_{mn}.$$

By making use of the Levi-Civita density one can also express ϵ_{mn} in terms of $d\Omega_l$ according to

$$\epsilon_{mn} = \sum_{l} \delta_{lmn}\, d\Omega_l,$$

and the expression for $d\Omega'$ in terms of $d\Omega$ then becomes

$$d\Omega_i' = \frac{1}{2} \sum_{\substack{j,k,l\\m,n}} \delta_{ijk}\, \delta_{lmn} b_{jm} b_{kn}\, d\Omega_l.$$

It is now asserted that the summation over j, k, m, and n reduces to the simple result

$$\frac{1}{2} \sum_{\substack{j,k\\m,n}} \delta_{ijk}\, \delta_{lmn} b_{jm} b_{kn} = b_{il}\, |\mathsf{B}|. \tag{4–97}$$

The proof is based on the following expression for the value of a determinant:

$$\sum_{l,m,n} \delta_{lmn} b_{il} b_{jm} b_{kn} = |\mathsf{B}|,$$

where i, j, k form an even, or cyclic, permutation of 123. However if the order of i, j, k is an odd permutation of 123, corresponding to an odd number of interchanges of the rows, then the sum becomes the negative of the determinant. It therefore follows that

$$\frac{1}{2} \sum_{\substack{j,k\\l,m,n}} \delta_{ijk}\, \delta_{lmn} b_{il} b_{jm} b_{kn} = |\mathsf{B}|.$$

The orthogonality of B furnishes the identity

$$\sum_l b_{il}^2 = 1,$$

which can be introduced into the above equation to put it in the form

$$\sum_l b_{il} \frac{1}{2} \sum_{\substack{j,k\\m,n}} \delta_{ijk}\, \delta_{lmn} b_{jm} b_{kn} = \sum_l b_{il}(b_{il}\, |\mathsf{B}|).$$

Finally, since the quantities b_{il} are elements of any arbitrary orthogonal matrix the validity of the identity (4–97) is established. With this result the transformation equations for $d\Omega_i$ become

$$d\Omega_i' = |\mathsf{B}| \sum_l b_{il}\, d\Omega_l. \tag{4–98}$$

Eqs. (4–98) are almost identical with the desired vector transformation Eqs. (4–95), the difference consisting in the factor $|\mathsf{B}|$. For proper rotations the two sets of transformation equations agree exactly, but if B contains an inversion the determinant introduces an additional minus sign. These conclusions are in complete agreement with the results of the previous less rigorous discussion. Quantities transforming as in Eq. (4–98) are known as *pseudovectors*, but the more familiar designation is *axial vector*.

It will be noticed that the infinitesimal nature of $\boldsymbol{\epsilon}$ has not been used, only its antisymmetry. Hence the elements of every 3×3 antisymmetrical matrix form the components of a pseudovector.

While $d\mathbf{\Omega}$ in this sense is not quite a vector, the difference for most purposes is unimportant. Indeed many quantities usually thought of as vectors bear this bar sinister, so to speak. Thus any cross product of two vectors must be a pseudovector, for the components of $\mathbf{C} = \mathbf{A} \times \mathbf{B}$ are defined as

$$C_i = A_j B_k - A_k B_j, \qquad i, j, k \text{ in cyclic order.}$$

Under inversion the components both of $\mathbf{A}$ and $\mathbf{B}$ change sign, hence those of $\mathbf{C}$ do not, indicating that it must be a pseudovector. The angular momentum $\mathbf{L} = \mathbf{r} \times \mathbf{p}$, and the magnetic intensity are therefore examples of pseudovectors. The dot product of a pseudovector and a vector is called a *pseudoscalar*. Whereas a true scalar will be completely invariant under an orthogonal transformation, a pseudoscalar changes sign under an improper rotation.

When the question of associating a vector with a rotation was first considered it was suggested that the obvious direction for such a vector would be along the axis of rotation, and that the natural value for the magnitude would be the angle of rotation. It was found impossible to construct a vector to represent a finite rotation, but this difficulty did not occur for infinitesimal rotations. In fact, the matrix description of the infinitesimal rotation automatically led to a vector, $d\mathbf{\Omega}$, specifying the rotation. It can now be shown that the direction and magnitude of $d\mathbf{\Omega}$ agree with the original suggestions made for a finite rotation vector.

From Eq. (4–94) it follows that the only vectors whose components are unaffected by the infinitesimal transformation must be parallel to $d\mathbf{\Omega}$. However, it has been established that vectors unchanged by a rotation must lie along the axis of rotation, which is therefore in the direction of $d\mathbf{\Omega}$. The magnitude of $d\mathbf{\Omega}$ can be evaluated by examining the form of the $\boldsymbol{\epsilon}$ matrix when the z axis of the coordinate system is aligned parallel to the axis of rotation. A comparison of Eq. (4–91) with Eq. (4–90) shows that the magnitude of $d\mathbf{\Omega}$ is then $d\phi$, the angle of rotation. By definition, the magnitude of a vector (or pseudovector) is invariant under orthogonal transformation, and hence the magnitude of $d\mathbf{\Omega}$ must be given by the angle of rotation in all coordinate systems. Indeed one can work backward. If $d\mathbf{\Omega}$ is defined to be along the axis of rotation and with magnitude $d\Omega$ equal to the angle of rotation then Eq. (4–94) can be derived by an elementary demonstration. Thus, consider the change in a vector $\mathbf{r}$ upon rotating it clockwise through a small angle $d\Omega$ about the z-axis (the operation corre-

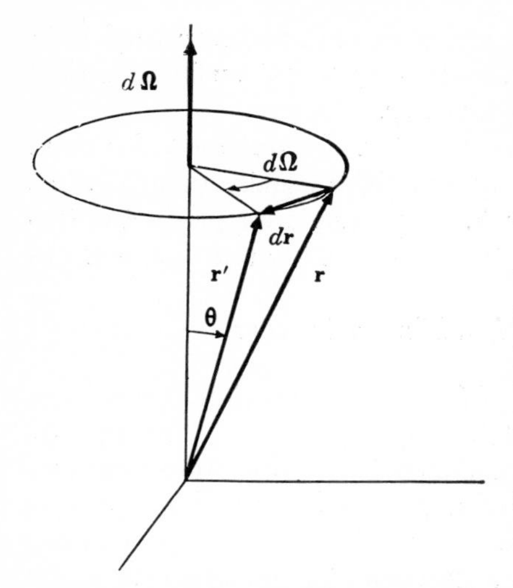

FIG. 4–12. Change in a vector produced by an infinitesimal rotation.

sponding to a counterclockwise rotation of the coordinates). From Fig. 4–12, the magnitude of $d\mathbf{r}$, to first order in $d\Omega$, is

$$dr = r \sin\theta \, d\Omega,$$

which agrees with the magnitude of $\mathbf{r} \times d\mathbf{\Omega}$. Further, $d\mathbf{r}$ must be perpendicular to $d\mathbf{\Omega}$ and $\mathbf{r}$. Finally, the sense is correct, as is shown by Fig. 4–12.

Conversely, the fact that an infinitesimal orthogonal transformation can be written in the form of Eq. (4–94) constitutes an independent proof of Euler's theorem. Any finite displacement of a rigid body with one point fixed can be built up as a succession of infinitesimal displacements. If an infinitesimal transformation is a rotation then the finite transformation must likewise correspond to a rotation.

4–8 Rate of change of a vector. The concept of an infinitesimal rotation provides a powerful tool for describing the motion of a rigid body in time. Consider some arbitrary vector $\mathbf{G}$ involved in the mechanical problem, such as the position vector of a point in the body, or the total angular momentum. Usually such a vector will vary in time as the body moves, but the change will often depend on the coordinate system to which the observations are referred. For example, if the vector happens to be the radius vector from the origin of the body set of axes to a point in the rigid body then, clearly, such a vector appears constant when measured by the body set of axes. However, to an observer fixed in the space set of axes the components of the vector, as measured on *his* axes will vary in time, if the body is in motion.

The change in a time dt of the components of a general vector $\mathbf{G}$ with respect to the body axes will differ from the corresponding change with respect to the space axes only by the effects of the rotation of the body axes. Expressed in symbols we have:

$$(d\mathbf{G})_{\text{body}} = (d\mathbf{G})_{\text{space}} + (d\mathbf{G})_{\text{rot}}.$$

However, the change in the components of a vector, arising solely from an infinitesimal coordinate rotation, is exactly what is given by Eq. (4–94):

$$(d\mathbf{G})_{\text{rot}} = \mathbf{G} \times d\boldsymbol{\Omega}.$$

Transposing, the differential $d\mathbf{G}$, as observed in the space set, is related to the $d\mathbf{G}$, as observed in the body set, by the equation:

$$(d\mathbf{G})_{\text{space}} = (d\mathbf{G})_{\text{body}} + d\boldsymbol{\Omega} \times \mathbf{G}. \tag{4–99}$$

The time *rate of change* of the vector $\mathbf{G}$ is then obtained by dividing Eq. (4–99) by the differential time element dt under consideration:

$$\left(\frac{d\mathbf{G}}{dt}\right)_{\text{space}} = \left(\frac{d\mathbf{G}}{dt}\right)_{\text{body}} + \boldsymbol{\omega} \times \mathbf{G}. \tag{4–100}$$

Here $\boldsymbol{\omega}$ is the *angular velocity* of the body, defined as

$$\boldsymbol{\omega} = \frac{d\boldsymbol{\Omega}}{dt}, \tag{4–101}$$

the instantaneous angular rate of rotation of the body. The vector $\boldsymbol{\omega}$ lies along the axis of the infinitesimal rotation occurring at the time t, a direction which is known as the *instantaneous* axis of rotation.

Eq. (4–100) is not so much an equation about the particular vector $\mathbf{G}$ as it is a statement of the transformation of the time derivative between the two coordinate systems. No conditions were imposed upon the vector in the course of the derivation; and the arbitrary nature of $\mathbf{G}$ can be emphasized by writing Eq. (4–100) as an operator equation acting on some given vector:

$$\left(\frac{d}{dt}\right)_{\text{space}} = \left(\frac{d}{dt}\right)_{\text{body}} + \boldsymbol{\omega} \times . \tag{4–102}$$

As with any vector equation, one may take components of Eq. (4–100) in any desired system of coordinates. Thus

$$\mathbf{F} = \left(\frac{d\mathbf{G}}{dt}\right)_{\text{space}}$$

is the vector time rate of change of $\mathbf{G}$ *as observed in the space set of axes.* But once the vector $\mathbf{F}$ has been obtained, its components may be given with respect to any axes, even the moving axes, as often proves convenient. However, a word of caution is in order: when resolved along the moving axes F_x is *not* the same as

$$\left(\frac{dG_x}{dt}\right)_{\text{space}}$$

For example, the time rate of change of a rotating vector **r** as observed in the fixed set of axes is some definite vector **v**, whose direction and magnitude is fixed by Eq. (4–94). The component of **v** along the x-axis of a system rotating with **r** will not, in general, be zero. On the other hand, the x component of **r** in the same system will be constant in time and its time derivative will be found to vanish, no matter in what system the observer is located. When a time derivative of a vector is with respect to one coordinate system, components may be taken along another set of coordinate axes only *after* the differentiation has been carried out.

It is often convenient to express the angular velocity vector in terms of the Euler angles and their time derivatives. The general infinitesimal rotation associated with $\boldsymbol{\omega}$ can be considered as consisting of three successive infinitesimal rotations with angular velocities $\omega_\phi = \dot{\phi}$, $\omega_\theta = \dot{\theta}$, $\omega_\psi = \dot{\psi}$. In consequence of the vector property of infinitesimal rotations, the vector $\boldsymbol{\omega}$ can be obtained as the sum of the three separate angular velocity vectors. Unfortunately, the directions $\boldsymbol{\omega}_\phi$, $\boldsymbol{\omega}_\theta$, and $\boldsymbol{\omega}_\psi$ are not symmetrically placed: $\boldsymbol{\omega}_\phi$ is along the space z-axis, $\boldsymbol{\omega}_\theta$ is along the line of nodes, while $\boldsymbol{\omega}_\psi$ alone is along the body z'-axis. However, the orthogonal transformations B, C, D of Section 4–4 may be used to furnish the components of these vectors along any desired set of axes.

The body set of axes proves to be the most useful for discussing the equations of motion, and we shall therefore obtain the components of $\boldsymbol{\omega}$ for such a coordinate system. Since $\boldsymbol{\omega}_\phi$ is parallel to the space z-axis, its components along the body axes are given by applying the complete orthogonal transformation $\mathsf{A} = \mathsf{BCD}$ (4–46):

$$(\boldsymbol{\omega}_\phi)_{x'} = \dot{\phi} \sin\theta \sin\psi, \quad (\boldsymbol{\omega}_\phi)_{y'} = \dot{\phi} \sin\theta \cos\psi, \quad (\boldsymbol{\omega}_\phi)_{z'} = \dot{\phi} \cos\theta.$$

The line of nodes, which is the direction of $\boldsymbol{\omega}_\theta$, coincides with the ξ'-axis, so that the components of $\boldsymbol{\omega}_\theta$ with respect to the body axes are furnished by applying only the final orthogonal transformation B (4–45):

$$(\boldsymbol{\omega}_\theta)_{x'} = \dot{\theta} \cos\psi, \quad (\boldsymbol{\omega}_\theta)_{y'} = -\dot{\theta} \sin\psi, \quad (\boldsymbol{\omega}_\theta)_{z'} = 0.$$

No transformation is necessary for the components of $\boldsymbol{\omega}_\psi$ which lies along the z'-axis. Adding these components of the separate angular velocities, the components of $\boldsymbol{\omega}$ with respect to the body axes are:

$$\begin{aligned} \omega_{x'} &= \dot{\phi} \sin\theta \sin\psi + \dot{\theta} \cos\psi \\ \omega_{y'} &= \dot{\phi} \sin\theta \cos\psi - \dot{\theta} \sin\psi \\ \omega_{z'} &= \dot{\phi} \cos\theta + \dot{\psi}. \end{aligned} \tag{4–103}$$

Similar techniques may be used to express the components of $\boldsymbol{\omega}$ along the space set of axes in terms of the Euler angles.

4-9 The Coriolis force. Eq. (4-102) is the basic kinematical law upon which the dynamical equations of motion for a rigid body are founded. But its validity is not restricted solely to rigid body motion. It may be used whenever we wish to discuss the motion of a particle, or system of particles, relative to a rotating coordinate system. A most important problem in this latter category is the description of particle motion relative to coordinate axes rotating with the earth. In classical mechanics it is postulated that Newton's second law of motion, Eq. (1-1), is valid in a reference frame fixed at the center of the sun, and called an *inertial system.*

Terrestrial measurements are usually made with respect to a coordinate system fixed in the earth, which therefore rotates uniformly with a constant angular velocity $\boldsymbol{\omega}$ relative to the inertial system. Eq. (4-102) then provides the modifications of the equations of motion in this noninertial frame.

The initial step is to apply Eq. (4-102) to the radius vector, $\mathbf{r}$, from the origin of the terrestrial system to the given particle:

$$\mathbf{v}_s = \mathbf{v}_r + \boldsymbol{\omega} \times \mathbf{r}, \tag{4-104}$$

where $\mathbf{v}_s$ and $\mathbf{v}_r$ are the velocities of the particle relative to the space and rotating set of axes respectively. In the second step Eq. (4-102) is used to obtain the time rate of change of $\mathbf{v}_s$:

$$\begin{aligned}\left(\frac{d\mathbf{v}_s}{dt}\right)_s = \mathbf{a}_s &= \left(\frac{d\mathbf{v}_s}{dt}\right)_r + \boldsymbol{\omega} \times \mathbf{v}_s \\ &= \mathbf{a}_r + 2(\boldsymbol{\omega} \times \mathbf{v}_r) + \boldsymbol{\omega} \times (\boldsymbol{\omega} \times \mathbf{r}),\end{aligned} \tag{4-105}$$

where $\mathbf{v}_s$ has been substituted from Eq. (4-104), and where $\mathbf{a}_s$ and $\mathbf{a}_r$ are the accelerations of the particle in the two systems. Finally, the equation of motion, which in the inertial system is simply

$$\mathbf{F} = m\mathbf{a}_s,$$

expands, when expressed in the rotating coordinates, into the equation

$$\mathbf{F} - 2m(\boldsymbol{\omega} \times \mathbf{v}_r) - m\boldsymbol{\omega} \times (\boldsymbol{\omega} \times \mathbf{r}) = m\mathbf{a}_r. \tag{4-106}$$

To an observer in the rotating system it therefore appears as if the particle is moving under the influence of an effective force $\mathbf{F}_{\text{eff}}$:

$$\mathbf{F}_{\text{eff}} = \mathbf{F} - 2m(\boldsymbol{\omega} \times \mathbf{v}_r) - m\boldsymbol{\omega} \times (\boldsymbol{\omega} \times \mathbf{r}). \tag{4-107}$$

Let us examine the nature of the terms occurring in Eq. (4-107). The last term is a vector normal to $\boldsymbol{\omega}$ and pointing outward. Further, its magnitude is $m\omega^2 r \sin\theta$. It will therefore be recognized that this term is simply the familiar centrifugal force. When the particle is stationary in the moving system the centrifugal force is the only added term in the effective force. However, when the particle is moving the middle term, known as the *Coriolis* force, comes into play. The order of magnitude of both of these forces may easily be calculated for a particle on the earth's surface. The earth rotates counterclockwise about the North Pole with an angular velocity

$$\omega = \frac{2\pi}{24 \times 3600} = 7.29 \times 10^{-5}\ \text{sec}^{-1}.$$

With this value for ω, and with r equal to the radius of the earth, the maximum centripetal acceleration is

$$\omega^2 r = 3.38\ \text{cm/sec}^2,$$

or about 0.3% of the acceleration of gravity. While small, this acceleration is not completely negligible. The centrifugal force is always repulsive, and at the Equator it will be parallel to the radius vector $\mathbf{r}$. However, at all other latitudes the force will not be along $\mathbf{r}$. Now, the equilibrium position of a plumb bob pendulum is the resultant of the force of gravity and the centrifugal force. Hence, except at the Equator, the plumb bob will not be exactly along the radius vector, although the deviation is quite small. No correction is made for this phenomenon in determining the vertical by means of a plumb bob, because, actually, the vertical is defined as the direction of a plumb bob, rather than as the direction of the radius vector.*

Since the total apparent force of gravity acting on a pendulum is the sum of the actual gravitational force and the centrifugal force, g will vary with latitude, being least at the Equator and greatest at the poles. The flattening of the earth increases this tendency.

We have neglected here the centrifugal force due to revolution about the sun, since it is small compared with the effect of rotation. The angular frequency of revolution is smaller than the angular rotation frequency by the factor 1/365 or about 2.7×10^{-3}. On the other hand, r is larger by the ratio of the radius of the orbit around the sun to the radius of the earth, roughly $10^8\ \text{mi}/4 \times 10^3\ \text{mi} = \frac{1}{4} \times 10^5$. Hence the ratio of the centrifugal force arising from the earth's revolution, to the centrifugal force due to its rotation, is approximately

$$\tfrac{1}{4} \times 10^5 \times (2.7 \times 10^{-3})^2 \simeq 0.2,$$

which is not large enough to be important.

* The vertical may be defined equivalently as the normal to the surface of a liquid in equilibrium.

The Coriolis force on a moving particle is perpendicular to both $\boldsymbol{\omega}$ and $\mathbf{v}$.* In the Northern Hemisphere, where $\boldsymbol{\omega}$ points out of the ground, the Coriolis force, $2m(\mathbf{v} \times \boldsymbol{\omega})$ tends to deflect a projectile shot along the earth's surface, to the right of its direction of travel, cf. Fig. 4-13. The Coriolis deflection reverses direction in the Southern Hemisphere, and is zero at the Equator, where $\boldsymbol{\omega}$ is horizontal. The magnitude of the Coriolis acceleration is always less than

$$2\omega v \simeq 1.5 \times 10^{-4} v,$$

which for a velocity of 10^5 cm/sec (roughly 2,000 mph) is 15 cm/sec^2, or about 0.015 g. Normally, such an acceleration is extremely small but there are instances where it becomes important. To take an artificial illustration, suppose a projectile were fired horizontally from a battleship at the North Pole. The Coriolis acceleration would then have the magnitude $2\omega v$, so that the linear deflection after a time t is $\omega v t^2$; and the angular deflection would be the linear deflection divided by the distance of travel:

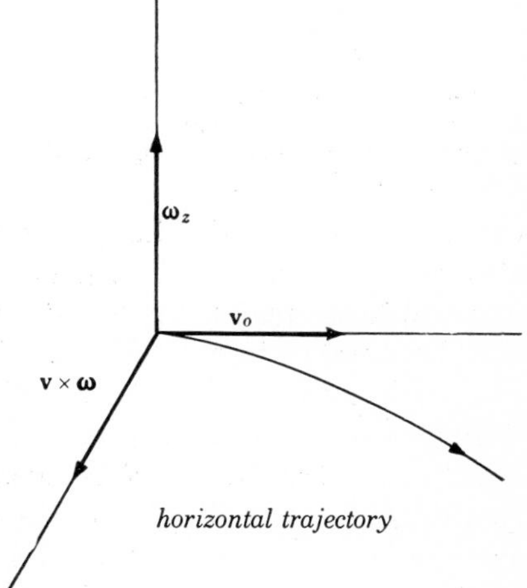

FIG. 4-13. Direction of Coriolis deflection in the Northern Hemisphere.

$$\theta = \frac{\omega v t^2}{vt} = \omega t, \tag{4-108}$$

which is the angle the earth rotates in the time t. Physically, this result means that a projectile shot off at the North Pole has no initial rotational motion and hence its trajectory in the inertial space is a straight line, the apparent deflection being due to the earth rotating beneath it. Some idea of the magnitude of the effect can be obtained by substituting a time of flight of 100 seconds — not unusual for large projectiles — in Eq. (4-108). The angular deflection is then of the order of 7×10^{-3} radians, about 0.4°, which is not inconsiderable. Clearly, the effect is even more important for guided missiles, such as the V-2, which have a much longer time of flight.

The Coriolis force is of still greater significance in the meteorological problem of wind circulation, for then the effective "time of flight" appear-

* From here on, the subscript r will be dropped from v as all velocities will be taken with respect to the rotating coordinate axes only.

ing in Eq. (4–108) is much larger than for projectiles. Wind is simply an air mass in motion, and in the absence of Coriolis forces the direction of motion would be along the pressure gradient, from high to low pressure, and therefore perpendicular to the isobars. However, in the Northern Hemisphere, Coriolis forces deflect the wind to the right of this direction, as shown in Fig. 4–14. At equilibrium the wind patterns are stationary, with the velocity neither increasing nor decreasing, and the net forces on the air mass must vanish. In such case the Coriolis force must be equal and opposite to the pressure gradient force, which requires that the wind direction be parallel to the isobars. A region of low pressure with roughly concentric isobars is known technically as a *cyclone*. As a result of the Coriolis forces the winds around a cyclone circulate in a counterclockwise direction in the Northern Hemisphere, and clockwise in the Southern Hemisphere. Actually, in addition to pressure and Coriolis forces, there are viscosity forces so that at equilibrium the winds are not exactly along the isobars. In northern latitudes the angle between the isobars and the wind direction is about 20°–30°, cf. Fig. 4–15.

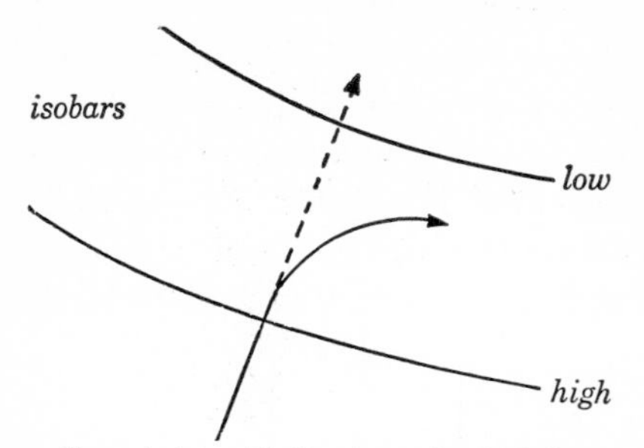

Fig. 4–14. Deflection of wind from the direction of the pressure gradient by the Coriolis force (shown for the Northern Hemisphere).

Another classical experiment where Coriolis force plays an important role is in the deflection from the vertical of a freely falling particle. Since the particle velocity is almost vertical, and $\boldsymbol{\omega}$ lies in the north-south vertical plane, the deflecting force, $2m(\mathbf{v} \times \boldsymbol{\omega})$ is in the east-west direction. Thus, in the Northern Hemisphere, a body falling freely will be deflected to the east. Calculation of the deflection is greatly simplified by choosing the

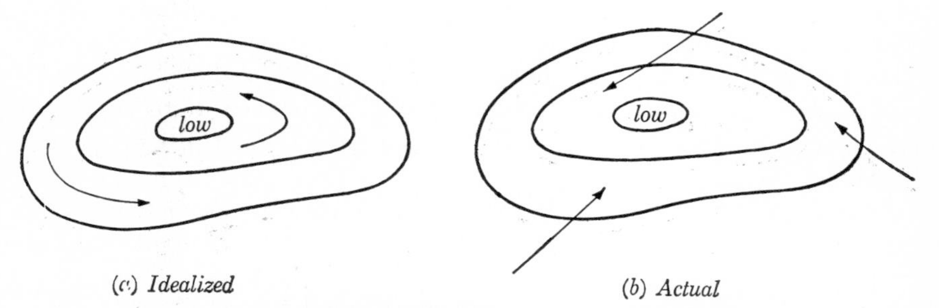

(a) *Idealized* (b) *Actual*

Fig. 4–15. Cyclone pattern in the Northern Hemisphere.

z-axis of the terrestrial coordinate system to be along the direction of the vertical as previously defined. With this choice the centrifugal force term appears only as a slight modification in the value of g. If the yz plane is taken as the north-south vertical plane, then the equation of motion in the x direction is

$$m \frac{d^2x}{dt^2} = -2m(\boldsymbol{\omega} \times \mathbf{v})_x$$
$$= -2m\omega v_z \sin\theta, \tag{4-109}$$

where θ is the colatitude. The effect of the Coriolis force on v_z would constitute a small correction to the deflection, which itself is very small. Hence the vertical velocity appearing in (4-109) may be computed as if Coriolis forces were absent:

$$v_z = gt$$

and

$$t = \sqrt{\frac{2z}{g}}.$$

With these values, Eq. (4-109) may be easily integrated to give the deflection as

$$x = -\frac{\omega g}{3} t^3 \sin\theta$$

or

$$x = -\frac{\omega}{3}\sqrt{\frac{(2z)^3}{g}} \sin\theta.$$

An order of magnitude of the deflection can be obtained by assuming $\theta = \pi/2$ (corresponding to the Equator) and $z = 100$ ft. The deflection is then, roughly,

$$x \simeq 0.15 \text{ inch}.$$

The actual experiment is difficult to perform, as the small deflection may often be masked by the effects of wind currents, viscosity, or other disturbing influences.

More easily observable is the well-known experiment of the Foucault pendulum. If a pendulum is set swinging at the North Pole in a given plane in space, then its linear momentum perpendicular to the plane is zero, and it will continue to swing in this invariable plane while the earth rotates beneath it. To an observer on the earth the plane of oscillation appears to rotate once a day. At other latitudes the result is more complicated, but the phenomenon is qualitatively the same and detailed calculation will be left as an exercise.

Effects due to the Coriolis terms also appear in atomic physics. Thus, two types of motion may occur simultaneously in polyatomic molecules: The molecule *rotates* as a rigid whole, and the atoms *vibrate* about their equilibrium positions. As a result of the vibrations the atoms are in motion relative to the rotating coordinate system of the molecule. The Coriolis term will then be different from zero, and will cause the atoms to move in a direction perpendicular to the original oscillations. Perturbations in molecular spectra due to Coriolis forces thus appear as interactions between the rotational and vibrational levels of the molecule.

SUGGESTED REFERENCES

H. MARGENAU AND G. M. MURPHY, *The Mathematics of Physics and Chemistry.* Many elaborate treatises exist on the theory of matrices, but for our purposes Chapter 10 of this reference forms a more than adequate survey of the mathematics involved. In addition Sections 15.15 and 15.16 contain material on the Cayley-Klein parameters and the Pauli spin matrices (although in a form thickly surrounded with fearsome notation).

H. JEFFREYS AND B. S. JEFFREYS, *Methods of Mathematical Physics.* Though covering much of the same ground as Margenau and Murphy, the emphasis in the discussion of matrices is more on the physical applications. Many of the topics of the present chapter will be found discussed in Chapters 3 and 4, including a treatment of the Pauli spin matrices and their connection with the three-dimensional rotation matrices. The section on Euler angles is practically unintelligible, chiefly due to a poor diagram. Heading each chapter are apt and witty quotations which add an unexpected flavor to the book!

M. BÔCHER, *Introduction to Higher Algebra.* One of the standard older treatises, this book will be found especially valuable for material on determinants, conditions for independence of sets of equations, and solutions of linear independent equations.

R. COURANT AND D. HILBERT, *Methoden der Mathematischen Physik*, Vol. I (not to be confused with the Jeffreys' work above). Courant and Hilbert has long been the classic introduction to the mathematical techniques of theoretical physics. Chapter 1 is on "The algebra of linear transformations and quadratic forms" and discusses, in a powerful and lucid style, the eigenvalue problem for linear transformations, along with many other related topics. Infinitesimal transformations are briefly mentioned in the appendix to Chapter 1.

H. O. NEWBOULT, *Analytical Method in Dynamics.* The use of matrix algebra for the systematic discussion of spatial rotations is rare in texts on classical mechanics, although common practice in quantum mechanics. However, the treatment of the kinematics of rigid body motion given in Chapter II of this slim volume does follow the matrix description to some extent. In particular, the rotation matrix is evaluated in terms of the Euler angles by multiplication of the three elementary rotations. Chapter III also gives a short discussion of motion relative to moving axes.

L. Brillouin, *Les Tenseurs en Mécanique et en Elasticité*. This charmingly written book contains much information on a wide variety of topics, from differential geometry to the quantum mechanics of solids. It is a practically unique source on pseudo quantities, i.e., those which do not have the proper transformation properties under reflection. Chapter III is devoted entirely to such subjects.

E. T. Whittaker, *Analytical Dynamics*. Chapter I contains the material pertinent to our purposes. The section on Eulerian angles is difficult to follow because of the lack of *any* diagram. Reference should be made to our footnote in Section 4–4 in comparing his results with our equations. Section 12 discusses the relation of the Cayley-Klein parameters to the so-called "homographic" transformation.

W. F. Osgood, *Mechanics*. Chapter IX briefly discusses motion relative to moving axes and the effects of centrifugal and Coriolis forces.

G. Herzberg, *Infrared and Raman Spectra*. The effect of Coriolis forces on the spectra of polyatomic molecules is exhaustively discussed in this book, especially in Chapter IV, Sections 1 and 2, although a grounding in small oscillation theory (see our Chapter 10) and quantum mechanics is necessary for a full understanding. See, however, pp. 372–375 for a brief classical treatment.

EXERCISES

1. Prove that matrix multiplication is associative. Show that the product of two orthogonal matrices is also orthogonal.

2. Prove the following properties of the transposed and adjoint matrices:

$$\widetilde{\mathsf{AB}} = \tilde{\mathsf{B}}\tilde{\mathsf{A}},$$
$$(\mathsf{AB})^\dagger = \mathsf{B}^\dagger\mathsf{A}^\dagger.$$

3. Show that the trace of a matrix is invariant under any similarity transformation. Show also that the antisymmetry property of a matrix is preserved under an orthogonal similarity transformation, while the hermitean property is invariant under any unitary similarity transformation.

4. Obtain the matrix elements of the general rotation matrix in terms of the Euler angles, Eq. (4–46), by performing the multiplications of the successive component rotation matrices. Verify directly that the matrix elements obey the orthogonality conditions.

5. Show that the components of the angular velocity along the space set of axes are given in terms of the Euler angles by

$$\omega_x = \dot\theta \cos\phi + \dot\psi \sin\theta \sin\phi,$$
$$\omega_y = \dot\theta \sin\phi - \dot\psi \sin\theta \cos\phi,$$
$$\omega_z = \dot\psi \cos\theta + \dot\phi.$$

6. Express the "rolling" constraint of a sphere on a plane surface in terms of the Euler angles. Show that the conditions are nonintegrable and that the constraint is therefore nonholonomic.

7. Show that the two complex eigenvalues of an orthogonal matrix representing a proper rotation are $e^{\pm i\Phi}$, where Φ is the angle of the rotation.

8. Verify that the angle of rotation Φ is given in terms of the Euler angles by the expression

$$\cos\frac{\Phi}{2} = \cos\frac{\phi + \psi}{2}\cos\frac{\theta}{2}.$$

9. Show that the three Pauli spin matrices anticommute with each other, i.e., that

$$\boldsymbol{\sigma}_i\boldsymbol{\sigma}_j = -\boldsymbol{\sigma}_j\boldsymbol{\sigma}_i, \qquad i \neq j$$

and that

$$\boldsymbol{\sigma}_i\boldsymbol{\sigma}_j - \boldsymbol{\sigma}_j\boldsymbol{\sigma}_i = 2i\boldsymbol{\sigma}_k. \qquad i, j, k \text{ in cyclic order.}$$

Prove also that $\boldsymbol{\sigma}_i^2 = \mathbf{1}$, for all values of i.

10. Show that $\mathbf{Q}_\theta$ may be written symbolically in the form

$$\mathbf{Q}_\theta = e^{i\sigma_x\frac{\theta}{2}},$$

where the exponential is taken as standing for its series expansion, whose first term is $\mathbf{1}$.

11. A projectile is fired horizontally along the earth's surface. Show that to a first approximation the angular deviation from the direction of fire resulting from the Coriolis force varies linearly with time at a rate

$$\omega \cos\theta,$$

where ω is the angular frequency of the earth's rotation and θ is the colatitude, the direction of deviation being to the right in the northern hemisphere.

12. The Foucault pendulum experiment consists in setting a long pendulum in motion at a point on the surface of the rotating earth with its momentum originally in the vertical plane containing the pendulum bob and the point of suspension. Show that its subsequent motion may be described by saying that the plane of oscillation rotates uniformly $2\pi \cos\theta$ radians per day, where θ is the colatitude. What is the direction of rotation? The approximation of small oscillations may be used, if desired.

CHAPTER 5

THE RIGID BODY EQUATIONS OF MOTION

Chapter 4 has presented all the kinematical tools needed in the discussion of rigid body motion. In the Euler angles we have a set of three coordinates, defined rather unsymmetrically it is true, yet suitable for use as the generalized coordinates describing the orientation of the rigid body. In addition, the method of orthogonal transformations, and the associated matrix algebra, furnish a powerful and elegant technique for investigating the characteristics of rigid body motion. We have already had one application of the technique in deriving Eq. (4–100), the relation between the rates of change of a vector as viewed in the space system and in the body system. These tools will now be applied to obtain the dynamical equations of motion of the rigid body in their most convenient form. With the help of the equations of motion, some simple but highly important problems of rigid body motion can be discussed.

5–1 Angular momentum and kinetic energy of motion about a point. Chasle's theorem states that any general displacement of a rigid body can be represented by a translation plus a rotation. The theorem suggests that it ought to be possible to split the problem of rigid body motion into two separate phases, one concerned solely with the translational motion of the body, the other, with its rotational motion. Of course, if one point of the body is fixed the separation is obvious, for then there is only a rotational motion about the fixed point, without any translation. But even for a general type of motion such a separation is often possible. The six coordinates needed to describe the motion have already been formed into two sets in accordance with such a division: the three cartesian coordinates, of a point fixed in the rigid body to describe the translational motion and, say, the three Euler angles for the motion about the point. If, further, the origin of the body system is chosen to be the center of mass, then by Eq. (1–26) the total angular momentum divides naturally into contributions from the translation of the center of mass and from the rotation about the center of mass. The former term will involve only the cartesian coordinates of the center of mass, the latter only the angle coordinates. By Eq. (1–29) a similar division holds for the total kinetic energy T, which can be written in the form

$$T = \tfrac{1}{2}Mv^2 + T'(\phi, \theta, \psi),$$

as the sum of the kinetic energy of the entire body as if concentrated at the center of mass, plus the kinetic energy of motion about the center of mass.

Often the potential energy can be similarly divided, each term involving only one of the coordinate sets, either the translational or rotational. Thus, the gravitational potential energy will depend only upon the cartesian vertical coordinate of the center of gravity.* Or if the force on a body is due to a uniform magnetic field, **B**, acting on its magnetic dipole moment, **M**, then the potential is proportional to $\mathbf{M} \cdot \mathbf{B}$, which involves only the orientation of the body. Certainly, almost all problems soluble in practice will allow of such a separation. In such case the entire mechanical problem does indeed split into two, for the Lagrangian, $L = T - V$, divides into two parts, one involving only the translational coordinates, the other only the angle coordinates. These two groups of coordinates will then be completely separated, and the translational and rotational problems can be solved independently of each other. It is of obvious importance, therefore, to obtain expressions for the angular momentum and kinetic energy of the motion about some point fixed in the body.

When a rigid body moves with one point stationary, the total angular momentum about that point is

$$\mathbf{L} = \sum_i m_i(\mathbf{r}_i \times \mathbf{v}_i), \tag{5–1}$$

where $\mathbf{r}_i$ and $\mathbf{v}_i$ are the radius vector and velocity, respectively, of the ith particle relative to the given point. Since $\mathbf{r}_i$ is a fixed vector relative to the body, the velocity $\mathbf{v}_i$ with respect to the space set of axes arises solely from the rotational motion of the rigid body about the fixed point. From Eq. (4–100), $\mathbf{v}_i$ is then

$$\mathbf{v}_i = \boldsymbol{\omega} \times \mathbf{r}_i. \tag{5–2}$$

Hence Eq. (5–1) can be written as

$$\mathbf{L} = \sum_i m_i(\mathbf{r}_i \times (\boldsymbol{\omega} \times \mathbf{r}_i)),$$

or, expanding the triple cross product:

$$\mathbf{L} = \sum_i m_i(\boldsymbol{\omega} r_i^2 - \mathbf{r}_i(\mathbf{r}_i \cdot \boldsymbol{\omega})). \tag{5–3}$$

Again expanding, the x component of the angular momentum becomes

$$L_x = \omega_x \sum_i m_i(r_i^2 - x_i^2) - \omega_y \sum_i m_i x_i y_i - \omega_z \sum_i m_i x_i z_i, \tag{5–4}$$

* The center of gravity, of course, coincides with the center of mass in a uniform gravitational field.

with similar equations for the other components of **L**. Thus each of the components of the angular momentum is a linear function of all the components of the angular velocity. *The angular momentum vector is related to the angular velocity by a linear transformation.* To emphasize the similarity of (5–4) with the equations of a linear transformation, (4–12), we may write L_x as

$$L_x = I_{xx}\omega_x + I_{xy}\omega_y + I_{xz}\omega_z. \tag{5–5}$$

Analogously, for L_y and L_z we have:

$$\begin{aligned} L_y &= I_{yx}\omega_x + I_{yy}\omega_y + I_{yz}\omega_z \\ L_z &= I_{zx}\omega_x + I_{zy}\omega_y + I_{zz}\omega_z. \end{aligned} \tag{5–5}$$

The nine coefficients I_{xx}, I_{xy}, etc. are the nine elements of the transformation matrix. The diagonal elements are known as *moment of inertia coefficients,* and have the form illustrated by

$$I_{xx} = \sum_i m_i(r_i^2 - x_i^2), \tag{5–6}$$

while the off-diagonal elements are designated as *products of inertia,* a typical one being

$$I_{xy} = -\sum_i m_i x_i y_i. \tag{5–7}$$

In Eqs. (5–6) and (5–7) the matrix elements appear in the form suitable if the rigid body is composed of discrete particles. For continuous bodies the summation is replaced by a volume integration, with the particle mass becoming a mass density. Thus, the diagonal element I_{xx} appears as

$$I_{xx} = \int_V \rho(\mathbf{r})(r^2 - x^2)\, dV. \tag{5–6$'$}$$

Up to the present, the coordinate system used in resolving the components of **L** has not been specified. From now on, it will be convenient to take it to be a system fixed in the body.* The various distances x_i, y_i, z_i are then constant in time, so that the matrix elements are likewise constants, peculiar to the body involved, and dependent on the origin and orientation of the particular body set of axes in which they are expressed.

The equations (5–5) relating the components of **L** and $\boldsymbol{\omega}$ can be summarized by a single operator equation,

$$\mathbf{L} = \mathbf{I}\boldsymbol{\omega}, \tag{5–8}$$

* In Chapter 4, such a system was denoted by primes. As components along spatial axes will rarely be used here, this convention will be dropped from now on to simplify the notation.

where the symbol I stands for the operator whose matrix elements are the inertia coefficients appearing in (5–5). Of the two interpretations which have been given to the operator of a linear transformation (cf. Section 4–2), it is clear that here I must be thought of as acting upon the vector $\boldsymbol{\omega}$, and not upon the coordinate system. The vectors $\mathbf{L}$ and $\boldsymbol{\omega}$ are two physically different vectors, having different dimensions, and are not merely the same vector expressed in two different coordinate systems. Unlike the operator of rotation, I will have dimensions — mass times length squared — and it is not restricted by any orthogonality conditions. Eq. (5–8) is to be read as saying that the operator I acting upon the vector $\boldsymbol{\omega}$ results in the physically new vector $\mathbf{L}$. While full use will be made of the matrix algebra techniques developed in the discussion of the rotation operator, more attention must be paid here to the nature and physical character of the operator per se.

5–2 Tensors and dyadics. The quantity I may be considered as defining the quotient of $\mathbf{L}$ and $\boldsymbol{\omega}$:

$$\mathsf{I} = \frac{\mathbf{L}}{\boldsymbol{\omega}},$$

for the product of I and $\boldsymbol{\omega}$ gives $\mathbf{L}$. Now, the quotient of two quantities is often not a member of the same class as the dividing factors, but may belong to a more complicated class. Thus, the quotient of two integers is in general not an integer but rather a rational number. Similarly, the quotient of two vectors, as is well known, cannot be defined consistently within the class of vectors. It is not surprising therefore to find that I is a new type of quantity, a *tensor of the second rank.*

In three-dimensional space, a tensor T of the Nth rank may be defined for our purposes as a quantity having 3^N components $T_{ijk\ldots}$ (with N indices) which transform under an orthogonal transformation of coordinates, A, according to the following scheme: *

$$T'_{ijk\ldots} = \sum_{l,m,n,\ldots} a_{il}a_{jm}a_{kn}\ldots T_{lmn\ldots}. \tag{5–9}$$

By this definition, a tensor of the zero rank has one component, which is invariant under orthogonal transformation. Hence, *a scalar is a tensor of zero rank.* A tensor of the first rank has three components transforming as

$$T'_i = \sum_j a_{ij}T_j.$$

* The distinction between covariant and contravariant indices will be disregarded here, as it is of no significance for cartesian coordinate systems.

Comparison with the transformation equations for a vector, (4–14), shows that a *tensor of the first rank is completely equivalent to a vector.*

Finally, the nine components of a tensor of the second rank transform as

$$T'_{ij} = \sum_{k,l} a_{ik}a_{jl}T_{kl}. \tag{5–10}$$

The transformation properties of the components of I are determined by the fact that the matrix of I transforms under A by a similarity transformation:

$$\mathsf{I}' = \mathsf{A}\mathsf{I}\mathsf{A}^{-1},$$

or, since A is orthogonal,

$$\mathsf{I}' = \mathsf{A}\mathsf{I}\tilde{\mathsf{A}}.$$

The ijth element of the transformed matrix is then

$$I'_{ij} = \sum_{k,l} a_{ik}I_{kl}\tilde{a}_{lj} = \sum_{k,l} a_{ik}a_{jl}I_{kl}, \tag{5–11}$$

which agrees in form with (5–10), confirming the identification of I as a tensor of the second rank.

Rigorously speaking, one must distinguish between the tensor I and the square matrix formed from its components. A tensor is defined only in terms of its transformation properties under orthogonal coordinate transformations. On the other hand, a matrix is in no way restricted in the types of transformations it may undergo, and indeed may be considered entirely independently of its properties under some particular class of transformation. Nevertheless, the distinction must not be stressed unduly. Within the restricted domain of orthogonal transformations there is a practical identity. The tensor components and the matrix elements are manipulated in the same fashion; and for every tensor equation there will be a corresponding matrix equation, and vice versa. The equivalence between the tensor and the matrix is not restricted to tensors of the second rank. For example, we already know that the components of a vector, which is a tensor of the first rank, form a column matrix, and vector manipulation may be treated completely in terms of these associated matrices.

Still another useful representation of the operator I is as a dyadic. A *dyad* is simply a pair of vectors, written in a definite order $\mathbf{AB}$, $\mathbf{A}$ being known as the *antecedent* and $\mathbf{B}$ the *consequent.* The scalar dot product of a dyad with a vector $\mathbf{C}$ can be performed in two ways, either as

$$\mathbf{AB} \cdot \mathbf{C} = \mathbf{A}(\mathbf{B} \cdot \mathbf{C})$$

or as

$$\mathbf{C} \cdot \mathbf{AB} = \mathbf{B}(\mathbf{A} \cdot \mathbf{C}).$$

In the first case **C** is called the *postfactor*, in the second the *prefactor*. The two products, in general, will not be equal — dyad scalar multiplication is not commutative. In both cases, it should be noted, the result of the dot product is to produce a vector having a direction and magnitude in general different from **C**. One can also define the double dot product of two dyads as the scalar given by

$$\mathbf{AB} : \mathbf{CD} = (\mathbf{A} \cdot \mathbf{C})(\mathbf{B} \cdot \mathbf{D}).$$

A more convenient notation is to write the double dot product as

$$\mathbf{AB} : \mathbf{CD} \equiv \mathbf{C} \cdot \mathbf{AB} \cdot \mathbf{D}.$$

A *dyadic* is defined as a linear polynomial of dyads:

$$\mathbf{AB} + \mathbf{CD} + \cdots.$$

Actually, any dyad **AB** can be expressed as a dyadic by writing the vectors **A** and **B** in component form in terms of the unit vectors **i**, **j**, and **k**. When expanded in this manner, the dyad appears as

$$\begin{aligned} \mathbf{AB} = \; & A_xB_x\mathbf{ii} + A_xB_y\mathbf{ij} + A_xB_z\mathbf{ik} \\ & + A_yB_x\mathbf{ji} + A_yB_y\mathbf{jj} + A_yB_z\mathbf{jk} \\ & + A_zB_x\mathbf{ki} + A_zB_y\mathbf{kj} + A_zB_z\mathbf{kk}. \end{aligned} \tag{5–12}$$

Eq. (5–12) is the *nonion* form of the dyad, so called from the nine coefficients involved. Obviously dyadics in like fashion can always be reduced to a nonion form. Since the coefficients in the nonion representation of a dyadic are homogeneous quadratic functions of vector components, they will clearly have the transformation properties characteristic of second rank tensor components, cf. Eq. (5–10). Conversely, a dyadic can always be formed from a tensor of the second rank, by using the tensor components as the nonion coefficients. There is thus a complete formal identity between a dyadic and a tensor of the second rank. They are also equivalent in the effect they produce as operators acting on vectors, for it has been seen that the dot product of a dyad or dyadic with a vector results in a new vector. Indeed, one may write the operator **I** so as to show its dyadic form explicitly by introducing the unit dyadic **1**:

$$\mathbf{1} = \mathbf{ii} + \mathbf{jj} + \mathbf{kk}. \tag{5–13}$$

The designation is certainly well merited for the matrix of **1** is exactly the unit matrix, and direct multiplication shows that

$$\mathbf{1} \cdot \mathbf{A} = \mathbf{A} \cdot \mathbf{1} = \mathbf{A}.$$

In dyadic notation I may be written as

$$\mathsf{I} = \sum_i m_i(r_i^2 \mathsf{1} - \mathbf{r}_i\mathbf{r}_i),$$

for

$$\mathsf{I} \cdot \boldsymbol{\omega} = \sum_i m_i(r_i^2\boldsymbol{\omega} - \mathbf{r}_i(\mathbf{r}_i \cdot \boldsymbol{\omega})) = \mathbf{L}, \tag{5–14}$$

in agreement with Eq. (5–3).

5–3 The inertia tensor and the moment of inertia. The quantity I may therefore be designated equally well as a second rank tensor or as a dyadic; it is usually called the *moment of inertia tensor* (or *dyadic*), or more briefly as the *inertia tensor*. The advantage of using the dyadic form for I is that the familiar methods of vector manipulation can still be employed. Thus, one is led in a natural fashion to express the kinetic energy of rotation in terms of the dyadic I. The kinetic energy of motion about a point is

$$T = \frac{1}{2}\sum_i m_i v_i^2,$$

where $\mathbf{v}_i$ is the velocity relative to the fixed point. By Eq. (5–2), T may also be written as

$$T = \frac{1}{2}\sum_i m_i \mathbf{v}_i \cdot (\boldsymbol{\omega} \times \mathbf{r}_i),$$

which, upon permuting the vectors in the triple dot product, becomes

$$T = \frac{\boldsymbol{\omega}}{2} \cdot \sum_i m_i(\mathbf{r}_i \times \mathbf{v}_i).$$

The summation will be recognized as the angular momentum of the body about the origin, and in consequence the kinetic energy can be written in the form

$$T = \frac{\boldsymbol{\omega} \cdot \mathbf{L}}{2} = \frac{\boldsymbol{\omega} \cdot \mathsf{I} \cdot \boldsymbol{\omega}}{2}. \tag{5–15}$$

Let $\mathbf{n}$ be a unit vector in the direction of $\boldsymbol{\omega}$ so that $\boldsymbol{\omega} = \omega\mathbf{n}$. Then an alternative form for the kinetic energy is

$$T = \frac{\omega^2}{2}\mathbf{n} \cdot \mathsf{I} \cdot \mathbf{n} = \frac{1}{2} I\omega^2, \tag{5–16}$$

where I is a scalar, defined by

$$I = \mathbf{n} \cdot \mathsf{I} \cdot \mathbf{n} = \sum_i m_i(r_i^2 - (\mathbf{r}_i \cdot \mathbf{n})^2), \tag{5–17}$$

and known as the *moment of inertia about the axis of rotation.*

In the usual elementary discussions the moment of inertia about an axis is defined as the sum, over the particles of the body, of the product of the particle mass and the square of the perpendicular distance from the axis. It must be shown that this definition is in accord with the expression given in Eq. (5–17). The perpendicular distance is equal to the magnitude of the vector $\mathbf{r}_i \times \mathbf{n}$ (cf. Fig. 5–1). Therefore, the customary definition of I may be written as

$$I = \sum_i m_i(\mathbf{r}_i \times \mathbf{n}) \cdot (\mathbf{r}_i \times \mathbf{n}).$$

Considering the first cross product as a vector by itself, one may permute the vectors of the triple dot product, and write I as

$$I = \sum_i m_i \mathbf{r}_i \cdot \mathbf{n} \times (\mathbf{r}_i \times \mathbf{n}).$$

Finally, upon expanding the triple cross product the expression for I becomes

$$I = \sum_i m_i \mathbf{r}_i \cdot (\mathbf{r}_i - \mathbf{n}(\mathbf{r}_i \cdot \mathbf{n}))$$
$$= \sum_i m_i(r_i^2 - (\mathbf{r}_i \cdot \mathbf{n})^2),$$

in agreement with Eq. (5–17).

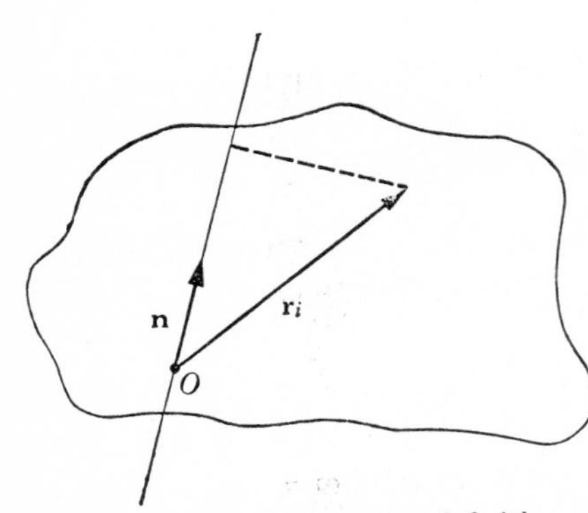

FIG. 5–1. Illustrating the definition of the moment of inertia.

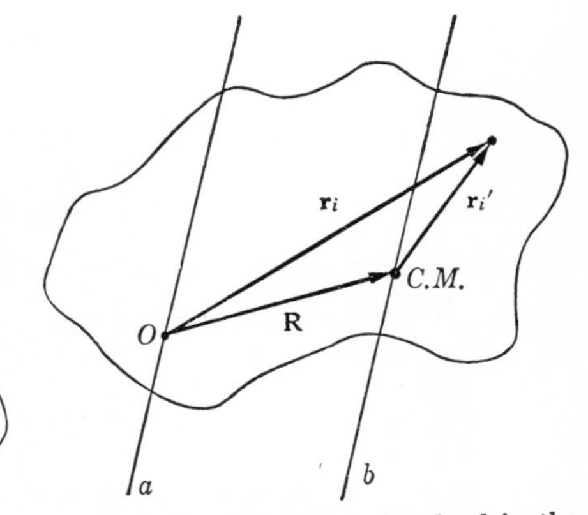

FIG. 5–2. The vectors involved in the relation between moments of inertia about parallel axes.

The value of the moment of inertia depends upon the direction of the axis of rotation. As $\boldsymbol{\omega}$ usually changes its direction with respect to the body in the course of time, the moment of inertia must also be considered a function of time. When the body is constrained so as to rotate only about a fixed axis then the moment of inertia is a constant. In such case the kinetic energy in (5-16) is almost in the form required to fashion the Lagrangian and the equations of motion. The one further step needed is to express ω as the time derivative of some angle, which can usually be done without difficulty.

Along with the inertia tensor, the moment of inertia depends also upon the choice of origin of the body set of axes. However, the moment of inertia about some given axis is related simply to the moment about a parallel axis through the center of mass. Let the vector from the given origin O to the center of mass be $\mathbf{R}$, and let the radii vectors from O and the center of mass to the ith particle be $\mathbf{r}_i$ and $\mathbf{r}'_i$, respectively. The three vectors so defined are connected by the relation (cf. Fig. 5-2)

$$\mathbf{r}_i = \mathbf{R} + \mathbf{r}'_i.$$

The moment of inertia about the axis a is therefore

$$I_a = \sum_i m_i(\mathbf{r}_i \times \mathbf{n})^2 = \sum_i m_i[(\mathbf{r}'_i + \mathbf{R}) \times \mathbf{n}]^2$$

or

$$I_a = \sum_i m_i(\mathbf{R} \times \mathbf{n})^2 + \sum_i m_i(\mathbf{r}'_i \times \mathbf{n})^2 + 2 \sum_i m_i(\mathbf{R} \times \mathbf{n}) \cdot (\mathbf{r}'_i \times \mathbf{n}).$$

The last term in this expression can be rearranged as

$$-2(\mathbf{R} \times \mathbf{n}) \cdot (\mathbf{n} \times \sum_i m_i \mathbf{r}'_i).$$

By the definition of center of mass, the summation $\sum_i m_i \mathbf{r}'_i$ vanishes. Hence I_a can be expressed in terms of moment about the parallel axis b as

$$I_a = I_b + M(\mathbf{R} \times \mathbf{n})^2. \tag{5-18}$$

The magnitude of $\mathbf{R} \times \mathbf{n}$ is the perpendicular distance of the center of mass from the axis passing through O.

Consequently the moment of inertia about a given axis is equal to the moment of inertia about a parallel axis through the center of mass plus the moment of inertia of the body, as if concentrated at the center of mass, with respect to the original axis. One is thus led to a type of resolution for the moment of inertia very similar to that found for the linear and angular momentum and the kinetic energy in Section (1-2).

5-4 The eigenvalues of the inertia tensor and the principal axis transformation. The preceding discussion has served to emphasize the important role the inertia tensor plays in the discussion of the motion of rigid bodies. An examination, at this point, of the properties of this tensor and its associated matrix will therefore prove of considerable interest. From the defining equation, (5–7), it is seen that the components of the tensor are symmetrical, that is

$$I_{xy} = I_{yx}.$$

Since the components are real it follows that the tensor is equal to its adjoint (cf. Eq. 4–38), which is to say $\mathbf{I}$ is *self-adjoint* or *hermitean*. Thus, while the inertia tensor will in general have nine components, only six of them will be independent — the three along the diagonal plus three of the off-diagonal elements.

The inertia coefficients depend both upon the location of the origin of the body set of axes and upon the orientation of these axes with respect to the body. It would be very convenient, of course, if for a given origin one could find a particular orientation of the body axes for which the inertia tensor is diagonal, so that the dyadic could be written as

$$\mathbf{I}' = I_1\mathbf{ii} + I_2\mathbf{jj} + I_3\mathbf{kk}. \tag{5–19}$$

With respect to such a set of axes each of the components of $\mathbf{L}$ would involve only the corresponding component of $\boldsymbol{\omega}$, thus:

$$L_x = I_1\omega_x, \qquad L_y = I_2\omega_y, \qquad L_z = I_3\omega_z. \tag{5–20}$$

A similar simplification would also occur in the form of the kinetic energy:

$$T = \frac{\boldsymbol{\omega}\cdot\mathbf{I}\cdot\boldsymbol{\omega}}{2} = \frac{1}{2}I_1\omega_x^2 + \frac{1}{2}I_2\omega_y^2 + \frac{1}{2}I_3\omega_z^2. \tag{5–21}$$

One can show that it is always possible to find such axes, and the proof is based essentially on the hermitean nature of the inertia tensor.

It has been remarked, in Section 4–6, that the eigenvalue equation of a matrix can be solved by transforming the matrix to a diagonal form, the elements of which are then the desired eigenvalues. Hence the problem of finding a set of axes in which $\mathbf{I}$ is diagonal is equivalent to the eigenvalue problem for the matrix of $\mathbf{I}$, I_1, I_2 and I_3 being the three eigenvalues. It also follows that for the coordinate system in which $\mathbf{I}$ is diagonal the direction of the axes coincides with the direction of the eigenvectors. For example, suppose $\boldsymbol{\omega}$ is along one of the axes, say the x-axis. Then, by (5–20) the angular momentum $\mathbf{L} = \mathbf{I}\cdot\boldsymbol{\omega}$ is also along the x-axis. The effect of $\mathbf{I}$ on any vector parallel to the three coordinate axes is thus to produce another vector in the same direction. By definition such a vector must therefore be one of the eigenvectors of $\mathbf{I}$.

In Section 4–6 we outlined the scheme for diagonalizing any matrix and finding its eigenvalues. By itself, however, this procedure does not constitute a proof for the existence of a *real cartesian* coordinate system in which **I** is diagonal. Thus, it will be recalled that an orthogonal matrix, except for trivial cases, has only one real eigenvalue and therefore only one real direction corresponding to an eigenvector (namely, the axis of rotation). In contrast, what we seek to prove here is that *all* the eigenvalues of **I** are real, and that the three real directions of the eigenvectors are mutually orthogonal.*

Let X_{kj} be the kth component of the jth eigenvector of **I**, following the notation of Section 4-6. The eigenvalue equation is then

$$\sum_k I_{ik}X_{kj} = I_jX_{ij}. \tag{5-22}$$

Multiply Eq. (5-22) by X^*_{il} and sum over the index i, obtaining:

$$\sum_{i,k} X^*_{il}I_{ik}X_{kj} = I_j \sum_i X^*_{il}X_{ij}. \tag{5-23}$$

The summation on the right will be recognized as $\mathbf{R}^*_l \cdot \mathbf{R}_j$, the dot product of the complex conjugate of the lth eigenvector with the jth eigenvector. The complex conjugate of the similar eigenvalue equation for I_l is:

$$\sum_i I^*_{ki}X^*_{il} = I^*_lX^*_{kl}.$$

Multiplying this time by X_{kj} and summing over k we obtain

$$\sum_{i,k} X^*_{il}I^*_{ki}X_{kj} = I^*_l \sum_k X^*_{kl}\, X_{kj}. \tag{5-24}$$

The summation on the right is again $\mathbf{R}^*_l \cdot \mathbf{R}_j$. Further, the hermitean nature of **I** states that $I_{ik} = I^*_{ki}$. Hence the left-hand sides of Eqs. (5-23) and (5-24) are the same, so that the difference of the two equations can be reduced to

$$(I_j - I^*_l)\mathbf{R}^*_l \cdot \mathbf{R}_j = 0. \tag{5-25}$$

If l is equal to j, then $\mathbf{R}^*_l \cdot \mathbf{R}_j = |\mathbf{R}_j|^2$ which must be positive definite. Hence Eq. (5-25) can vanish in that case only if $I_j = I^*_j$; which proves one of the desired results. Notice that the proof has used only the hermitean property of **I** — *the eigenvalues of any hermitean matrix are real.* Since **I** is also real, the direction cosines of the eigenvectors $\mathbf{R}_j$ must likewise be real.

If l is chosen different from j, and the eigenvalues are distinct, then Eq. (5-25) can be satisfied only if $\mathbf{R}_l \cdot \mathbf{R}_j$ vanishes, which verifies the second

---

* In terms of the matrix **X** which diagonalizes **I** by means of a similarity transformation, these conditions state that **X** must be a *real orthogonal* matrix.

requirement that the eigenvectors be orthogonal.* If the eigenvalues are not all distinct this orthogonality proof falls through, but it may be mended with little difficulty. When two of the eigenvalues are equal the corresponding eigenvectors need not be orthogonal. Further, any linear combination of these eigenvectors must also be an eigenvector of $\mathbf{I}$ with the same eigenvalue. Hence *all* of the vectors in the plane defined by the original two eigenvectors are also eigenvectors. The eigenvector corresponding to the third distinct eigenvalue must therefore be normal to this plane. One may therefore choose arbitrarily two perpendicular vectors lying in the plane, and these, together with the third vector, will give the three orthogonal axes desired. Similarly if all eigenvalues are the same then all directions in space are eigenvectors. But then $\mathbf{I}$ will be diagonal to start with, and no diagonalization is necessary.

The methods of matrix algebra thus enable us to show that for any point in a rigid body one can find a set of cartesian axes for which the inertia tensor will be diagonal. The axes are called the *principal axes* and the corresponding diagonal elements I_1, I_2, I_3 are known as the *principal moments of inertia.* Given some initial body set of coordinates one can transform to the principal axes by a particular orthogonal transformation, which is therefore known as the *principal axis transformation.* In practice, of course, the principal moments of inertia, being the eigenvalues of $\mathbf{I}$, are found as the roots of the secular equation. To recall the steps leading to the secular equation, it will be noted that with $i = 1, 2, 3$ Eqs. (5–22) form a set of three homogeneous linear equations for the components of the eigenvector. These equations are consistent only if the determinant of the coefficients vanishes:

$$\begin{vmatrix} I_{xx} - I & I_{xy} & I_{zx} \\ I_{xy} & I_{yy} - I & I_{yz} \\ I_{zx} & I_{yz} & I_{zz} - I \end{vmatrix} = 0, \tag{5–26}$$

where the symmetry of $\mathbf{I}$ has been displayed explicitly. Eq. (5–26) is the secular equation, in the form of a cubic in I, whose three roots are the de-

* The proof given here has a close analog in the theory of the Sturm-Liouville problem in partial differential equations. There one shows that the eigenvalues of a hermitean differential operator are real and the corresponding eigenfunction solutions of the differential equation are orthogonal. The similarity is not accidental, one can always construct matrix quantities to correspond to any given problem involving a partial differential equation, and this is exactly what happens in the correspondence between the matrix mechanics and wave mechanics formulations of quantum theory.

sired principal moments. For each of these roots the Eqs. (5–22) can be solved to obtain the direction of the corresponding principal axis. In most of the easily soluble problems in rigid dynamics the principal axes can be determined by inspection. For example, one almost always has to deal with rigid bodies that are solids of revolution about some axis, with the origin of the body system on the symmetry axis. All directions perpendicular to the axis of symmetry are then alike, which is the mark of a double root to the secular equation. The principal axes are then the symmetry axis and any two perpendicular axes in the plane normal to the symmetry axis.

One may also be led to the concept of principal axes through some geometrical considerations which historically formed the first approach to the subject. The moment of inertia about a given axis has been defined as $I = \mathbf{n} \cdot \mathbf{I} \cdot \mathbf{n}$. Let the direction cosines of the axis be α, β, and γ so that

$$\mathbf{n} = \alpha\mathbf{i} + \beta\mathbf{j} + \gamma\mathbf{k};$$

I then can be written as

$$I = I_{xx}\alpha^2 + I_{yy}\beta^2 + I_{zz}\gamma^2 + 2I_{xy}\alpha\beta + 2I_{yz}\beta\gamma + 2I_{zx}\gamma\alpha, \qquad (5\text{–}27)$$

using explicitly the symmetry of $\mathbf{I}$. It is convenient to define a vector $\boldsymbol{\rho}$ by the equation:

$$\boldsymbol{\rho} = \frac{\mathbf{n}}{\sqrt{I}}.$$

The magnitude of $\boldsymbol{\rho}$ is thus related to the moment of inertia about the axis whose direction is given by $\mathbf{n}$. In terms of the components of this new vector Eq. (5–27) takes on the form

$$1 = I_{xx}\rho_1^2 + I_{yy}\rho_2^2 + I_{zz}\rho_3^2 + 2I_{xy}\rho_1\rho_2 + 2I_{yz}\rho_2\rho_3 + 2I_{zx}\rho_3\rho_1. \qquad (5\text{–}28)$$

Considered as a function of the three variables ρ_1, ρ_2, ρ_3, Eq. (5–28) is the equation of some surface in ρ space. In particular Eq. (5–28) is the equation of an ellipsoid designated as the *inertia ellipsoid*. It is well known that one can always transform to a set of cartesian axes in which the equation of an ellipsoid takes on its normal form:

$$1 = I_1\rho_1'^2 + I_2\rho_2'^2 + I_3\rho_3'^2, \qquad (5\text{–}29)$$

with the principal axes of the ellipsoid along the new coordinate axes. But (5–29) is just the form Eq. (5–28) has in a system of coordinates in which the inertia tensor $\mathbf{I}$ is diagonal. Hence the coordinate transformation which puts the equation of ellipsoid into its normal form is exactly the principal axis transformation previously discussed. The principal moments of inertia determine the lengths of the axes of the inertia ellipsoid. If two of the roots of the secular equation are equal the inertia ellipsoid thus has two equal axes and is an ellipsoid of revolution. If all three principal moments are equal the inertia ellipsoid is a sphere.

A quantity closely related to the moment of inertia is the *radius of gyration*, R_0, defined by the equation

$$I = MR_0^2. \tag{5-30}$$

In terms of the radius of gyration the vector $\boldsymbol{\rho}$ can be written as

$$\boldsymbol{\rho} = \frac{\mathbf{n}}{R_0\sqrt{M}}.$$

The radius vector to a point on the inertia ellipsoid is thus inversely proportional to the radius of gyration about the direction of the vector.

5–5 Methods of solving rigid body problems and the Euler equations of motion. Practically all the tools necessary for setting up and solving problems in rigid body dynamics have by now been assembled. If nonholonomic constraints are present then special means must be taken to include the effects of these constraints in the equations of motion. For example, if there are "rolling constraints" these must be introduced into the equations of motion by the method of Lagrange undetermined multipliers, as in Section 2–4. However, with the exception of such unusual cases, most problems dealing with rigid body motion involve only holonomic constraints and conservative forces, and the Lagrangian by itself is then sufficient to tell the whole story. If the motion of the body is completely unconstrained, the full set of six generalized coordinates will be needed: three cartesian coordinates to describe the translational motion and the three Euler angles to describe the rotational motion. In particular the kinetic energy T can always be written as the translational energy of the center of mass, $\frac{1}{2}Mv^2$, plus the rotational energy about the center of mass, $\frac{1}{2}I\omega^2$. For convenience in expressing this last term one almost always works in terms of the principal axes so that the rotational energy has the simple form shown in Eq. (5–21), with the components of $\boldsymbol{\omega}$ expressed in terms of the Euler angles. Of course, if one point in the body is fixed by the constraints of the problem, then the kinetic energy contains only the rotational terms and the solution is greatly simplified. Often, too, the motion is effectively only in two dimensions, as in the motion of a rigid lamina in a plane. The axis of rotation is then fixed in the direction perpendicular to the plane; only one angle of rotation is necessary and one may dispense with the cumbersome machinery of the Euler angles.

While formally the Lagrange equations of motion are sufficient for the solution of the problem, in the case of motion with one point fixed it is often convenient to use a different set, known as *Euler's equations of motion*. For conservative forces the Lagrangian in such case can be written as

$$L = T - V = \tfrac{1}{2}(I_1\omega_x^2 + I_2\omega_y^2 + I_3\omega_z^2) - V(\theta, \phi, \psi), \tag{5-31}$$

where I_1, I_2, I_3 are the principal moments of inertia for the fixed point. The components of the angular velocity along the principal axes, ω_x, ω_y, ω_z, can be expressed, through Eqs. (4–103), in terms of the Euler angles which relate the principal axis system to some space set of axes. By this means the entire Lagrangian can be written as a function of the three rotation angles θ, ϕ, ψ. It will be remembered that the generalized force corresponding to a generalized coordinate of rotation is the component of the impressed torque along the axis of rotation (cf. Section 2–6). The generalized forces corresponding to θ, ϕ, ψ, are therefore components of the torque, not along the principal axes, but rather along the line of nodes, the space z-axis, and the body z-axis respectively. Hence, of these three only the generalized force corresponding to ψ, namely $-\dfrac{\partial V}{\partial \psi}$, directly furnishes a component of the total torque along one of the principal axes, in this case along the z-axis. The Lagrange equation for ψ can therefore be written

$$\frac{d}{dt}\left(\frac{\partial T}{\partial \dot{\psi}}\right) - \frac{\partial T}{\partial \psi} = N_z. \tag{5-32}$$

An examination of Eqs. (4–103) reveals that only ω_z explicitly involves $\dot{\psi}$, while ψ itself occurs only in ω_x and ω_y; and in fact we have the relations:

$$\frac{\partial \omega_z}{\partial \dot{\psi}} = 1, \qquad \frac{\partial \omega_x}{\partial \psi} = \omega_y, \qquad \frac{\partial \omega_y}{\partial \psi} = -\omega_x.$$

With these formulae, and with the form for the kinetic energy given in Eq. (5–21) the partial derivatives occurring in (5–32) can be written as

$$\frac{\partial T}{\partial \dot{\psi}} = I_3\omega_z,$$

$$\frac{\partial T}{\partial \psi} = I_1\omega_x\omega_y - I_2\omega_y\omega_x.$$

The Lagrange equation for the coordinate ψ therefore has the form

$$I_3\dot{\omega}_z - \omega_x\omega_y(I_1 - I_2) = N_z. \tag{5-33}$$

The identification of some one of the principal axes as the z-axis is entirely arbitrary. Clearly we can permute the indices and write an equation similar to (5–33) for the component of the total torque along any of the other principal axes. Without further ado, the complete set of equations for all the axes is

$$\begin{aligned} I_1\dot{\omega}_x - \omega_y\omega_z(I_2 - I_3) &= N_x \\ I_2\dot{\omega}_y - \omega_z\omega_x(I_3 - I_1) &= N_y \\ I_3\dot{\omega}_z - \omega_x\omega_y(I_1 - I_2) &= N_z. \end{aligned} \tag{5–34}$$

These are the so-called Euler equations of motion for a rigid body with one point fixed. Although the third of Eqs. (5–34) happens to be the Lagrange equation for ψ, it must be emphasized that the other two are *not* the Lagrange equations corresponding to the θ or ϕ coordinates. Thus $-\dfrac{\partial V}{\partial \theta}$ corresponds neither to N_x nor N_y, but is rather the component of $\mathbf{N}$ along the line of nodes.

An alternative derivation of Euler's equations goes back to the fundamental equation of motion for the total angular momentum, Eq. (1–24):

$$\frac{d\mathbf{L}}{dt} = \mathbf{N}.$$

Here the time derivative clearly refers to the space axes, for the equation only holds in an inertial system. On the other hand the equations (5–34) involve the time derivatives of components with respect to moving axes. By Eq. (4–100), however, the two time derivatives are related by:

$$\left(\frac{d\mathbf{L}}{dt}\right)_{\text{space}} = \left(\frac{d\mathbf{L}}{dt}\right)_{\text{body}} + \boldsymbol{\omega} \times \mathbf{L}$$

and the component of the equation of motion along the x principal axis is

$$\frac{dL_x}{dt} + \omega_y L_z - \omega_z L_y = N_x. \tag{5–35}$$

But the components of the angular momentum along the principal axes are proportional to the corresponding components of the angular velocity, the factors of proportionality being the principal moments of inertia. Hence Eq. (5–35) reduces to

$$I_1\dot{\omega}_x - \omega_y\omega_z(I_2 - I_3) = N_x,$$

which agrees with the first of Eqs. (5–34), and the other equations follow by cyclic permutation of indices.

5-6 Force-free motion of a rigid body. One problem in rigid dynamics where Euler's equations are applicable is in the motion of a rigid body not subject to any net forces or torques. The center of mass is then either at rest or moving uniformly, and it does not decrease the generality of the solution to discuss the rotational motion in a reference frame in which the center of mass is stationary. In such case the angular momentum arises only from rotation about the center of mass and Euler's equations are the equations of motion for the system. In the absence of any net torques they reduce to

$$\begin{aligned} I_1\dot{\omega}_x &= \omega_y\omega_z(I_2 - I_3) \\ I_2\dot{\omega}_y &= \omega_z\omega_x(I_3 - I_1) \\ I_3\dot{\omega}_z &= \omega_x\omega_y(I_1 - I_2). \end{aligned} \tag{5-36}$$

The same equations, of course, will also describe the motion of a rigid body when one point is fixed and there are no net applied torques. We know two immediate integrals of the motion, for both the kinetic energy and the total angular momentum vector must be constant in time. With these two integrals it is possible to integrate (5-36) completely in terms of elliptic functions, but such a treatment is not very illuminating. However, it is possible to derive an elegant geometrical description of the motion, known as Poinsot's construction, without requiring a complete solution to the problem.

Consider a coordinate system oriented along the principal axes of the body but whose axes measure the components of the vector $\boldsymbol{\rho}$ defined in Section 5-4, rather than the components of the position vector $\mathbf{r}$. In this space we define a function

$$F(\rho) = \boldsymbol{\rho} \cdot \mathbf{I} \cdot \boldsymbol{\rho},$$

where the surfaces of constant F are ellipsoids, the particular surface $F = 1$ being the inertia ellipsoid. As the direction of the axis of rotation changes in time the parallel vector $\boldsymbol{\rho}$ moves accordingly, its tip always defining a point on the inertia ellipsoid. The gradient of F, evaluated at this point, furnishes the direction of the corresponding normal to the inertia ellipsoid. Now, from the definition of the function F, remembering the diagonal form of $\mathbf{I}$ in the principal axes, the partial derivative of F with respect to ρ_1 is

$$\frac{\partial F}{\partial \rho_1} = (\nabla F)_1 = 2I_1\rho_1.$$

Since the vector $\boldsymbol{\rho}$ may be defined also as

$$\boldsymbol{\rho} = \frac{\boldsymbol{\omega}}{\omega\sqrt{I}},$$

this component of the ρ gradient of F can be written

$$(\nabla F)_1 = \frac{2}{\omega\sqrt{I}} I_1\omega_x$$

or

$$(\nabla F)_1 = \frac{2}{\omega\sqrt{I}} L_x.$$

Similarly, the other components of ∇F are

$$(\nabla F)_2 = \frac{2}{\omega\sqrt{I}} L_y,$$

$$(\nabla F)_3 = \frac{2}{\omega\sqrt{I}} L_z.$$

Thus the $\boldsymbol{\omega}$ vector will always move such that the corresponding normal to the inertia ellipsoid is in the direction of the angular momentum. In the particular case under discussion the direction of **L** is fixed in space and it is the inertia ellipsoid (fixed with respect to the body) which must move in space in order to preserve this connection between $\boldsymbol{\omega}$ and **L** (cf. Fig. 5–3).

It can also be shown that the distance between the origin of the ellipsoid and the plane tangent to it at the point $\boldsymbol{\rho}$ must similarly be constant in time. This distance is equal to the projection of $\boldsymbol{\rho}$ on **L**, and is given by

$$\frac{\boldsymbol{\rho}\cdot\mathbf{L}}{L} = \frac{\boldsymbol{\omega}\cdot\mathbf{L}}{\omega L\sqrt{I}} = \frac{2T}{L\sqrt{I\omega^2}}$$

or

$$\frac{\boldsymbol{\rho}\cdot\mathbf{L}}{L} = \frac{\sqrt{2T}}{L}. \tag{5–37}$$

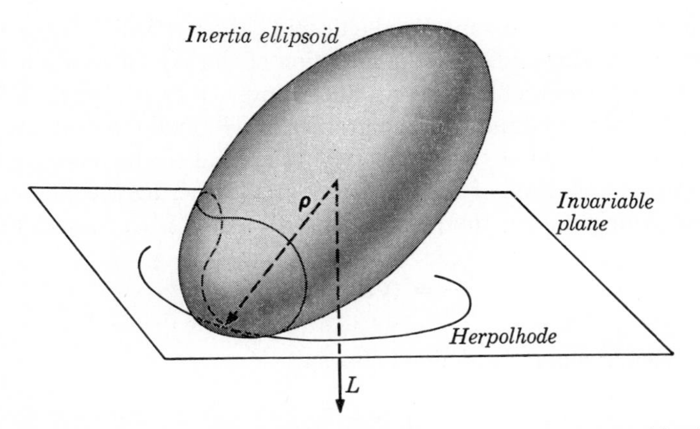

FIG. 5–3. The motion of the inertia ellipsoid relative to the invariable plane.

Both T, the kinetic energy, and L, the angular momentum, are constants of the motion and the tangent plane is therefore always a fixed distance from the origin of the ellipsoid. Since the normal to the plane, being along $\mathbf{L}$, also has a fixed direction, the tangent plane is known as the *invariable plane.* We can picture the force-free motion of the rigid body as being such that the inertia ellipsoid rolls, without slipping, on the invariable plane, with the center of the ellipsoid a constant height above the plane. The rolling occurs without slipping because the point of contact is defined by the position of $\boldsymbol{\rho}$ which, being along the instantaneous axis of rotation, is the one direction in the body momentarily at rest. The curve traced out by the point of contact on the inertia ellipsoid is known as the *polhode,* while the similar curve on the invariable plane is called the *herpolhode.**

Poinsot's geometrical discussion is quite adequate to describe completely the force-free motion of the body. The direction of the invariable plane and the height of the inertia ellipsoid above it are determined by the values of T and $\mathbf{L}$, which are among the initial conditions of the problem. It is then a matter of geometry to trace out the polhode and the herpolhode. The direction of the angular velocity in space is given by the direction of $\boldsymbol{\rho}$, while the instantaneous orientation of the body is provided by the orientation of the inertia ellipsoid, which is fixed in the body. Elaborate descriptions of force-free motion obtained in this fashion are to be found frequently in the literature.† In the special case of a symmetrical body, the inertia ellipsoid is an ellipsoid of revolution, so that the polhode on the ellipsoid is clearly a circle about the symmetry axis. Correspondingly the angular velocity vector moves on the surface of a cone. Physically this means that the direction of $\boldsymbol{\omega}$ *precesses* in time about the axis of symmetry of the body.

For a symmetrical rigid body the analytical solution for the force-free motion is not difficult to obtain, and one may directly confirm the precessing motion predicted by the Poinsot construction. Let the symmetry axis be taken as the z principal axis so that $I_1 = I_2$. Euler's equations (5-36) reduce then to

$$\begin{aligned} I_1\dot{\omega}_x &= (I_1 - I_3)\omega_z\omega_y \\ I_1\dot{\omega}_y &= -(I_1 - I_3)\omega_z\omega_x \\ I_3\dot{\omega}_z &= 0. \end{aligned} \tag{5-38}$$

* Hence the jabberwockian sounding statement: the polhode rolls without slipping on the herpolhode lying in the invariable plane.

† See especially Webster, *Dynamics of Particles and Rigid Bodies,* the article by Winkelmann and Grammel in Vol. V of the *Handbuch der Physik,* and the treatise by F. Klein and A. Sommerfeld, *Theorie des Kreisels.*

The last of these equations states that ω_z is a constant and it can therefore be treated as one of the known initial conditions of the problem. Either ω_x or ω_y can be eliminated from the remaining two equations. To solve for ω_x take the time derivative of the first of Eqs. (5–38):

$$I_1\ddot{\omega}_x = (I_1 - I_3)\omega_z\dot{\omega}_y,$$

and substitute for $\dot{\omega}_y$ the expression given by the second equation, obtaining finally

$$\ddot{\omega}_x = -\left[\frac{(I_1 - I_3)\omega_z}{I_1}\right]^2 \omega_x. \tag{5–39}$$

Eq. (5–39) describes a simple harmonic motion with an angular frequency

$$\Omega = \frac{I_1 - I_3}{I_1}\,\omega_z, \tag{5–40}$$

so that a typical solution for ω_x could be written as

$$\omega_x = A \sin \Omega t, \tag{5–41}$$

where A is some constant. The corresponding solution for ω_y can be found by substituting (5–41) in the first of Eqs. (5–38) and solving for ω_y:

$$\omega_y = A \cos \Omega t. \tag{5–42}$$

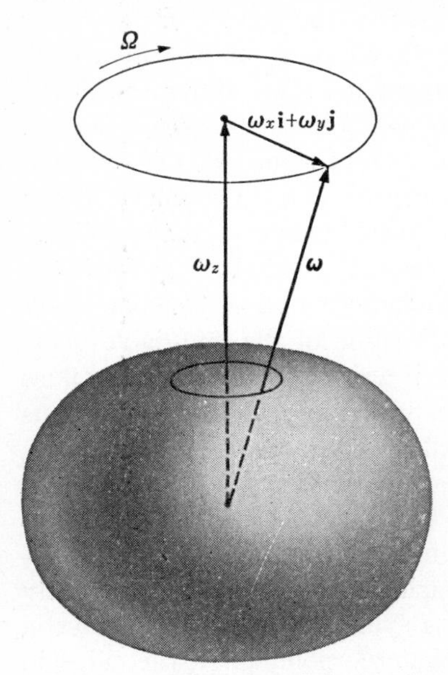

FIG. 5–4. Precession of the angular velocity about the axis of symmetry in the force-free motion of a symmetrical rigid body.

The solutions (5–41 and 5–42) show that the vector $\omega_x\mathbf{i} + \omega_y\mathbf{j}$ has a constant magnitude and rotates uniformly about the z-axis of the body with the angular frequency Ω (cf. Fig. 5–4). Hence the total angular velocity $\boldsymbol{\omega} = \omega_x\mathbf{i} + \omega_y\mathbf{j} + \omega_z\mathbf{k}$ is also constant in magnitude and *precesses* about the z-axis with the same frequency, exactly as predicted by the Poinsot construction.* It should be remembered that the precession described here is relative to the body axes, which are themselves rotating in space with the larger frequency $\boldsymbol{\omega}$. From Eq. (5–40) it is seen that the closer I_1 is to

* The precession can be demonstrated in another fashion by defining a vector $\boldsymbol{\Omega}$ lying along the z-axis with magnitude given by (5–40). Eqs. (5–38) are then essentially equivalent to the vector equation

$$\dot{\boldsymbol{\omega}} = \boldsymbol{\omega} \times \boldsymbol{\Omega},$$

which immediately reveals the precession of $\boldsymbol{\omega}$ with the frequency Ω.

I_3 the slower will be the precession frequency Ω compared to the rotation frequency ω. The constants A (the amplitude of the precession) and ω_z can be evaluated in terms of the more usual constants of the motion, namely the kinetic energy and the magnitude of the angular momentum. Both T and L^2 can be written as functions of A and ω_z:

$$T = \frac{1}{2} I_1 A^2 + \frac{1}{2} I_3 \omega_z^2,$$
$$L^2 = I_1^2 A^2 + I_3^2 \omega_z^2,$$

and these relations in turn may be solved for A and ω_z in terms of T and L.

It would be expected that the axis of rotation of the earth should exhibit this precession, for the external torques acting on the earth are so weak that the *rotational* motion may be considered as force-free. The earth is symmetrical about the polar axis and slightly flattened at the poles so that I_1 is less than I_3. Numerically the ratio of the moments is such that

$$\frac{I_1 - I_3}{I_1} = -.0033$$

and the magnitude of the precession angular frequency should therefore be

$$\Omega = \frac{\omega_z}{300}.$$

Since ω_z is practically the same as the magnitude of $\boldsymbol{\omega}$ this result predicts a period of precession of 300 days or about 10 months. An observer on the earth should therefore find that the axis of rotation traces out a circle about the North Pole once every 10 months. Something vaguely resembling such a phenomenon has actually been observed. The amplitude of the precession is quite small, the axis of rotation never wandering more than about 15 ft from the North Pole. But the orbit is quite irregular, and the fundamental period seems to be about 427 days rather than the 300 days predicted. The fluctuations are ascribed to small shifts in the mass distribution of the earth such as are caused by atmospheric motion, while the difference in the period arises from the fact that the earth is not completely rigid but has the elastic properties of a material like steel.*

---

* The force-free precession of the earth's axis is not to be confused with its slow precession about the normal to the ecliptic. This *astronomical* precession of the equinoxes is due to the gravitational torques of the sun and the moon which were considered negligible in the above discussion. That the assumption is justified is shown by the long period of the precession of the equinoxes (26,000 years) compared to a period of roughly one year for the force-free precession. The astronomical precession is discussed further in Section 5-7 and in the exercises.

5–7 The heavy symmetrical top with one point fixed. As a further and more complicated example of the application of the methods of rigid dynamics we shall consider the motion of a symmetrical body in a gravitational field when one point on the symmetry axis is fixed in space. A wide variety of physical systems, ranging from a child's top to complicated gyroscopic navigational instruments, are approximated by such a *heavy symmetrical top*. Both for its practical applications and as an illustration of many of the techniques previously developed, the motion of the heavy symmetrical top deserves a detailed exposition.

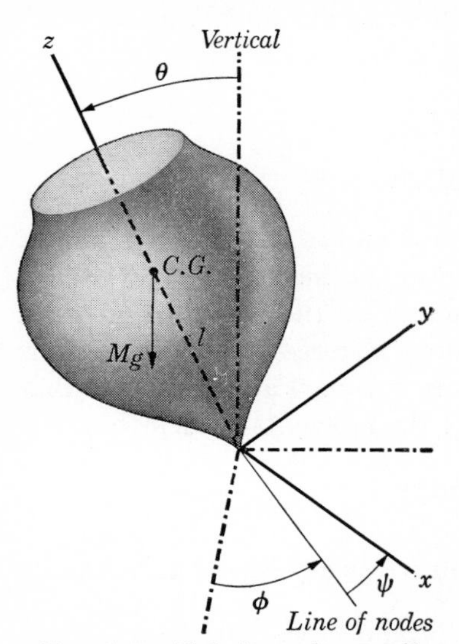

Fig. 5–5. Euler's angles specifying the orientation of a symmetrical top.

The symmetry axis is of course one of the principal axes and will be chosen as the z-axis of the coordinate system fixed in the body. Since one point is stationary the configuration of the top is completely specified by the three Euler angles: θ gives the inclination of the z-axis from the vertical, ϕ measures the azimuth of the top about the vertical while ψ is the rotation angle of the top about its own z-axis (cf. Fig. 5–5). The distance of the center of gravity (located on the symmetry axis) from the fixed point will be denoted by l.

The Lagrangian procedure, rather than Euler's equations, will be used to obtain a solution for the motion of the top. Since the body is symmetrical the kinetic energy can be written as

$$T = \frac{1}{2} I_1(\omega_x^2 + \omega_y^2) + \frac{1}{2} I_3\omega_z^2,$$

or, in terms of Euler's angles, and using Eqs. (4–103), as:

$$T = \frac{I_1}{2}(\dot{\theta}^2 + \dot{\phi}^2 \sin^2\theta) + \frac{I_3}{2}(\dot{\psi} + \dot{\phi}\cos\theta)^2, \tag{5–43}$$

where the cross terms in ω_x^2 and ω_y^2 cancel. The potential energy is simply

$$V = Mgl\cos\theta \tag{5–44}$$

so that the Lagrangian is

$$L = \frac{I_1}{2}(\dot{\theta}^2 + \dot{\phi}^2 \sin^2\theta) + \frac{I_3}{2}(\dot{\psi} + \dot{\phi}\cos\theta)^2 - Mgl\cos\theta. \tag{5-45}$$

It will be noticed that ϕ and ψ do not appear explicitly in the Lagrangian; they are therefore cyclic coordinates, indicating that the corresponding generalized momenta are constant in time. Now, we have seen that the momentum conjugate to a rotation angle is the component of the total angular momentum along the axis of rotation, which for ϕ is the vertical axis, and for ψ, the z-axis in the body. One can in fact show from elementary principles that these components of the angular momentum must be unvarying. Since the torque of gravity is along the line of nodes, there is no component of the torque along either the vertical or the body z-axis, for by definition both of these axes are perpendicular to the line of nodes. Hence the components of the angular momentum along these two axes must be constant in time.

We therefore have two immediate first integrals of the motion:

$$p_\psi = \frac{\partial L}{\partial \dot{\psi}} = I_3(\dot{\psi} + \dot{\phi}\cos\theta) = I_3\omega_z = I_1 a \tag{5-46}$$

and

$$p_\phi = \frac{\partial L}{\partial \dot{\phi}} = (I_1 \sin^2\theta + I_3\cos^2\theta)\dot{\phi} + I_3\dot{\psi}\cos\theta = I_1 b. \tag{5-47}$$

Here the two constants of the motion are expressed in terms of new constants a and b. There is one further first integral available; since the system is conservative the total energy E is constant in time:

$$E = T + V = \frac{I_1}{2}(\dot{\theta}^2 + \dot{\phi}^2\sin^2\theta) + \frac{I_3}{2}\omega_z^2 + Mgl\cos\theta. \tag{5-48}$$

Only three additional quadratures are needed to solve the problem, and they can easily be obtained from these three first integrals without directly using the Lagrange equations. From Eq. (5-46) $\dot{\psi}$ is given in terms of $\dot{\phi}$ by

$$I_3\dot{\psi} = I_1 a - I_3\dot{\phi}\cos\theta, \tag{5-49}$$

and this result can be substituted in (5-47) to eliminate $\dot{\psi}$:

$$I_1\dot{\phi}\sin^2\theta + I_1 a\cos\theta = I_1 b,$$

or

$$\dot{\phi} = \frac{b - a\cos\theta}{\sin^2\theta}. \tag{5-50}$$

Thus if θ were known as a function of time Eq. (5–50) could be integrated to furnish the dependence of ϕ on time. Substituting Eq. (5–50) back in Eq. (5–49) results in a corresponding expression for $\dot{\psi}$:

$$\dot{\psi} = \frac{I_1 a}{I_3} - \cos\theta \frac{b - a\cos\theta}{\sin^2\theta}, \tag{5–51}$$

which furnishes $\dot{\psi}$ if θ is known. Finally, Eqs. (5–50) and (5–51) can be used to eliminate $\dot{\phi}$ and $\dot{\psi}$ from the energy equation, resulting in a differential equation involving θ alone. First notice that Eq. (5–46) says ω_z is constant in time and equal to $\frac{I_1}{I_3} a$. Therefore $E - \frac{1}{2} I_3 \omega_z^2$ is a constant of the motion, which we shall designate as E'. The energy equation can thus be written in the form

$$E' = \frac{I_1}{2}(\dot{\theta}^2 + \dot{\phi}^2 \sin^2\theta) + Mgl\cos\theta, \tag{5–52}$$

or, after substituting Eq. (5–50) and rearranging the terms, as

$$\sin^2\theta\ \dot{\theta}^2 = \sin^2\theta\,(\alpha - \beta\cos\theta) - (b - a\cos\theta)^2 \tag{5–53}$$

where α and β are two constants,

$$\alpha = \frac{2E'}{I_1}, \quad \beta = \frac{2Mgl}{I_1}. \tag{5–54}$$

If the variable is now changed to $u = \cos\theta$, Eq. (5–53) appears as:

$$\dot{u}^2 = (1 - u^2)(\alpha - \beta u) - (b - au)^2 \tag{5–55}$$

which can be reduced immediately to a quadrature:

$$t = \int_{u(o)}^{u(t)} \frac{du}{\sqrt{(1 - u^2)(\alpha - \beta u) - (b - au)^2}}. \tag{5–56}$$

With this result, and Eqs. (5–50) and (5–51), ϕ and ψ can also be reduced to quadratures. However, the polynomial in the radical is a cubic so that we have to deal with elliptic integrals. Extensive discussions of these solutions involving elliptic functions are to be found in the literature,* but, as in the case of the force-free motion, the physics tends to be obscured in the profusion of mathematics. Fortunately, the general nature of the motion can be discovered without actually performing the integrations.

---

* See, for example, the treatise by F. Klein and A. Sommerfeld, Whittaker, loc. cit., or the very detailed treatment in Macmillan, *Dynamics of Rigid Bodies*.

Let the right-hand side of Eq. (5–55) be denoted by $f(u)$. The roots of this cubic polynomial furnish the angles at which $\dot{\theta}$ changes sign, i.e., the "turning angles" in θ. For u large, the dominant term in $f(u)$ is βu^3. Since β (cf. Eq. (5–54)) is always greater than zero, $f(u)$ is positive for large positive u and negative for large negative u. At the points $u = \pm 1$, $f(u)$ becomes equal to $-(b \mp a)^2$ and is therefore always negative, except for the unusual case where $u = \pm 1$ is a root (corresponding to a vertical top). Hence at least one of the roots must lie in the region $u > 1$, a region which does not correspond to real angles. Indeed, physical motion of the top can occur only when $\dot{u}^2$ is positive somewhere in the interval between $u = -1$ and $u = +1$, i.e., θ between 0 and $+\pi$. We must conclude, therefore, that for any actual top $f(u)$ will have two roots, u_1 and u_2, between -1 and $+1$ (cf. Fig. 5–6), and that the top moves such that $\cos\theta$ always remains between these two roots. The location of these roots, and the behavior of $\dot{\phi}$ and $\dot{\psi}$ for values of θ between them, provide much qualitative information about the motion of the top.*

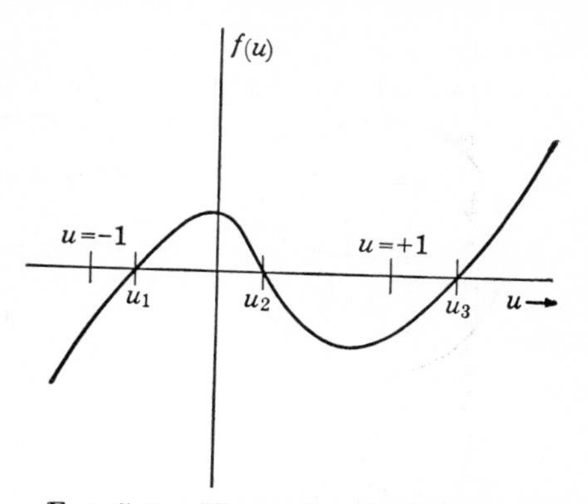

FIG. 5–6. Illustrating the location of the turning angles of θ in the motion of a heavy symmetrical top.

It is customary to depict the motion of the top by tracing the curve of the intersection of the figure axis on a sphere of unit radius about the fixed point. This curve will be known as the *locus* of the figure axis. The polar coordinates of a point on the locus are identical with the Euler angles θ, ϕ for the body system. From the discussion of the above paragraph it is seen that the locus lies between the two bounding circles of colatitude $\theta_1 = \arccos u_1$ and $\theta_2 = \arccos u_2$, with $\dot{\theta}$ vanishing at both circles. The shape of the locus curve is in large measure determined by the value of the root of $b - au$, which will be denoted by u':

$$u' = \frac{b}{a}. \tag{5–57}$$

* There is an obvious similarity between this method of discussing the top motion and the "effective potential" method used in Chapter 3 for central forces. Indeed the modified energy Eq. (5–52) can be considered as representing the one-dimensional motion of a particle of mass I_1 with an equivalent potential energy

$$Mgl\cos\theta + \frac{I_1}{2}\frac{(b - a\cos\theta)^2}{\sin^2\theta}.$$

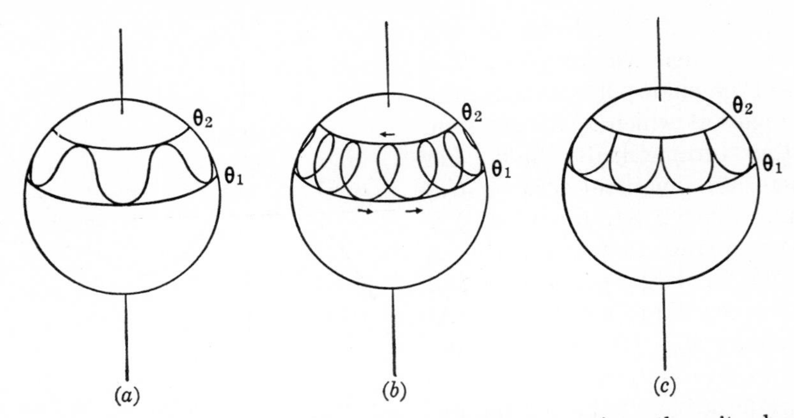

FIG. 5–7. The possible shapes for the locus of the figure axis on the unit sphere.

Suppose, for example, the initial conditions are such that u' is larger than u_2. Then, by Eq. (5–50), $\dot{\phi}$ will always have the same sign for the allowed inclination angles between θ_1 and θ_2. Hence the locus of the figure axis must be tangent to the bounding circles in such a manner that $\dot{\phi}$ is in the same direction at both θ_1 and θ_2, as is shown in Fig. 5–7a. Since ϕ therefore increases secularly in one direction or the other, the axis of the top may be said to *precess* about the vertical axis. But it is not the regular precession encountered in force-free motion, for as the figure axis goes around it nods up and down between the bounding angles θ_1 and θ_2 — the top *nutates* during the precession.

Should b/a be such that u' lies between u_1 and u_2 the direction of the precession will be different at the two bounding circles and the locus of the figure axis exhibits loops, as shown in Fig. 5–7b. The average of $\dot{\phi}$ will not vanish, however, so that there is always a net precession in one direction or the other. It can also happen that u' coincides with one of the roots of $f(u)$. At the corresponding bounding circles both $\dot{\theta}$ and $\dot{\phi}$ must then vanish, which requires that the locus have cusps touching the circle, as shown in Fig. 5–7c. This last case is not as exceptional as it sounds; it corresponds in fact to the initial conditions usually stipulated in elementary discussions of tops: one assumes that initially the symmetrical top is spinning about its figure axis, which is fixed in some direction θ_0. At time $t = 0$ the figure axis is released and the problem is to describe the subsequent motion. Explicitly these initial conditions are that at $t = 0$, $\theta = \theta_0$ and $\dot{\theta} = \dot{\phi} = 0$. The angle θ_0 must therefore be one of the roots of $f(u)$, in fact it corresponds to the upper circle. For proof, note that E' initially is equal to $Mgl \cos \theta_0$ and that the terms involving $\dot{\theta}$ and $\dot{\phi}$ are always positive.

Hence as $\dot{\theta}$ and $\dot{\phi}$ begin to differ from their initial zero values, energy can be conserved only by a decrease in the potential energy, that is, by an increase in θ. The initial θ_0 is therefore the same as θ_2, the minimum value θ can have. When released in this manner, *the top always starts to fall*, and continues to fall until the other bounding angle θ_1 is reached, precessing the meanwhile. The figure axis then begins to rise again to θ_2, the complete motion being as shown in Fig. 5–7c.

Some quantitative predictions can be made about the motion of the top under these initial conditions of vanishing $\dot{\theta}$ and $\dot{\phi}$, provided that the initial kinetic energy of rotation is assumed large compared to the maximum change in potential energy:

$$\frac{1}{2}I_3\omega_z^2 >> 2Mgl. \tag{5–58}$$

The effects of the gravitational torques, namely the precession and accompanying nutation, will then be only small perturbations on the dominant rotation of the top about its figure axis. In this situation, we speak of the top as being a "fast top." With this assumption we can obtain expressions for the extent of the nutation, the nutation frequency, and the average frequency of precession.

Under the given initial conditions θ_0 corresponds to the upper bounding circle, so that u_2 is identical with u_0. The extent of the nutation therefore depends on the location of the other real angle which is a root of $f(u)$. From the requirement $\dot{\phi} = 0$ at $u = u_0$ it follows that the two constants of the motion a and b are related to u_0 by

$$b = au_0.$$

Since $f(u_0)$ vanishes, from Eq. (5–55) one must also have the connection

$$\alpha = \beta u_0,$$

which states merely that E' is given by $Mgl \cos \theta_0$. With these relations $f(u)$ may be rewritten more simply as

$$f(u) = (u_0 - u)\{\beta(1 - u^2) - a^2(u_0 - u)\}. \tag{5–59}$$

The roots of $f(u)$ other than u_0 are given by the roots of the quadratic expression in the curly brackets, and the desired root u_1 therefore satisfies the equation

$$(1 - u_1^2) - \frac{a^2}{\beta}(u_0 - u_1) = 0. \tag{5–60}$$

Denoting $u_0 - u$ by x and $u_0 - u_1$ by x_1, Eq. (5–60) can be rewritten as

$$x_1^2 + px_1 - q = 0, \tag{5–61}$$

where

$$p = \frac{a^2}{\beta} - 2\cos\theta_0, \quad q = \sin^2\theta_0.$$

The term $2\cos\theta_0$ in p may be neglected, as the ratio a^2/β can be written in the form:

$$\frac{a^2}{\beta} = \left(\frac{I_3}{I_1}\right)\frac{I_3\omega_z^2}{2Mgl}.$$

From the assumption of a "fast" top, (5–58), unless I_3 is much less than I_1 (which would correspond to a top in the unusual shape of a cigar) this ratio is much greater than 2. By the same token p^2 is much greater than $4q$ so that the solutions of (5–61) are approximately

$$x_1 = \frac{q}{p}, \quad -p - \frac{q}{p}.$$

The second solution corresponds to $u > 1$, hence the desired root is

$$x_1 = \frac{\beta\sin^2\theta_0}{a^2} = \frac{I_1}{I_3}\frac{2Mgl}{I_3\omega_z^2}\sin^2\theta_0. \tag{5–62}$$

Thus the extent of the nutation, as measured by $x_1 = u_0 - u_1$, goes down as $1/\omega_z^2$; the faster the top is spun the less is the nutation.

The *frequency* of nutation likewise can easily be found for the "fast" top. Since the amount of nutation is small the term $(1 - u^2)$ in Eq. (5–59) can be replaced by its initial value, $\sin^2\theta_0$. Eq. (5–59) then reads:

$$f(u) = \dot{x}^2 = x(\beta\sin^2\theta_0 - a^2x). \tag{5–63}$$

Eq. (5–63) constitutes a differential equation for the variation of x with time. It may be integrated directly without difficulty, the solution for the initial condition $x = 0$ at $t = 0$ being

$$x = \frac{x_1}{2}(1 - \cos at), \tag{5–64}$$

where x_1 is given by (5–62). The angular frequency of nutation of the figure axis between θ_0 and θ_1 is therefore

$$a = \frac{I_3}{I_1}\omega_z, \tag{5–65}$$

which *increases* the faster the top is spun initially.

Finally, the angular velocity of precession, from (5–50), is given by

$$\dot{\phi} = \frac{a(u_0 - u)}{\sin^2 \theta} \approx \frac{ax}{\sin^2 \theta_0},$$

or, substituting Eqs. (5–64) and (5–62):

$$\dot{\phi} = \frac{\beta}{2a}(1 - \cos at). \tag{5–66}$$

The rate of precession is therefore not uniform but varies harmonically with time, with the same frequency as the nutation. The *average* precession frequency, however, is

$$\bar{\dot{\phi}} = \frac{\beta}{2a} = \frac{Mgl}{I_3\omega_z}, \tag{5–67}$$

which indicates that the rate of precession decreases as the initial rotational velocity of the top is increased.

We are now in a position to present a complete picture of the motion of the fast top when the figure axis initially has zero velocity. Immediately after the figure axis is released, the initial motion of the top is always to fall under the influence of gravity. But as it falls the top picks up a precession velocity, directly proportional to the extent of its fall, which starts the figure axis moving sideways about the vertical. The initial fall results in a periodic nutation of the figure axis in addition to the precession. As the top is spun faster and faster initially, the extent of the nutation decreases rapidly, although the frequency of nutation increases, while at the same time the precession about the vertical becomes slower. In practice, for a sufficiently fast top the nutation is damped out by the friction at the pivot and becomes unobservable. The top then *appears* to precess uniformly about the vertical axis. Because the precession is regular only in appearance, Klein and Sommerfeld have dubbed it a *pseudoregular* precession. In most of the elementary discussions of precession the phenomenon of nutation is neglected. In consequence such derivations seem to lead to the paradoxical conclusion that upon release the top *immediately* begins to precess uniformly; a motion that is *normal* to the forces of gravity which are the ultimate cause of the precession. Our discussion of pseudoregular precession serves to resolve the paradox; the precession builds up continuously from rest without any infinite accelerations, and the initial tendency of the top *is* to move in the direction of the forces of gravity.

It is of interest to determine exactly what initial conditions will result in a true regular precession. In such case the angle θ remains constant at its initial value θ_0, which means that $\theta_1 = \theta_2 = \theta_0$, or in other words, $f(u)$ must have a double root at u_0 (cf. Fig. 5–8). Equivalently, we can require that at θ_0 both $\dot{\theta}$ and $\ddot{\theta}$ shall be zero.* Eq. (5–53), the modified energy equation, can also be written as

$$\dot{\theta}^2 = (\alpha - \beta \cos \theta) - \frac{(b - a \cos \theta)^2}{\sin^2 \theta}, \tag{5–68}$$

and its derivative with respect to time is

$$2\dot{\theta}\ddot{\theta} = \beta \sin \theta \, \dot{\theta} + 2 \cos \theta \frac{(b - a \cos \theta)^2}{\sin^3 \theta} \dot{\theta} - \frac{2a(b - a \cos \theta)}{\sin \theta} \dot{\theta}.$$

There is the same factor of $\dot{\theta}$ on both sides of the equation, and the condition that $\ddot{\theta} = 0$ at $\theta = \theta_0$ can therefore be written as

$$\frac{\beta}{2} = (a\dot{\phi} - \dot{\phi}^2 \cos \theta), \tag{5–69}$$

using Eq. (5–50) for $\dot{\phi}$. With the definitions of β, Eq. (5–54), and of a, Eq. (5–46), this condition becomes

$$Mgl = \dot{\phi}(I_3\dot{\psi} - (I_1 - I_3)\dot{\phi} \cos \theta). \tag{5–70}$$

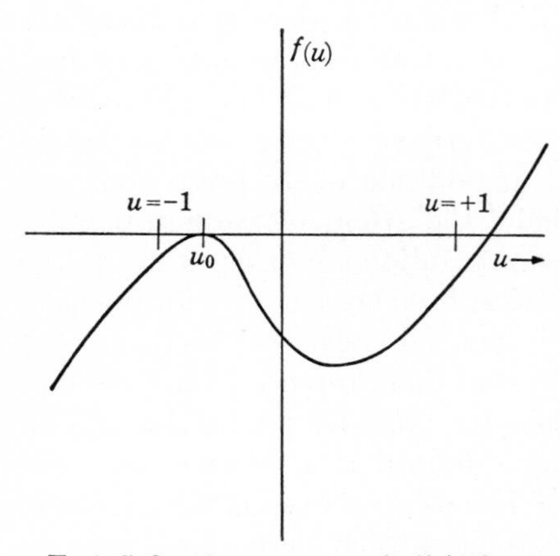

FIG. 5–8. Appearance of $f(u)$ for a regular precession.

The initial conditions for the problem of the heavy top require the specification of $\dot{\theta}$, $\dot{\phi}$, $\dot{\psi}$, θ, ϕ and ψ at the time $t = 0$. Because they are cyclic the initial values of ϕ and ψ are largely irrelevant and in general we can choose any desired value for each of the four others. But if in addition we require that the motion of the figure axis be one of uniform precession without nutation then our choice of these four initial values is no longer completely unrestricted, instead they must satisfy Eq. (5–70). One may still choose initial values

* The condition that $\ddot{\theta}$ vanish is actually equivalent to the requirement that $f(u)$ shall have a double root at $\theta = \theta_0$, for the latter condition can be expressed as

$$\frac{d\dot{\theta}^2}{d\theta} = 0 = 2\dot{\theta}\frac{d\dot{\theta}}{d\theta} = 2\ddot{\theta},$$

since $\dfrac{d\dot{\theta}}{d\theta} = \ddot{\theta}/\dot{\theta}$. The expression obtained here for $\ddot{\theta}$ is of course the same as the Lagrange equation for the coordinate θ.

of $\dot{\theta}$ and say, θ and $\dot{\psi}$, almost arbitrarily, but the value of $\dot{\phi}$ is then determined. The phrase "almost arbitrarily" is used because Eq. (5–70) is quadratic, and in order that $\dot{\phi}$ be real the discriminant of the equation must be positive:

$$I_3\dot{\psi}^2 - 4Mgl(I_1 - I_3)\cos\theta \geq 0,$$

which limits the allowable range of initial values of θ and $\dot{\psi}$. As a result of the quadratic nature of Eq. (5–70) there will in general be two solutions for $\dot{\phi}$, known as the "fast" and "slow" precession. It will also be noticed that (5–70) can never be satisfied by $\dot{\phi} = 0$ for finite $\dot{\psi}$; to obtain uniform precession we must always give the top a shove to start it on its way. Without this correct initial precessional velocity one can obtain at best only a pseudoregular precession.

If the precession is slow, so that $\dot{\phi}\cos\theta$ may be neglected compared to a, then an approximate solution for $\dot{\phi}$ is

$$\dot{\phi} \approx \frac{\beta}{2a},$$

which agrees with the average rate of pseudoregular precession for a fast top. This result is to be expected, of course; if the rate of precession is slow there is little difference between starting the gyroscope off with a little shove or with no shove at all.

One further case deserves some attention, namely, when $u = 1$ corresponds to one of the roots of $f(u)$. Suppose, for instance, a top is set spinning with its figure axis initially vertical. Clearly then $b = a$, for I_1b and I_1a are the constant components of the angular momentum about the vertical axis and the figure axis respectively, and these axes are initially coincident. Since the initial angular velocity is only about the figure axis the energy equation (5–52) evaluated at time $t = 0$ states that

$$E' = E - \frac{1}{2}I_3\omega_z^2 = Mgl.$$

By the definitions of α and β (Eq. (5–54)) it follows that $\alpha = \beta$.

The energy equation at any angle may therefore be written as

$$\dot{u}^2 = (1 - u^2)\beta(1 - u) - a^2(1 - u)^2$$

or

$$\dot{u}^2 = (1 - u)^2\{\beta(1 + u) - a^2\}.$$

The form of the equation indicates that $u = 1$ is always a double root, with the third root given by

$$u_3 = \frac{a^2}{\beta} - 1.$$

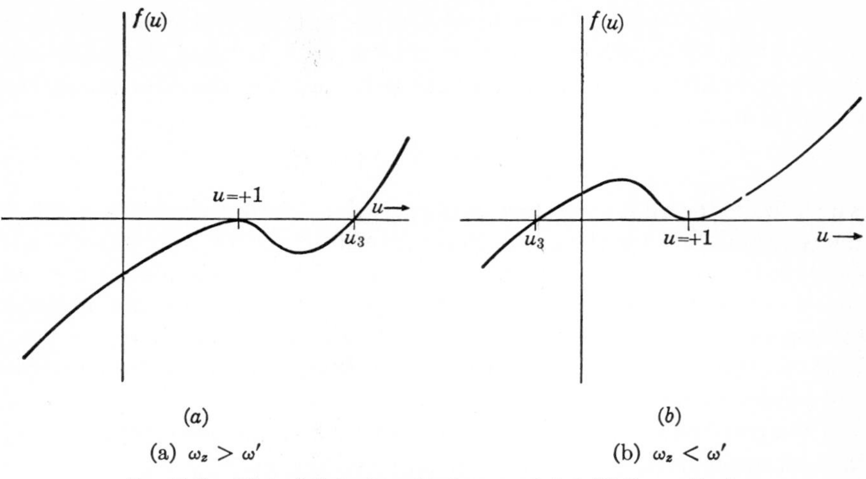

(a) $\omega_z > \omega'$ (b) $\omega_z < \omega'$

FIG. 5–9. Plot of $f(u)$ when the figure axis is initially vertical.

If $a^2/\beta > 2$ (which corresponds to the condition for a "fast" top) u_3 is larger than 1 and the only possible motion is for $u = 1$; the top merely continues to spin about the vertical. For this state of affairs the plot of $f(u)$ appears as shown in Fig. 5–9a. On the other hand if $a^2/\beta < 2$, u_3 is then less than 1, $f(u)$ takes on the form shown in Fig. 5–9b, and the top will nutate between $\theta = 0$ and $\theta = \theta_3$. There is thus a critical angular velocity, ω', above which only vertical motion is possible, whose value is given by

$$\frac{a^2}{\beta} = \left(\frac{I_3}{I_1}\right)\frac{I_3\omega'^2}{2Mgl} = 2$$

or

$$\omega'^2 = 4\,\frac{MglI_1}{I_3^2}. \tag{5–71}$$

In practice, if a top is started spinning with its axis vertical and with ω_z greater than the critical angular velocity, it will continue to spin quietly for a while about the vertical (hence the designation as a "sleeping" top). However, friction gradually reduces the frequency of rotation below the critical value and the top then begins to wobble in ever larger amounts as it slows down.

It has been mentioned previously that the earth is a top whose figure axis is precessing about the normal to the ecliptic, a motion known astronomically as the precession of the equinoxes. Were the earth completely spherical none of the other members of the solar system could exert a gravitational torque on it. But the earth is slightly flat at the poles and therefore "bulges" to a small extent at the equator. It is just the net torque on these bulges due to gravitational attraction, chiefly of the sun and moon, that sets the earth's axis precessing in space. The torque is quite small so that the precession is extremely slow — the period being 26,000 years compared to the rotational period of one day. The total applied torque is not constant in time, because the torques of the sun and moon have slightly different directions to the ecliptic and vary as the three bodies move around each other. As a result there are irregularities in the precession, often designated as *astronomical nutation*, but this must not be confused with the true nutation, discussed above, which is present even if the source of the torques is unvarying. Klein and Sommerfeld have pointed out that the true nu ation appears as the force-free precession of the earth's rotation axis about its figure axis, as derived in the previous section. The earth was apparently set spinning with an initial value of $\dot{\phi}$ far larger than needed for uniform precession, so that the nutation appears in space roughly as shown in Fig. 5–7b One can show that under these conditions the period of nutation as observed on the rotating body approaches the force-free precession period given by Eq. (5–40).

Some short mention should also be made of the many technical applications of gyroscopes, though space limitations forbid any extended discussion. By the term "gyroscope" is usually meant a symmetrical top mounted in gimbal rings in such a manner that the motion of the figure axis is unrestricted, while the center of gravity remains stationary. No net gravitational torque is exerted about the center of gravity and the angular momentum vector is constant. If the gyroscope is set spinning along the figure axis (which is therefore in coincidence with the angular momentum direction), the figure axis will always preserve its original direction. One can therefore use it as a "directional gyro" to serve as a reference direction independent of the motion of the vehicle carrying the gyroscope.

The action of a gyrocompass is much subtler. In this instrument the axis of the top is constrained to move only in the horizontal plane. But due to the rotation of the earth the horizontal plane is changing its direction relative to an inertial space, so that in effect the constraint forces the gyroscope to precess, with a period of one day, about the earth's axis. The gyroscope axis tends to remain stationary, the mounting is such as to constrain it to precess; as a result the bearings exert forces on the gyroscope. It can be shown that these forces act always to line up the gyroscope axis with the precession axis, here the direction of rotation of the earth. Such an arrangement therefore serves to indicate the direction of the meridian, it is a "gyrocompass."

5–8 Precession of charged bodies in a magnetic field. The involved discussions of the previous section show that the motion of a symmetrical top in the gravitational field may be quite complicated. In contrast, a charged spinning body in a uniform *magnetic field* has a relatively simple motion, but one which is of the greatest importance for atomic physics. Instead of using the Lagrangian formulation it will prove simpler to go back to first principles and start from the statement that the time rate of change of total angular momentum is equal to the total impressed torque:

$$\frac{d\mathbf{L}}{dt} = \mathbf{N}. \tag{1–24}$$

It will be assumed that the body is composed of particles all having the same e/m ratio. The motion of the charges as a result of the rotation of the body constitutes an electric current distribution which can interact with a magnetic field. If the field is uniform there will be no net force on the body * but there will be a net torque, approximately given by: †

$$\mathbf{N} = \mathbf{M} \times \mathbf{B}, \tag{5–72}$$

where $\mathbf{M}$ is the magnetic moment of the current distribution and $\mathbf{B}$ the magnetic intensity. The equation of motion under such a torque is therefore

$$\frac{d\mathbf{L}}{dt} = \mathbf{M} \times \mathbf{B}. \tag{5–73}$$

For any volume distribution of current the magnetic moment is defined, in Gaussian units, as

$$\mathbf{M} = \frac{1}{2c}\int \mathbf{r} \times \mathbf{j}\, dV, \tag{5–74}$$

where $\mathbf{j}$ is the current density.§ As an illustration of this definition consider a plane loop of thin wire in which a current i is flowing. The combination $\mathbf{j}\, dV$ can then be written as

$$\mathbf{j}\, dV = \mathbf{j}\, dS\, dl = i\, d\mathbf{l},$$

where dS is the cross-sectional area of the wire and $d\mathbf{l}$ an element of length along the direction of current flow. With this rearrangement the magnetic moment can be written as

$$\mathbf{M} = \frac{i}{c}\frac{1}{2}\oint \mathbf{r} \times d\mathbf{l}.$$

* In such case the center of mass can be taken as stationary and we are dealing with motion about a fixed point. It is then not necessary to specify the reference point for either the torque or angular momentum (cf. Sec. 1–2).

† See, for example, J. A. Stratton, *Electromagnetic Theory*, N.Y., 1941, 176, 242.

§ See R. Becker, *Theorie der Elektrizität*, Vol. II, 6th ed., p. 98, or J. A. Stratton, op. cit., p. 235 (where MKS units are used).

But $\frac{1}{2}\mathbf{r} \times d\mathbf{l}$ is an element of area swept out by the radius vector (cf. the discussion of areal velocity in Section 3–2) and the integral is proportional to the area of the loop, designated by A. If $\mathbf{n}$ denotes a unit vector normal to the plane of the loop then the magnetic moment is simply

$$\mathbf{M} = \frac{Ai}{c}\,\mathbf{n},$$

which is perhaps a more familiar form.

Returning to the definition Eq. (5–74), one can substitute for $\mathbf{j}$ the product of the charge density and the velocity. But the charge density is also the product of the e/m ratio and the mass density ρ, and one can therefore write

$$\mathbf{j} = \frac{e}{m}\,\rho\mathbf{v}.$$

In consequence of this relation the magnetic moment can be written

$$\mathbf{M} = \frac{e}{2mc}\int \mathbf{r} \times \rho\mathbf{v}\, dV. \tag{5–75}$$

The integral will be recognized as the total angular momentum of the body; there is thus a unique relationship, in classical physics at least, between the angular momentum of a body and its magnetic moment: *

$$\mathbf{M} = \frac{e}{2mc}\,\mathbf{L}. \tag{5–76}$$

The equation of motion (5–73) can now be written as

$$\frac{d\mathbf{L}}{dt} = \mathbf{L} \times \frac{e\mathbf{B}}{2mc}. \tag{5–77}$$

But this is exactly the equation of motion for a vector of constant magnitude which is rotating in space about the direction of $\mathbf{B}$ with an angular velocity

$$\boldsymbol{\omega}_l = -\frac{e\mathbf{B}}{2mc}. \tag{5–78}$$

It must be concluded therefore that the effect of a uniform magnetic field on a charged body obeying Eq. (5–72), is to cause the angular momentum vector to *precess uniformly* with the angular velocity (5–78), known as the *Larmor frequency*. For electrons, where e is negative, the precession is counterclockwise around the direction of $\mathbf{B}$.

The uniform precession of a charged body in a magnetic field is encountered constantly in atomic physics and is customarily referred to as the *Larmor precession*. It will be noted that we haven't actually required the body to be rigid. The basic equation of motion, Eq. (1–24), holds true

* It is characteristic of the quantum nature of the "spin" angular momentum of the electron that (5–76) is not obeyed in that case, the coefficient being e/mc instead.

no matter what the nature of the system, and the integral in (5–75) is the angular momentum about any point for all systems, provided the center of mass is at rest. For any system of charged particles, therefore, the total angular momentum will undergo a Larmor precession in a magnetic field, as given by (5–78), the only requirement being that all particles have the same e/m ratio.

The torque law of Eq. (5–72) is strictly true only for a permanent magnetic dipole whose magnitude is independent of the body's orientation. Actually, when the magnetic dipole arises from the rotation of a charged body the magnitude of **M** depends on the angular momentum, which is not necessarily independent of the orientation. It can be shown that one must then add to Eq. (5–72) a small term proportional to ω_l, which produces a small nutation in **L**. So long as ω_l is small compared with the rotation frequency of the body, the average precession frequency is still given by the Larmor frequency.*

SUGGESTED REFERENCES

A. P. WILLS, *Vector and Tensor Analysis.* Many of the references cited in the previous chapter discuss tensors, dyadics, and principal axis transformations. Additional material, excellently presented, will be found in Chapters VI, VIII, and IX of Wills' text. J. W. Gibbs introduced dyadics, and the fullest exposition of this subject is in the classic work, *Vector Analysis* by Gibbs and Wilson (1901).

A. G. WEBSTER, *Dynamics.* The literature on the dynamics of rigid bodies is very extensive; there are many books dealing with this subject alone, and every general treatise on mechanics devotes some space to it. Most such sources either date from the end of the 19th century or closely follow the traditional exposition of rigid dynamics developed at that time. One of the best of the earlier works is this general treatise of Webster (1st edition, 1904). Compared to Whittaker's text, Webster covers more ground (he includes potential theory, elasticity, and hydrodynamics), but the general level is more elementary. Many of the advanced topics are barely touched upon. The style is pleasantly discursive and the writing is less formal and more physical than in Whittaker, and consequently more intelligible. Vector notation is not used, for it was practically in its infancy at the time the book was written. Part II is on rigid body dynamics, and contains a particularly elaborate discussion of the force-free motion of the symmetrical top. The treatment of the heavy top is similar to the one given here, but is more extensive.

E. A. MILNE, *Vectorial Mechanics.* In contrast to Webster, vector and tensor notations are used here throughout, but in a manner that often obscures rather than reveals the physics of the subject. Chapter XIV contains a concise discussion of the inertia tensor and its properties. The unique appendix to this chapter is concerned with the calculation of the inertia tensor itself (rather than with the individual components) for a wide variety of bodies. Chapter XVII, on gyrostatic problems, is also of interest.

* For further details, see Herbert Goldstein, *American Journal of Physics* **19**, **100** (1951).

W. D. MACMILLAN, *Dynamics of Rigid Bodies.* While not recommendea for a systematic study of rigid body dynamics, this work contains much material not readily available elsewhere. Chapter VII, in particular, has long and elaborate discussions of Poinsot motion, and of the motion of the heavy symmetrical top, including the explicit solutions in terms of elliptic functions. The chapter on the complex problems of rolling rigid bodies is also worthy of note.

M. WINKELMANN AND R. GRAMMEL, *Kinetik der Starren Körper*, Vol. V of the *Handbuch der Physik.* This article is primarily of interest for the detailed yet clear discussions of the force-free top, the heavy symmetrical top with one point fixed, motion of billiard balls, the spinning coin, and similar problems. The extensive section on gyroscopic motion in rotating coordinate axes (as on the earth's surface) will unfortunately be inaccessible to most readers because the authors make use of the unfamiliar vector concept called a "motor."

J. L. SYNGE AND B. A. GRIFFITH, *Principles of Mechanics.* Although based on elementary principles, this text has a remarkably complete discussion of Poinsot motion and of the heavy symmetrical top, even including some of the explicit description of the motion in terms of elliptic functions. There are also some short sections on rolling motion and on the technical applications of gyroscopes (chiefly the gyrocompass).

F. KLEIN AND A. SOMMERFELD, *Theorie des Kreisels.* This monumental work on the theory of the top, in four volumes, has all the external appearances of the typical stolid and turgid German "Handbuch." Appearances are deceiving, however, for it is remarkably readable, despite the handicap of being written in the German language. The graceful, informal style has the fluency and attention to pedagogic details characteristic of all of Sommerfeld's later writings. Although the treatment becomes highly mathematical at times, the physical world is never lost sight of, and one does not founder in a maze of formula. Although limited by the title to tops and gyroscopes, the treatise actually provides a liberal education in all of rigid body mechanics, with excursions into other branches of physics and mathematics. Thus Chapter I discusses, among other items, Euler angles, infinitesimal rotations, and the Cayley-Klein parameters and their connections with the homographic transformation and with the theory of quaternions. The later notes to this chapter (in Vol. IV) discuss also the connections with electrodynamics and special relativity (quantum mechanics was still far in the future). By and large, Vol. I lays the necessary foundations in rigid dynamics and gives a physical description of top motion with little mathematics.

Vol. II is devoted to the detailed exposition of the heavy symmetrical top, although there is also much on Poinsot motion, and it contains a summary of what was then known about the asymmetric top. The distinction between regular and pseudoregular precession was first introduced here and the authors spend much time in examining the two motions, and the approach to regular precession. Many pages are given to a thorough demolishing of the popular or elementary "derivations" of gyroscopic precession. (The authors remark that it was the unsatisfactory nature of these derivations that led them to write the treatise!) There is a long discussion on questions of the stability of motion. Most of the treatment is based on the solution in terms of elliptic integrals and not merely on the approximate small nutation, as was done here.

Vol. III is mainly on perturbing forces (chiefly friction) and astronomical applications (nutation of the earth, precession of the equinoxes, etc.). The discussion of the wandering of the earth's poles is especially complete, including an estimation of the effects of the earth's elasticity and the transport of atmospheric masses by the wind circulation. Vol. IV is on technical applications, and is rather out of date by now.

F. KLEIN, *The Mathematical Theory of the Top.* In 1896 Felix Klein gave a series of lectures at Princeton, the notes for which constitute this slim volume. Most of the book is concerned with highly abstract mathematical details of the theory, but the first Lecture provides a readable account of Cayley-Klein parameters. It is interesting to note that both in this work and in the larger treatise with Sommerfeld use was made of a four-dimensional non-Euclidean space in which time is the fourth dimension — anticipating the use in special relativity by many years (see next chapter). However, the space was solely for mathematical convenience and no physical significance was intended.

A. SOMMERFELD, *Mechanik.* Sommerfeld's work with Klein on the top was one of his first publications, while this text, published more than forty years later, is one of the latest of his writings. His interest in the top has apparently not diminished in that time and he devotes considerable space to qualitative discussions of a wide range of gyroscopic and top phenomena — even to a page or two on the asymmetrical top. The thirty or so pages on the entire subject occupy almost all of the chapter on rigid bodies, and practically forms an abstract of the larger work! The treatment is extensive, rather than intensive, and there is little detailed discussion.

A. GRAY, *Treatise on Gyrostatics and Rotational Motion.* This is another detailed treatise, somewhat on the pattern of Klein and Sommerfeld. The treatment is much less systematic, however, and is not as readable or as informative. Of particular interest are the sections devoted to the more complicated gyroscopic systems — ship stabilizers, chains of gyroscopes, even boomerangs!

M. DAVIDSON, *The Gyroscope and Its Applications.* This reference is included because it presents an up-to-date (1947) presentation of the many technical applications of gyroscopes to navigational instruments, automatic pilots, and predicters. The theory is elementary and insufficient.

S. TIMOSHENKO AND D. H. YOUNG, *Advanced Dynamics.* The last chapter of this recent engineering text discusses the theory of some technical applications of gyroscopes, including the modern gyrocompass (which differs considerably from the simple Foucault gyrocompass described in the text).

EXERCISES

1. Calculate the change in the inertia tensor if the point of reference in the rigid body is shifted by the vector $\mathbf{r}_0$. Show that the new inertia tensor will have the same set of principal axes as the old if the original reference point was the center of mass *and* $\mathbf{r}_0$ is along one of the principal axes. What is the change in the principal moments of inertia produced by such a shift?

2. Find the principal moments of inertia about the center of mass of a flat rigid body in the shape of a 45° right triangle with uniform mass density. What are the principal axes?

3. Three equal mass points are located at $(a, 0, 0)$, $(0, a, 2a)$, and $(0, 2a, a)$. Find the principal moments of inertia about the origin and a set of principal axes.

4. A compound pendulum consists of a rigid body in the shape of a lamina suspended in the vertical plane at a point other than the center of gravity. Compute the period for small oscillations in terms of the radius of gyration about the center of gravity and the separation of the point of suspension from the center of gravity. Show that if the pendulum has the same period for two points of suspension at unequal distances from the center of gravity, then the sum of these distances is equal to the length of the equivalent simple pendulum.

5. A uniform bar of mass M and length $2l$ is suspended from one end by a spring of force constant k. The bar can swing freely only in one vertical plane, and the spring is constrained to move only in the vertical direction. Set up the equations of motion in the Lagrangian formulation.

6. A uniform rod slides with its ends on a smooth vertical circle. If the rod subtends an angle of 120° at the center of the circle, show that the equivalent simple pendulum has a length equal to the radius of the circle.

7. An automobile is started from rest with one of its doors initially at right angles. If the hinges of the door are toward the front of the car, the door will slam shut as the automobile picks up speed. Obtain a formula for the time needed for the door to close if the acceleration f is constant, the radius of gyration of the door about the axis of rotation is r_0, and the center of mass is at a distance a from the hinges. Show that if f is 1 ft/sec^2 and the door is a uniform rectangle 4 ft wide, the time will be approximately 3.04 seconds.

8. A wheel rolls down a flat inclined surface which makes an angle α with the horizontal. The wheel is constrained so that its plane is always perpendicular to the inclined plane, but it may rotate about the axis normal to the surface. Obtain the solution for the two-dimensional motion of the wheel, using Lagrange's equations and the method of undetermined multipliers.

9. (a) Show that the angular momentum of the force-free symmetrical top rotates in the body coordinates about the symmetry axis with an angular frequency Ω. Show also that the symmetry axis rotates in space about the fixed direction of the angular momentum with the angular frequency

$$\dot{\phi} = \frac{I_3\omega_z}{I_1 \cos\theta},$$

where ϕ is the Euler angle of the line of nodes with respect to the angular momentum as the space z-axis.

(b) Using the results of Exercise 5, Chapter 4, show that $\boldsymbol{\omega}$ rotates in space about the angular momentum with the same frequency $\dot{\phi}$, but that the angle θ' between $\boldsymbol{\omega}$ and $\mathbf{L}$ is given by

$$\sin\theta' = \frac{\Omega}{\dot{\phi}} \sin\theta'',$$

where θ'' is the inclination of $\boldsymbol{\omega}$ to the symmetry axis. Using the figures given in Section 5–6, show therefore that the earth's rotation axis and the axis of angular momentum are never more than 0.6 inch apart on the surface of the earth.

(c) Show from parts (a) and (b) that the motion of the force-free symmetrical top can be described in terms of the rotation of a cone fixed in the body whose axis is the symmetry axis, rolling on a fixed cone in space whose axis is along the angular momentum. The angular velocity vector is along the line of contact of the two cones. Show that the same description follows immediately from the Poinsot construction in terms of the inertia ellipsoid.

10. When the rigid body is not symmetrical, an analytic solution to Euler's equation for the force-free motion cannot be given in terms of elementary functions. Show, however, that the conservation of energy and angular momentum can be used to obtain expressions for the body components of $\boldsymbol{\omega}$ in terms of elliptic integrals.

11. Obtain from Euler's equations of motion the condition (5–70) for the uniform precession of a symmetrical top in a gravitational field, by imposing the requirement that the motion be a uniform precession without nutation.

12. Show that the magnitude of the angular momentum for a heavy symmetrical top can be expressed as a function of θ and the constants of the motion only. Prove that as a result the angular momentum vector precesses uniformly only when there is uniform precession of the symmetry axis.

13. It was stated in the text that the precession of the equinoxes was due to the torque of the sun and moon on the earth. This torque is entirely due to the flattening of the earth, for there can be no gravitational torque exerted on a perfect sphere. As a first approximation, one can therefore replace the earth by a perfect sphere plus a "girdle" at the equator, choosing the mass of the girdle and moment of inertia of the sphere such that the combination has the same principal moments of inertia as the earth. The entire torque is then due only to the equatorial girdle. Since the precession is very slow compared to the revolution of the moon around the earth and of the earth around the sun, we can roughly replace the sun and moon by circular mass distributions about the earth as a center. It will also suffice for order of magnitude purposes to assume that the orbits of the sun and moon are in a plane, the so-called ecliptic plane. Calculate the gravitational potential

$$V = -G \int\int \frac{dm_e dm_s}{r}$$

between the earth and sun rings by expanding the distance r between two elements of the rings in powers of $\frac{a}{R_s}$, where a is the radius of the girdle and R_s is the mean distance to the sun. For convenience, take the ecliptic plane as the xy-plane and use spherical polar coordinates. By a similar calculation, obtain the potential between the earth and moon rings. The resulting total potential is a function of the inclination angle θ of the earth's axis to the plane of the ecliptic, and the negative derivative of V with respect to θ gives the torque exerted by the sun and moon on the earth. Show in this fashion that the first nonvanishing term in the torque is given by

$$N = \frac{3}{4} G(I_3 - I_1) \sin 2\theta \left[\frac{m_s}{r_s^3} + \frac{m_l}{r_l^3} \right],$$

where G is the universal constant of gravity, and the subscripts s and l refer to the sun and moon respectively. Find the frequency of regular precession for this torque (using, for example, the method of Exercise 11), assuming that the precession is very slow compared to the rotation frequency. Compare this result with the measured precession period of 25,800 years.

14. In Section 5–6 the precession of the earth's axis of rotation about the pole was calculated on the basis that there were no torques acting on the earth. The previous exercise, on the other hand, showed that the earth is undergoing a forced precession due to the torques of the sun and moon. Actually, both results are valid; the motion of the axis of rotation about the symmetry axis appears as the nutation of the earth in the course of its forced precession. To prove this statement, calculate θ and $\dot{\phi}$ as a function of time for a heavy symmetrical top which is given an initial velocity $\dot{\phi}_0$ that is large compared with the net precession velocity $\beta/2a$, but which is small compared with ω_z. Under these conditions, the bounding circles for the figure axis still lie close together, but the orbit of the figure axis appears as in Fig. 5–7(b), i.e., shows large loops which move only slowly around the vertical. Show for this case that (5–64) remains valid but now

$$x_1 = \left(\frac{\beta}{a^2} - \frac{2\dot{\phi}_0}{a}\right)\sin^2\theta_0.$$

From these values of θ and $\dot{\phi}$ obtain ω_x and ω_y, and show that for $\beta/2a$ small compared with $\dot{\phi}_0$, the vector $\boldsymbol{\omega}$ precesses around the figure axis with an angular velocity

$$\Omega = \frac{I_1 - I_3}{I_1}\omega_z,$$

in agreement with Eq. (5–40). Verify from the numbers given in Section 5–6 that $\dot{\phi}_0$ corresponds to a period of about 1600 years, so that $\dot{\phi}_0$ is certainly small compared with the daily rotation, and is sufficiently large compared with $\beta/2a$, which corresponds to the precession period of 26,000 years.

15. (Foucault Gyrocompass.) A gyroscope is mounted with its center of gravity at the center of the gimbals, so that there is no gravitational torque. However, the figure axis is constrained to move only in the horizontal plane. If the gyroscope is set spinning on the earth's surface, there will be an additional rotational motion due to the earth's rotation. Show from Euler's equations that if the angular frequency of the gyroscope is large compared with the earth's rotation, the figure axis will oscillate symmetrically about the meridian and can thus be used as a compass.

16. A system consists of charged particles, all having the same ratio e/m, with a potential energy depending only on the relative positions of the particles. A uniform magnetic field **B** is now imposed on the system; the vector potential **A** being then given by

$$\mathbf{A} = \frac{1}{2}(\mathbf{B} \times \mathbf{r}).$$

Obtain the Lagrangian for the system and show that when expressed in terms of coordinate axes that are rotating about **B** with the angular velocity $\boldsymbol{\omega}_l$ the Lagrangian reduces to the field-free form if terms in B^2 are small and can be neglected. This constitutes an independent proof of Larmor's theorem, which in this form states that the *sole* effect of a weak magnetic field is to cause a precession of the entire motion about **B**. As given in the text, Larmor's theorem merely referred to the effect on the angular momentum.

17. Show that the Hamiltonian for a charged symmetrical top in a uniform magnetic field is identical with the kinetic energy and is a constant of the motion. The field therefore does no work on the system, as can also be seen from the Lorentz force (1–56). This behavior is in contrast to the gravitational top, where the added kinetic energy of precession was obtained from the gravitational field. Show that in the magnetic top the energy of precession comes from a diminution of the rotational velocity about the figure axis, and that this produces a nutation of the figure axis.

CHAPTER 6

SPECIAL RELATIVITY IN CLASSICAL MECHANICS

Our development of classical mechanics has been based on a number of definitions and postulates which were presented in Chapter 1. However, when the velocities involved approach the speed of light these postulates, as is well known, no longer represent the experimental facts, and they must then be altered to conform to the so-called *special theory of relativity*. This is a modification of the structure of mechanics which must not be confused with the far more violent recasting required by quantum theory. There are many physical instances where quantum effects are important but relativistic corrections are negligible. And conversely, phenomena frequently occur involving relativistic velocities where the refinements of quantum mechanics do not affect the discussion. There is no inherent connection between special relativity and quantum mechanics, and the effects of one may be discussed without the other. It is therefore of considerable practical importance to examine the changes in the formulation of classical mechanics required by special relativity.

It is not intended, however, to present a comprehensive discussion of the theory of special relativity and its consequences. We shall not be greatly concerned with the events and experiments which led to the construction of the theory, far less with its philosophical implications, its apparent paradoxes "which gaily mock at common sense." The emphasis will be on how special relativity may be fitted into the framework of classical mechanics and only as much of the theory as is needed for this task will be presented.

6–1 The basic program of special relativity. In the discussions of the previous chapter, frequent use has been made of such phrases as "space system" or "system fixed in space." By these phrases we have meant no more than an *inertial* system, one in which Newton's law of motion,

$$\mathbf{F} = m\mathbf{a}, \tag{6–1}$$

is valid. A system fixed in a body rotating with respect to an inertial system does not satisfy this qualification, one must add to (6–1) terms describing the effect of the rotation. On the other hand, it would appear that a system moving uniformly with respect to a "space system" should itself be an inertial system. If $\mathbf{r}'$ represents a radius vector from the origin of the second system to a given point, and $\mathbf{r}$ the corresponding vector in

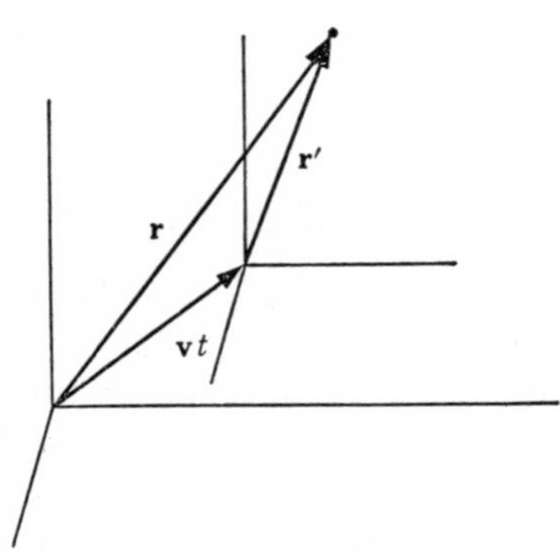

FIG. 6–1. Illustrating the Galilean transformation.

the first system, cf. Fig. 6–1, then it seems obvious that these two vectors are connected by the relation:

$$\mathbf{r}' = \mathbf{r} - \mathbf{v}t. \tag{6–2}$$

Since the relative velocity is constant, the first time derivative of Eq. (6–2) is

$$\dot{\mathbf{r}}' = \dot{\mathbf{r}} - \mathbf{v}, \tag{6–3}$$

and another differentiation gives:

$$\mathbf{a}' = \mathbf{a} \tag{6–4}$$

so that the acceleration is the same in both systems. If Newton's law, Eq. (6–1), holds in one system it should hold in the other.

On the other hand, the transformation represented by Eqs. (6–2, 4), known as the *Galilean transformation*, predicts that the velocity of light should be different in the two systems. Thus, suppose there is a source of light at the origin of the unprimed system emitting spherical waves traveling with the speed c. Let the radius vector $\mathbf{r}$ be the position vector of a point on some given wave surface. Then in the unprimed system the velocity of the point on the wave surface is $\dot{\mathbf{r}} = c\mathbf{n}$, where $\mathbf{n}$ is a unit vector along $\mathbf{r}$. According to (6–2), however, the corresponding wave velocity in the primed system is $\dot{\mathbf{r}}' = c\mathbf{n} - \mathbf{v}$. In the system moving with respect to the source of light the magnitude of the wave velocity will in general no longer be c; indeed, since it depends on direction, the waves will no longer be spherical.

A long series of investigations, especially the famous experiments of Michelson and Morley, have indicated that the velocity of light is always the same in all directions, and is independent of the relative uniform motions of the observer, the transmitting medium, and the source. The Galilean transformation thus cannot be correct, and must be replaced by another, the *Lorentz transformation*, which will preserve the velocity of light in all systems. Einstein showed that such a transformation requires revision of the usual concepts of time and simultaneity. But he went further; from the experimental fact that the speed of light is constant in all systems he generalized as a basic postulate that *all* phenomena of physics appear the same in all uniformly moving systems. This so-called *postulate of equivalence* states that it is impossible by means of any physical measurements to label a coordinate system as intrinsically "stationary" or "uniformly moving"; one can only infer that the two systems are moving *relative* to each other. Thus measurements made entirely *within* a given system must be incapable of distinguishing that system from all others moving uniformly with respect to it. The equivalence postulate requires that all physical laws must be phrased in an identical manner for all uniformly moving systems. For example, the statement that the speed of light is everywhere c means that a wave equation of the form

$$\nabla^2 \mathbf{E} - \frac{1}{c^2}\frac{\partial^2 \mathbf{E}}{\partial t^2} = 0$$

describes the propagation of light in all systems.

We have seen that Newton's equations of motion are invariant in form only under a Galilean transformation, which we know to be incorrect. It is highly probable a priori, therefore, that Newton's equations of motion, and perhaps other commonly accepted laws of physics, will not preserve their form under the correct Lorentz transformation. The equivalence postulate states that such laws are inaccurate representations of experimental phenomena, and must be suitably generalized into forms which do have the correct transformation properties. Of course the generalizations must be such as to reduce to the more customary forms for velocities much smaller than that of light, when the Galilean transformation is approximately correct.

The program of the theory of special relativity is therefore twofold. First, there must be obtained a transformation between two uniformly moving systems which will preserve the velocity of light. Second, the laws of physics must be examined as to their transformation properties under this Lorentz transformation. Those laws which do not keep their form invariant are to be generalized so as to obey the equivalence postulate. Abundant experimental verification has by now been obtained for the physical picture resulting from this program, and in the last analysis this is the only justification needed for Einstein's fundamental assumptions.

6–2 The Lorentz transformation. Consider two uniformly moving systems whose origins coincide at time $t = 0$, at which time a source of light fixed at the origin of the unprimed system emits a pulse of light. An experimenter fixed in this system will of course see a spreading spherical wave propagating with the speed c. The equation of the observed wave front will be

$$x^2 + y^2 + z^2 = c^2t^2. \tag{6–5}$$

But the experimental fact of the invariance of the speed of light indicates that an observer in the system moving with respect to the source will also see the light propagating as a spherical wave from *his* origin, with the equation of the wave front appearing as

$$x'^2 + y'^2 + z'^2 = c^2t'^2. \tag{6–6}$$

By putting a prime on the "t" explicit recognition is given to the possibility that the time scale will also transform in going from one system to another. Stated more fully in words, the desired transformation which yields Eq. (6–6) from Eq. (6–5) may require that the time interval between two events depend on the reference system of the observer. From Eqs. (6–5) and (6–6) the desired transformation must be such that

$$x^2 + y^2 + z^2 - c^2t^2 = x'^2 + y'^2 + z'^2 - c^2t'^2. \tag{6–7}$$

This condition is reminiscent of the definition of an orthogonal transformation, Eq. (4–13), especially if $x_1x_2x_3$ be written for xyz, so that Eq. (6–7) appears as

$$\sum_{i=1}^{3} x_i^2 - c^2t^2 = \sum_{i=1}^{3} x_i'^2 - c^2t'^2. \tag{6–7$'$}$$

Comparison of the form of Eq. (6–7′) with Eq. (4–13) suggests introducing formally a fourth, imaginary coordinate $ict = x_4$ to obtain an even closer resemblance to spatial orthogonal transformations:

$$\sum_{\mu=1}^{4} x_\mu^2 = \sum_{\mu=1}^{4} x_\mu'^2. \tag{6–8}$$

Eq. (6–8) shows that the transformation we seek corresponds to a rotation in a four-dimensional space consisting of the three dimensions of ordinary space plus a fourth imaginary dimension proportional to the time. This space is known as *world space* or *Minkowski space.* The Lorentz transformations are therefore simply the orthogonal transformations of Minkowski space. All the mathematical apparatus developed in Chapter 4 for spatial orthogonal transformations is automatically also applicable here in the discussion of Lorentz transformations.

Clearly a spatial rotation between two systems at rest relative to each other is included as a subclass of the Lorentz transformation. We shall speak of a pure Lorentz transformation as one which does not involve any spatial rotation but is concerned only with uniformly moving systems whose axes are parallel. It is also evident without detailed proof * that any general Lorentz transformation is a product of a space rotation and a pure Lorentz transformation. Further, it is no loss of generality to take

* See R. Becker, *Theorie der Elektrizität*, Vol. II, 6th ed., Leipzig, 1933, p. 287.

the direction of velocity **v** between the two systems to be along one of the axes, say the x_3 axis. No matter what the direction of **v** one can always perform an ordinary spatial rotation on the coordinate axes to align the x_3 axis with the direction of **v**. It will therefore be sufficient to obtain the matrix elements $a_{\mu\nu}$ of the transformation between x and x',

$$x'_\mu = \sum_{\nu=1}^{4} a_{\mu\nu} x_\nu, \tag{6-9}$$

only for a pure Lorentz transformation with the velocity **v** along the direction of x_3.* The matrix elements must of course satisfy the same sort of orthogonality conditions as for spatial rotations (cf. Eq. (4-37)):

$$\sum_\nu a_{\mu\nu} a_{\lambda\nu} = \delta_{\mu\lambda}. \tag{6-10}$$

However, unlike the spatial orthogonal transformation, the matrix elements are now not all real. Since the coordinates $x'_1 x'_2 x'_3$ must remain real, the elements a_{i4}, $i = 1, 2, 3$, are required to be imaginary. Similarly the imaginary nature of x'_4 imposes the condition that the a_{4i} elements be imaginary while a_{44} must be real.

The directions perpendicular to the motion are obviously left unaffected by the transformation,

$$x'_1 = x_1, \qquad x'_2 = x_2,$$

for they do not participate in the motion and are effectively at rest. Only the x_3 and x_4 coordinates require any change in transforming from one system to the other. It can also be seen from qualitative arguments that neither x'_3 nor x'_4 will involve the values of x_1 or x_2. No one position in the x_1x_2 plane can be singled out by the physics of the situation as necessarily the origin of the coordinate system, and one can clearly shift the origin to any spot on the x_1x_2 plane without affecting the transformed values of x'_3 and x'_4. But such a shift *will* affect the values of x_1 and x_2, and hence these coordinates cannot enter into the transformation equations for x'_3 and x'_4. As a result of these simplifications, the matrix for the pure Lorentz transformation can be written as

$$\begin{pmatrix} 1 & 0 & 0 & 0 \\ 0 & 1 & 0 & 0 \\ 0 & 0 & a_{33} & a_{34} \\ 0 & 0 & a_{43} & a_{44} \end{pmatrix}.$$

---

* It has become conventional to use Greek letters, $\mu\nu\lambda$, etc., for indices running from 1 to 4, and Roman letters, ijk, etc., when the range is from 1 to 3, and this practice will be adhered to from now on.

The orthogonality conditions furnish three relations connecting the four matrix elements:

$$\begin{aligned} a_{33}^2 + a_{34}^2 &= 1, \\ a_{43}^2 + a_{44}^2 &= 1, \\ a_{33}a_{43} + a_{34}a_{44} &= 0. \end{aligned} \tag{6–11}$$

A fourth condition is needed to determine the remaining elements uniquely. It can be supplied from the observation that the origin of the primed system ($x_3' = 0$) is moving uniformly along the x_3 axis, so that at time t its x_3 coordinate is vt:

$$x_3 = vt = -i\beta x_4,$$

where

$$\beta = v/c. \tag{6–12}$$

In terms of the matrix elements this condition can be written as saying that the origin of the primed system is given by:

$$x_3' = x_4(a_{34} - i\beta a_{33}) = 0,$$

or

$$a_{34} = i\beta a_{33}.$$

The first of the orthogonality conditions, Eqs. (6–11) can then be simplified to

$$a_{33}^2(1 - \beta^2) = 1$$

or

$$a_{33} = \frac{1}{\sqrt{1-\beta^2}}, \tag{6–13}$$

and consequently

$$a_{34} = \frac{i\beta}{\sqrt{1-\beta^2}}. \tag{6–14}$$

Notice that a_{33} is real and a_{34} is imaginary, as required by the reality conditions on the matrix elements.

The remaining two elements can be obtained by solving the last of Eqs. (6–11) for a_{43}:

$$a_{43} = -a_{44}\frac{a_{34}}{a_{33}} = -i\beta a_{44},$$

and by substituting the result in the second orthogonality condition and solving for a_{44}:

$$a_{44} = \frac{1}{\sqrt{1-\beta^2}}.$$

The fourth matrix element is then

$$a_{43} = \frac{-i\beta}{\sqrt{1-\beta^2}}.$$

With these values for the four elements, the matrix of the Lorentz transformation has the form:*

$$\begin{pmatrix} 1 & 0 & 0 & 0 \\ 0 & 1 & 0 & 0 \\ 0 & 0 & \frac{1}{\sqrt{1-\beta^2}} & \frac{i\beta}{\sqrt{1-\beta^2}} \\ 0 & 0 & \frac{-i\beta}{\sqrt{1-\beta^2}} & \frac{1}{\sqrt{1-\beta^2}} \end{pmatrix}. \tag{6-15}$$

Notice that the complete matrix contains a submatrix of the same form as the matrix for a plane rotation:

$$\begin{pmatrix} \cos\phi & \sin\phi \\ -\sin\phi & \cos\phi \end{pmatrix}.$$

Indeed the matrix (6–15) does represent a rotation in the x_3x_4 plane, but the angle of rotation is imaginary, for here

$$\cos\phi = \frac{1}{\sqrt{1-\beta^2}}, \tag{6-16}$$

which is greater than unity.

The Lorentz transformation equations can also be written as

$$\begin{aligned} x' &= x, \\ y' &= y, \\ z' &= \frac{z - vt}{\sqrt{1-\beta^2}}, \\ t' &= \frac{t - \frac{vz}{c^2}}{\sqrt{1-\beta^2}}. \end{aligned} \tag{6-17}$$

The inverse transformation from x'_μ back to x_μ can be obtained simply by transposing the matrix (6–15). From the form of the matrix it is seen that the inverse equations differ from (6–17) only by a change in the sign

* The positive square root is chosen for a_{44} and a_{33} in order that the matrix reduce to the unit matrix when $\beta \to 0$. We are interested only in *proper* Lorentz transformations, with determinant $+1$.

of v. This result is to be expected on purely physical grounds, since the unprimed system is moving relative to the primed with the velocity $-\mathbf{v}$.

The aspect of the transformation equations most paradoxical by commonsense standards is contained in the relation between t and t'. Two events occurring at the same time at two different space points in the unprimed system will not appear as simultaneous to observers in the primed system, as a result of the vz/c^2 term n the equation for t'. It is not within our purpose here to enter into a physical discussion of this and similar apparent paradoxes,* but mention must be made of two famous consequences of the Lorentz transformations — the Lorentz-Fitzgerald contraction of length and the dilatation of time scales.

Consider a rigid rod at rest in the unprimed system lying along the z-axis and having length $l = z_2 - z_1$. A moving observer measures the length of the rod by locating the position of both end points in his system, z_1' and z_2', at a time t'. From the inverse equations we have:

$$z_1 = \frac{z_1' + vt'}{\sqrt{1 - \beta^2}},$$

$$z_2 = \frac{z_2' + vt'}{\sqrt{1 - \beta^2}},$$

so that the apparent length is:

$$z_2' - z_1' = l\sqrt{1 - \beta^2}. \tag{6–18}$$

The rod will appear contracted to the moving observer by the factor $\sqrt{1 - \beta^2}$, and this result is the basis for the famous Lorentz-Fitzgerald contraction hypothesis. Note that it is not convenient to use here the direct Eqs. (6–17), for while both ends are measured at the same time t' these are not simultaneous events in the unprimed space since they are at different points z_1 and z_2.

Suppose a clock were located in the unprimed system at a point z_1. At time t_1 on this clock an observer at the point but fixed in the moving system notes a time:

$$t_1' = \frac{t_1 - \frac{vz_1}{c^2}}{\sqrt{1 - \beta^2}},$$

* For such discussions see P. Bergmann, *An Introduction to the Theory of Relativity*, 1942, New York, and R. Becker, op. cit.

and at time t_2 a similar observer finds the time in his system to be

$$t_2' = \frac{t_2 - \frac{vz_1}{c^2}}{\sqrt{1 - \beta^2}},$$

so that the apparent time interval is:

$$t_2' - t_1' = \frac{t_2 - t_1}{\sqrt{1 - \beta^2}}. \tag{6-19}$$

When the stationary clock has turned one hour the moving observer will find his clock has gone $\dfrac{1}{\sqrt{1 - \beta^2}}$ hours. He will say the stationary clock is slow, that it is losing time; hence the name "dilatation of time" given to this phenomenon. But it should be emphasized that observers in the unprimed system examining the rate of a clock fixed in the primed system likewise come to the conclusion that it is running slow compared to theirs. The same statement is true for Lorentz contraction; an observer in the unprimed system observes the same contraction (6-18) for objects fixed in the primed system. Thus no one system is singled out as the stationary one and the other the moving one — the motion is only relative; all (uniformly moving) systems are completely equivalent.

The Lorentz transformation also indicates that one cannot have a relative velocity greater than c. If a body did have such a velocity with reference to a given system, then it should be possible to transform from the reference system to the system in which the body is at rest by means of a Lorentz transformation. But no Lorentz transformation to real coordinate systems is possible when $\beta > 1$, indicating that speeds greater than the speed of light cannot occur.

It might be thought possible to obtain a speed greater than c by a succession of Lorentz transformations. Thus one might try to transform from one system to a second moving with a relative speed v_1 greater than $c/2$, and then transform from this to a third system moving parallel to $\mathbf{v}_1$ at a speed v_2, also greater than $c/2$, relative to the second. Unfortunately, the speed of the third system relative to the first is not given simply by $v_1 + v_2$. One may find the Lorentz transformation from the first to the third system directly by multiplying the matrices of the two separate transformations. It is then found that the total transformation corresponds to a velocity v_3 given by the so-called Einstein addition law for velocities:

$$v_3 = \frac{v_1 + v_2}{1 + \frac{v_1 v_2}{c^2}},$$

or equivalently

$$\beta_3 = \frac{\beta_1 + \beta_2}{1 + \beta_1\beta_2}. \tag{6-20}$$

It is seen that β_3 always remains less than unity. Proof of (6-20) will be left for the exercises.

6-3 Covariant four-dimensional formulations. Having obtained the Lorentz transformation to replace the incorrect Galilean transformation, we can now proceed to the second stage and require that the laws of mechanics, in common with all of physics, shall have the same form in all uniformly moving systems. The task of examining the laws of physics for invariance in form under Lorentz transformation is greatly facilitated by writing them in terms of the four-dimensional world introduced in the previous section. Indeed, as will be shown, it is then possible to verify the Lorentz invariance of a given equation merely by inspection.

Invariance of form under Lorentz transformation is not the only invariant property demanded of physical laws. Clearly the physical content of any given relation cannot be affected by the particular orientation chosen for the spatial axes; the laws of physics must also be invariant in form under rigid rotations, i.e., spatial orthogonal transformations. An examination of this more familiar invariance requirement will clarify the procedure to be followed in establishing invariance under Lorentz transformation.

Normally we do not worry about the invariance of our theories under spatial rotations. In constructing any equation it is always required that the terms of the equation be *all* scalars, or *all* vectors; in general all terms must be tensors of the same rank, and this requirement automatically insures the desired invariance under rotation. Thus a scalar relation will have the general form

$$a = b,$$

and since both sides of the equation, being scalars, are invariant under spatial rotations the relation obviously holds in all coordinate systems. A vector relation, of the form

$$\mathbf{F} = \mathbf{G},$$

really stands for three separate relations between the components of the vector:

$$F_i = G_i.$$

The values of these components, of course, are not invariant under the spatial rotation; rather they are transformed to new values F'_i, G'_i which are the components of the transformed vectors, $\mathbf{F}'$, $\mathbf{G}'$. But because both sides of the component relations transform in identical fashion, the same relation must hold between the transformed components:

$$F_i' = G_i'.$$

The relationship between the two vectors is thus undisturbed by the spatial rotation, in the new system we still have

$$\mathbf{F}' = \mathbf{G}'.$$

Note that the invariance in the form of the relationship is entirely in consequence of the fact that both sides transform as vectors. We speak of the terms of the equation as being *covariant.* Similarly, an equality between two tensors of the second rank

$$\mathsf{C} = \mathsf{D}$$

necessarily implies the same equality between the two transformed tensors

$$\mathsf{C}' = \mathsf{D}'$$

because the two tensors transform covariantly under a spatial rotation. On the other hand, an equation involving separately a component of a vector and, say, a component of a tensor obviously cannot remain invariant in form under a three-dimensional orthogonal transformation. *Invariance of a physical law under rotation of the spatial coordinate system requires covariance of the terms of the equation under three-dimensional orthogonal transformation.*

Now, the Lorentz transformation has been identified previously with the orthogonal transformations in world space. One can set up scalars, vectors, in general, tensors of any rank, in this four-dimensional world space, with transformation properties defined as obvious generalizations of those for the analogous spatial quantities. Thus, we will speak of *world scalars* and of *world vectors* (or *four-vectors*). The invariance of the form of any physical law under Lorentz transformation will then be immediately evident once it is expressed in a *covariant four-dimensional form,* all terms being world tensors of the same rank. A law failing to meet the requirements of the equivalence principle cannot be put into a covariant form. The four-dimensional transformation properties of the terms of a physical law thus act as the touchstone for examining its relativistic validity.

An important example of a four-vector is the position vector of a point in world space, with components $x_1x_2x_3x_4$. To avoid confusion with spatial vectors, a four-vector will be denoted by a typical one of its components; thus x_μ will stand for the position four-vector. It is often convenient to use the Einstein summation convention: wherever repeated indices occur in a term it is to be understood that the expression is to be summed over the index, even if the summation sign is not shown. For example, $x_\mu x_\mu$ would stand for

$$\sum_{\mu=1}^{4} x_\mu^2.$$

As a particle moves in ordinary space its corresponding point in four-dimensional space will describe a path known as the *world line*. The four-vector dx_μ represents the change in the position four-vector for a differential motion along the world line. From the dot product of dx_μ with itself we can form a world scalar (and hence a Lorentz invariant), denoted by $d\tau$ and defined by the equation

$$(d\tau)^2 = -\frac{1}{c^2}\sum_\mu (dx_\mu)^2. \tag{6–21}$$

The significance of $d\tau$ can be made clear by evaluating Eq. (6–21) in a system in which the particle is momentarily at rest. In such a system the components of the transformed vector dx'_μ are $(0, 0, 0, icdt')$ and the invariant $d\tau$ is given by

$$(d\tau)^2 = -\frac{1}{c^2}\sum_\mu (dx'_\mu)^2 = (dt')^2.$$

Thus $d\tau$ is the time interval as measured on a clock traveling with the particle,* hence it is referred to as an interval of the particle's *proper time* or *world time*.

The relation between $d\tau$ and an interval of time as measured in a given Lorentz system can be derived directly by expanding the defining equation (6–21):

$$(d\tau)^2 = -\frac{1}{c^2}\left((dx)^2 + (dy)^2 + (dz)^2 - c^2(dt)^2\right),$$

or

$$d\tau = dt\sqrt{1 - \frac{1}{c^2}\left[\left(\frac{dx}{dt}\right)^2 + \left(\frac{dy}{dt}\right)^2 + \left(\frac{dz}{dt}\right)^2\right]},$$

which is equivalent to the relation

$$\frac{d\tau}{\sqrt{1-\beta^2}} = dt. \tag{6–22}$$

---

* By definition $d\tau$ is taken as the positive square root of the expression given in Eq. (6–21).

Eq. (6-22) also follows from the time-dilatation formula, in consequence of the interpretation of $d\tau$ as the time interval on a clock fixed in the particle and dt as the corresponding time interval measured by observers moving with respect to the particle.

As one of the components of a four-vector is imaginary, the square of such a vector is no longer necessarily positive definite. Four-vectors for which the square of the magnitude is greater than or equal to zero are called *space-like;* when the magnitudes are negative they are known as *time-like* vectors. Since these characteristics arise from the magnitudes of the vectors, which are world scalars, the designations are obviously unaffected by any Lorentz transformation. The names stem from the fact that the square of a spatial vector is always positive definite, and a space-like four-vector can always be transformed so that its fourth component vanishes. On the other hand, a time-like four-vector must always have a fourth component, but it can be transformed so that the first three components vanish. As an example of these concepts it may be noted that the difference vector between two world points can be either space- or time-like. Let X_μ be the difference vector, defined as

$$X_\mu = x_{1\mu} - x_{2\mu},$$

the subscripts 1 and 2 denoting the two points. The magnitude of X_μ is given by

$$X_\mu X_\mu = |\mathbf{r}_1 - \mathbf{r}_2|^2 - c^2(t_1 - t_2)^2.$$

Thus, X_μ is space-like if the two world points are separated such that

$$|\mathbf{r}_1 - \mathbf{r}_2|^2 \geq c^2(t_1 - t_2)^2,$$

while it is time-like if

$$|\mathbf{r}_1 - \mathbf{r}_2|^2 < c^2(t_1 - t_2)^2.$$

The condition for a time-like difference vector is equivalent to stating that it is possible to bridge the distance between the two world points by a light signal, while if the points are separated by a space-like difference vector they cannot be connected by any wave traveling with the speed c.

The spatial axes can always be so oriented that the spatial difference vector $\mathbf{r}_1 - \mathbf{r}_2$ is along the x_3 axis, with $|\mathbf{r}_1 - \mathbf{r}_2|$ therefore equal to $z_1 - z_2$. Under a Lorentz transformation with velocity v parallel to the z-axis the fourth component of X_μ then transforms according to Eq. (6-17):

$$c(t_1' - t_2') = \frac{c(t_1 - t_2) - \dfrac{v}{c}(z_1 - z_2)}{\sqrt{1 - \beta^2}}.$$

If X_μ is space-like then

$$c(t_1 - t_2) < z_1 - z_2,$$

and it is therefore possible to find a velocity $v < c$ such that $ic(t'_1 - t'_2) \equiv X'_4$ vanishes, as was stated above. Now, a point in world space corresponds to something happening at a given point in space, $\mathbf{r}$, and at a given time, t; in a word, it describes an *event*. Physically the vanishing of X'_4 means that if the distance between two events is space-like then one can always find a Lorentz system in which the two events are simultaneous.

Examples of four vectors may be easily multiplied. Thus, the four-velocity u_ν is defined as the rate of change of the position vector of a particle with respect to its proper time:

$$u_\nu = \frac{dx_\nu}{d\tau}, \tag{6–23}$$

with space and time components

$$u_i = \frac{v_i}{\sqrt{1-\beta^2}} \quad \text{and} \quad u_4 = \frac{ic}{\sqrt{1-\beta^2}}. \tag{6–24}$$

The world velocity has a constant magnitude, for the sum $u_\nu u_\nu$ is given by

$$u_\nu u_\nu = \frac{v^2}{1-\beta^2} - \frac{c^2}{1-\beta^2} = -c^2, \tag{6–25}$$

and it is thus also time-like.

To illustrate the process of restating a physical law in a covariant four-dimensional formulation consider the scalar wave equation of the type

$$\nabla^2\psi - \frac{1}{c^2}\frac{\partial^2\psi}{\partial t^2} = 0. \tag{6–26}$$

In analogy to the three-dimensional gradient we may introduce a *four-gradient* differential operator, denoted by the symbol $\Box$, with components

$$\frac{\partial}{\partial x_1}, \quad \frac{\partial}{\partial x_2}, \quad \frac{\partial}{\partial x_3}, \quad \frac{\partial}{\partial x_4}.$$

It can readily be shown that the four-gradient transforms as a four-vector. From the laws of partial differentiation, the partial derivative with respect to the transformed coordinate x'_μ is given by

$$\frac{\partial}{\partial x'_\mu} = \sum_\nu \frac{\partial x_\nu}{\partial x'_\mu}\frac{\partial}{\partial x_\nu}. \tag{6–27}$$

But x_ν is related to x'_μ by the inverse transformation:

$$x_\nu = \sum_\mu a_{\mu\nu} x'_\mu,$$

so that the partial derivative of x_ν with respect to x'_μ is simply $a_{\mu\nu}$. Eq. (6-27) then takes the form

$$\frac{\partial}{\partial x'_\mu} = \sum_\nu a_{\mu\nu} \frac{\partial}{\partial x_\nu},$$

which is the transformation equation for the components of a four-vector. The dot product of $\square$ with itself, written $\square^2$ and known as the *D'Alembertian*, is therefore a world scalar differential operator:

$$\square^2 = \sum_\mu \frac{\partial^2}{\partial x_\mu^2} = \frac{\partial^2}{\partial x^2} + \frac{\partial^2}{\partial y^2} + \frac{\partial^2}{\partial z^2} - \frac{1}{c^2}\frac{\partial^2}{\partial t^2}.$$

Comparison with (6-26) shows that $\square^2$ is exactly the differential operator involved in the scalar wave equation, which may now be written as

$$\square^2 \psi = 0. \tag{6-28}$$

Providing that ψ is truly a world-scalar, this covariant formulation immediately indicates the invariance of the wave equation (6-26) under Lorentz transformation.

6-4 The force and energy equations in relativistic mechanics. We have seen that Newton's equations of motion, being invariant under a Galilean transformation, cannot be invariant under a Lorentz transformation; they must be suitably generalized to provide a law of force satisfying the equivalence principle. Of course, the generalization we seek must be such that for velocities small compared to c the new equations reduce to the familiar form

$$\frac{d}{dt}(mv_i) = F_i. \tag{6-29}$$

Now, the space components of a four-vector by themselves constitute a spatial vector, for a Lorentz transformation for which $a_{4i} = a_{i4} = 0$, $a_{44} = 1$, is merely an ordinary spatial rotation, and affects only the space components of a four-vector. The converse is not true, however; the components of a spatial vector do not necessarily transform as the spatial components of a four-vector. One may multiply the components of an ordinary vector by any function of β without changing their spatial rotation properties. But such a multiplication vitally affects the way in which they change under a Lorentz transformation. Thus the spatial components of the world velocity u_ν form a vector $\mathbf{v}/\sqrt{1-\beta^2}$, but note that $\mathbf{v}$ itself is not part of a four-vector, it must first be divided by $\sqrt{1-\beta^2}$.

While Eq. (6–29) is itself not Lorentz-invariant, we can expect, therefore, that its relativistic generalization will be a four-vector equation, whose spatial component reduces to (6–29) in the limit as $\beta \to 0$. It is not difficult to find a four-vector generalization of the left-hand side of the equation. The only four-vector whose space part reduces to $\mathbf{v}$ for small velocities is the world velocity u_ν. Further, while m can be taken as an invariant property of the particle, we know that the time t is not a Lorentz-invariant but it can obviously be replaced by the scalar proper time τ which approaches t as $\beta \to 0$. The desired generalization of Newton's equations of motion must therefore have the form

$$\frac{d}{d\tau}(mu_\nu) = K_\nu, \tag{6–30}$$

where K_ν is some four-vector, known as the *Minkowski force.*

It must not be thought that the spatial components of K_ν are to be identified with the components of the force. All that is required by Eq. (6–29) is that K_i reduce to F_i in the limit of small velocities. Thus K_i may be equal to the product of F_i with any function of β which reduces to unity as $\beta \to 0$; the exact relation clearly depends on the Lorentz transformation properties of the force components. Two types of approach have been used in the past to determine the behavior of $\mathbf{F}$ under Lorentz transformation.

One procedure begins by pointing out that, fundamentally, forces arise from only a few physical sources — forces are either gravitational, electromagnetic, or possibly nuclear. It is the duty of a correct theory of these physical phenomena to provide expressions for the forces involved, and these expressions, if stated in covariant form, automatically tell us the transformation properties of the force components. Unfortunately we don't have covariant theories of all the possible sources of force, indeed for nuclear forces we don't have any theory worth speaking about. Only classical electromagnetic theory can be expected to provide a covariant force equation, since the Lorentz transformation was expressly constructed so as to preserve the invariance of the theory. But this is sufficient for our purposes; the transformation properties must be the same for all forces no matter what their origin. The statement "a particle is in equilibrium under the influence of two forces" must hold true in all Lorentz systems which can only be the case if all forces transform in the same manner.

In Section 1–5 it was pointed out that the electromagnetic force on a particle is given by

$$F_i = -q\left(\frac{\partial}{\partial x_i}\left(\phi - \frac{1}{c}\mathbf{v}\cdot\mathbf{A}\right) + \frac{1}{c}\frac{dA_i}{dt}\right), \tag{6–31}$$

where ϕ and $\mathbf{A}$ are the scalar and vector electromagnetic potentials. It can also be shown that the invariance of the velocity of light requires that $\mathbf{A}$ and $i\phi$ transform as the spatial and time components, respectively, of a four-vector denoted by A_μ. Consequently the expression $\phi - \frac{1}{c}\mathbf{v} \cdot \mathbf{A}$ can be written covariantly as

$$\phi - \frac{1}{c}\mathbf{v} \cdot \mathbf{A} = -\frac{1}{c}\sqrt{1-\beta^2}u_\nu A_\nu, \tag{6-32}$$

and the force components F_i become

$$F_i = -\frac{q}{c}\sqrt{1-\beta^2}\left(-\frac{\partial}{\partial x_i}(u_\nu A_\nu) + \frac{dA_i}{d\tau}\right). \tag{6-33}$$

The expression in the parentheses transforms as the spatial component of a four-vector, so that F_i is equal to the product of $\sqrt{1-\beta^2}$ and the spatial component of a four-vector, which is to be identified as the Minkowski force K_μ. Hence the connection between the ordinary and Minkowski forces must be

$$F_i = K_i\sqrt{1-\beta^2}, \tag{6-34}$$

irrespective of the origin of the forces. A by-product to this derivation is the particular form of the Minkowski force on charged particles:

$$K_\mu = \frac{q}{c}\left(\frac{\partial}{\partial x_\mu}(u_\nu A_\nu) - \frac{dA_\mu}{d\tau}\right).$$

The alternative procedure attempts to avoid the necessity of using a physical theory beyond mechanics itself; it simply *defines* force as being the time rate of change of momentum, in all Lorentz systems:

$$\frac{dp_i}{dt} = F_i. \tag{6-35}$$

The momentum indicated in (6-35), however, is not mv_i, but rather some relativistic generalization which reduces to it in the limit of small velocities. Lewis and Tolman * have obtained an expression for the relativistic momentum, independent of the form (6-30) for the force law, by noting that a Lorentz-invariant consequence of the definition (6-35) is the conservation of momentum in the absence of external forces. They examine an elastic collision between two particles and find the form of p_i such that it is conserved in such a collision for all Lorentz systems. But having accepted (6-30) as the form of the force law it becomes possible to find the relativistic

* See P. Bergmann, op. cit., p. 87 f.

momentum and the meaning of K_i at once, by putting (6–30) in a form resembling (6–35) as closely as possible. From the relation between τ and t, and the definition of world velocity, we can write the spatial components of Eq. (6–30) as

$$\frac{d}{dt}\left(\frac{mv_i}{\sqrt{1-\beta^2}}\right) = K_i\sqrt{1-\beta^2}.$$

Comparison with (6–35) shows that the conservation of momentum theorem is invariant providing the momentum is defined by

$$p_i = \frac{mv_i}{\sqrt{1-\beta^2}}, \tag{6–36}$$

and that F_i and K_i are related according to Eq. (6–34). It will be noted that Eq. (6–36) properly reduces to mv_i as $\beta \to 0$. The two approaches thus lead to the same conclusion.

So far only the space part of the four-vector equation (6–30) has been discussed, nothing has been said about the physical significance of the fourth equation. The time-like part of the four-vector K_μ can be obtained directly from the dot product of (6–30) with the world velocity:

$$u_\nu \frac{d}{d\tau}(mu_\nu) = \frac{d}{d\tau}\left(\frac{m}{2}u_\nu u_\nu\right) = K_\nu u_\nu. \tag{6–37}$$

Since the square of the magnitude of u_ν is a constant, $-c^2$ (cf. Eq. (6–25)), and m is here likewise a constant, the left-hand side of (6–37) vanishes, leaving

$$K_\nu u_\nu \equiv \frac{\mathbf{F}\cdot\mathbf{v}}{1-\beta^2} + \frac{icK_4}{\sqrt{1-\beta^2}} = 0.$$

The fourth component of the Minkowski force is therefore

$$K_4 = \frac{i}{c}\frac{\mathbf{F}\cdot\mathbf{v}}{\sqrt{1-\beta^2}}, \tag{6–38}$$

and the corresponding fourth component of Eq. (6–30) appears as

$$\frac{d}{dt}\frac{mc^2}{\sqrt{1-\beta^2}} = \mathbf{F}\cdot\mathbf{v}. \tag{6–39}$$

Now, the kinetic energy T is defined in general to be such that $\mathbf{F}\cdot\mathbf{v}$, the rate at which the force does work on the particle, is the time rate of increase of T:

$$\frac{dT}{dt} = \mathbf{F}\cdot\mathbf{v}. \tag{6–40}$$

This is a definition of kinetic energy which agrees with the customary nonrelativistic form $\frac{1}{2}mv^2$; the relativistic expression, by comparison, is furnished by Eq. (6–39) which is thus seen to be the energy equation with

$$T = \frac{mc^2}{\sqrt{1-\beta^2}}. \tag{6–41}$$

In the limit as β^2 becomes much less than 1, Eq. (6–41) can be expanded as

$$T \rightarrow mc^2\left(1 + \frac{\beta^2}{2}\right) = mc^2 + \frac{mv^2}{2}. \tag{6–42}$$

This limiting value does not agree with the expected nonrelativistic form, there is here the added term mc^2. It would seem at first sight, however, that the term is of no importance, for clearly any constant of integration can be added to the right of Eq. (6–41) without affecting the validity of Eq. (6–40). In particular the constant could be $-mc^2$, which would bring T in line with the nonrelativistic value.

It is preferable, nevertheless, to define T as in (6–41) for then the momentum $\mathbf{p}$, as defined by (6–36), and iT/c form a four-vector *world momentum*, p_ν, defined as

$$p_\nu = mu_\nu. \tag{6–43}$$

It follows that if the momentum p_i is conserved, i.e., constant in time, then the energy T as defined in Eq. (6–41) is likewise constant. Otherwise one could transform to another system in which the transformed components p_i' are given in terms of p_i and T by the equations of the Lorentz transformation, and momentum would no longer be conserved. Thus the conservation laws for momentum and kinetic energy are no longer separate; they appear in special relativity as different aspects of the conservation law for a single four-vector. The term, mc^2, known as the *rest energy*, therefore has an important physical significance. In the nonrelativistic formulation of mechanics, such as given in Chapter 1, conservation of momentum could occur without conservation of kinetic energy. In such cases the relativistic kinetic energy as defined by (6–41) must still be conserved, which can only be brought about by changes in the rest energy, i.e., in the rest mass. The change in energy produced by a given change in the rest mass is furnished by the now famous Einstein relation

$$\Delta E = \Delta mc^2.$$

Many examples are quoted in the literature on relativity of the combined conservation of kinetic energy of motion plus rest energy. One of the first illustrations proposed was the inelastic collision of two bodies moving at nonrelativistic velocities. Here momentum is conserved, but kinetic energy, if defined as $\sum \frac{1}{2}mv^2$, is not. Ordinarily we say that the energy lost in the collision is converted into heat. In order that the relativistic kinetic energy shall nonetheless be conserved, there must be an increase in the rest mass, or inertia, of the system in proportion to the amount of heat produced. The mass changes involved in such a collision are, of course, quite small, since a joule of energy corresponds to a mass of only 1.1×10^{-14} gm. Modern physics presents a number of phenomena in which the mass changes involved are of much greater consequence, such as pair creation, in which two particles of finite mass are created from the energy of a massless photon. The most striking demonstration, to say the least, of conversion of mass into energy is provided by the explosion of an atomic bomb. Momentum is conserved in such an explosion, but a large amount of kinetic energy of motion is produced. The total energy T remains constant, but only by virtue of a decrease in the rest mass of the bomb contents. Despite the fantastic energies produced, the mass loss can at most amount only to 0.1% of the original mass.

Formally the connection between the energy T and the momentum is expressed in the statement that the magnitude of the momentum four-vector is constant:

$$p_\mu p_\mu = -m^2c^2 = p^2 - \frac{T^2}{c^2}$$

or

$$T^2 = p^2c^2 + m^2c^4. \tag{6–44}$$

Eq. (6–44) is the relativistic analog of the relation $T = p^2/2m$ in nonrelativistic mechanics, except that T here includes the rest energy.

In the present discussion the mass m has been treated as a scalar invariant property of the particle unaffected by Lorentz transformation, and as such is known as the rest mass. Occasionally another type of mass is introduced which we shall call the *relativistic mass* m_r defined such that

$$m_r = \frac{m}{\sqrt{1-\beta^2}}. \tag{6–45}$$

The sole occasion for introducing such a quantity is to preserve the momentum in the same form as in nonrelativistic mechanics:

$$\mathbf{p} = m_r\mathbf{v}.$$

Unlike the rest mass, the relativistic mass varies with velocity, becoming infinite as β approaches unity. Still other kinds of "masses" will be found in the literature, usually in connection with the form of the relativistic equation of motion:

$$F_i = \frac{d}{dt}\frac{mv_i}{\sqrt{1-\beta^2}}. \tag{6-46}$$

The force is no longer directly related to the acceleration, the two quantities being in general in different directions. Only if the acceleration is parallel or perpendicular to the instantaneous direction of **v** does it turn out that **F** is proportional to the acceleration (cf. the exercises at the end of the chapter). The coefficients of the acceleration for these two special cases are known as the longitudinal and transverse masses respectively:

$$\begin{aligned} m_l &= \frac{m}{(1-\beta^2)^{3/2}}, \\ m_t &= \frac{m}{(1-\beta^2)^{1/2}}. \end{aligned} \tag{6-47}$$

In general the use of these various "masses," m_r, m_l, and m_t is decreasing; they conceal the covariance of the formulation and obscure the physics of the situation more than they reveal.

6-5 The Lagrangian formulation of relativistic mechanics. Having established the appropriate generalization of Newton's equation of motion for special relativity we can now examine the Lagrangian formulation of the resulting relativistic mechanics. In a certain sense there is little difficulty in setting up a Lagrangian which will furnish the correct relativistic equations of motion. It is true that it would be difficult to determine the Lagrangian solely from D'Alembert's principle as was done in Chapter 1. While the principle itself,

$$\sum_i (\mathbf{F}_i^{(a)} - \dot{\mathbf{p}}_i) \cdot \delta \mathbf{r}_i = 0, \tag{1-42}$$

is still valid, the subsequent derivation used there no longer follows, as $\mathbf{p}_i$ is not now given by $m\mathbf{v}_i$. But one may also approach the Lagrangian formulation from the alternative route of Hamilton's principle (Section 2-1) and attempt simply to find a function L for which the Euler-Lagrange equations, as obtained from the variational principle

$$\delta I = \delta \int_{t_1}^{t_2} L\, dt = 0, \tag{6-48}$$

agree with the known relativistic equations of motion.

It is usually not difficult to find a function satisfying these requirements. For example, a suitable relativistic Lagrangian for a single particle acted on by conservative forces independent of velocity would be

$$L = -mc^2\sqrt{1-\beta^2} - V, \tag{6-49}$$

where V is the potential, depending only upon position. That this is the correct Lagrangian can be shown by demonstrating that the resultant Lagrange equations:

$$\frac{d}{dt}\left(\frac{\partial L}{\partial v_i}\right) - \frac{\partial L}{\partial x_i} = 0,$$

agree with Eqs. (6–46). Since the potential is velocity independent v_i occurs only in the first term of (6–49) and therefore

$$\frac{\partial L}{\partial v_i} = \frac{mv_i}{\sqrt{1-\beta^2}}. \tag{6-50}$$

The equations of motion derived from the Lagrangian (6–49) are then

$$\frac{d}{dt}\frac{mv_i}{\sqrt{1-\beta^2}} = -\frac{\partial V}{\partial x_i} = F_i,$$

which agree with (6–46). Note that the Lagrangian is no longer $L = T - V$ but that the partial derivative of L with velocity is still the momentum. Indeed it is just this last fact which insures the correctness of the Lagrange equations, and one could have worked backwards from Eq. (6–50) to supply at least the velocity dependence of the Lagrangian.

One can readily extend the Lagrangian (6–49) to systems of many particles, and change from cartesian coordinates to any desired set of generalized coordinates q_j. The canonical momenta will still be defined by

$$p_j = \frac{\partial L}{\partial \dot{q}_j},$$

so that the connection between cyclic coordinates and conservation of the corresponding momenta remains just as in the nonrelativistic theory. Only now the derivation of the energy conservation theorem requires some alteration. It will be recalled that it was shown in Section 2–6 that if L does not contain the time explicitly there exists a constant of the motion

$$H = \sum \dot{q}_j p_j - L.$$

This fact remains valid here too, for the derivation involved only the general form of the Lagrange equations and the definition of canonical momentum. The proof given for the rest of the theorem — that H is the total energy —

must be changed, for L is no longer $T - V$ nor is $\sum \dot{q}_j p_j$ equal to $2T$. The conclusion remains the same, nevertheless, for in the example of a single particle H is given by

$$H = \sum_i \frac{mv_i^2}{\sqrt{1-\beta^2}} + mc^2\sqrt{1-\beta^2} + V,$$

which, on collecting terms, reduces to

$$H = \frac{mc^2}{\sqrt{1-\beta^2}} + V = T + V = E.$$

The quantity H is thus again seen to be the total energy E, which is therefore a constant of the motion under these conditions.

The introduction of velocity-dependent potentials produces no particular difficulty here, and can be performed in exactly the same manner as in Section 1–5 for nonrelativistic mechanics. Thus the Lagrangian for a single particle in an electromagnetic field is

$$L = -mc^2\sqrt{1-\beta^2} - q\phi + \frac{q}{c}\mathbf{A}\cdot\mathbf{v}. \tag{6–51}$$

Note that the *canonical* momentum is no longer mu_i; there are now additional terms arising from the velocity-dependent part of the potential:

$$p_i = mu_i + \frac{q}{c}A_i. \tag{6–52}$$

This phenomenon is not a relativistic one, of course; exactly the same additional term was found in the earlier treatment (cf. Eq. (2–44)).

Almost all of the mechanism constructed previously for the solution of specific mechanical problems thus carries over into relativistic mechanics. For example, there is no difficulty in principle in obtaining the relativistic form of central force motion. The general physical characteristics of the orbits will remain the same as in Chapter 3, although details will naturally be affected by the new form of the Lagrangian.

6–6 Covariant Lagrangian formulations. The Lagrangian procedure as given above certainly predicts the correct relativistic equations of motion. Yet it is a relativistic formulation only "in a certain sense." No effort has been made to keep to the ideal of a covariant four-dimensional form for all the laws of mechanics. Thus the time t has been treated as a parameter entirely distinct from the spatial coordinates, while a covariant formulation would require that space and time be considered as entirely similar coordinates in world space. Clearly the invariant parameter used to trace the progress of the system point in configuration space should be the invariant proper time τ rather than t. Further, the examples of Lagrangian functions discussed in the previous section do not have any particular Lorentz transformation properties. Since the Lagrangian should be an invariant property of the system only, independent of the particular coordinate system used, we must expect it to be a world scalar, invariant under all Lorentz transformations. Finally, instead of being a function of x_i and $\dot{x}_i(\equiv v_i)$ with t as a parameter, a Lagrangian correctly expressed in four-dimensional language should be a function of the components of x_ν and $\frac{dx_\nu}{d\tau}$ $(\equiv u_\nu)$, with τ as the parameter. A covariant formulation of Hamilton's principle should therefore have the form

$$\delta I = \delta \int_{\tau_1}^{\tau_2} L'(x_\nu, u_\nu, \tau)\, d\tau = 0, \tag{6–53}$$

with both I and L', the covariant Lagrangian, as world scalars.

It is not always possible to set up a completely covariant formulation of a given problem in this manner. The explicit form of the potentials is determined by the nature of the forces involved, which must be provided by theoretical considerations lying outside the domain of mechanics. Only if these extramechanical theories have themselves been put in a covariant formulation, so that the forces have the correct transformation properties, can one find a covariant L' satisfying Eq. (6–53). It has already been noted that not all types of forces are available in a covariant form. Certainly the ordinary gravitational force, the mainspring of most mechanics problems, does not satisfy this requirement. Nor is the physical reason difficult to see. As usually stated, gravitational force is a static force, an example of "action-at-a-distance," implying infinite velocity of propagation. But the concept of action-at-a-distance makes no sense in relativity, where it is required that forces be transmitted with velocities less than or equal to c. On the other hand, it has already been noted that the electromagnetic forces automatically satisfy the relativity requirement. We shall therefore content ourselves with finding L' for a particle in only two situations, first when it is entirely free, and second when the particle is acted on by external electromagnetic forces.

The Euler-Lagrange equations for a single particle deduced from the covariant variational principle are obviously

$$\frac{d}{d\tau}\frac{\partial L'}{\partial u_\nu} - \frac{\partial L'}{\partial x_\nu} = 0, \tag{6–54}$$

and the left-hand sides of these four equations transform as the components of a four-vector. The Lagrangian L' for a free particle must be such that Eqs. (6–54) reduce to

$$\frac{d}{d\tau}mu_\nu = 0. \tag{6–55}$$

Now in form the equations (6–55) are similar to the nonrelativistic equations,

$$\frac{d}{dt}mv_i = 0,$$

which suggests obtaining L' by replacing v^2 in the nonrelativistic relation $L = \frac{1}{2}mv^2$ with the square of the world velocity:

$$L' = \frac{1}{2}mu_\nu u_\nu. \tag{6–56}$$

This choice is indeed the correct one, for we have

$$\frac{\partial L'}{\partial x_\nu} = 0, \qquad \frac{\partial L'}{\partial u_\nu} = mu_\nu = p_\nu,$$

so that the Lagrange Eqs. (6–54) agree with (6–55).*

If the forces on the particle are electromagnetic in origin, a suitable invariant Lagrangian is

$$L' = \frac{1}{2}mu_\nu u_\nu + \frac{q}{c}u_\lambda A_\lambda. \tag{6–57}$$

The canonical momenta are here

$$p_\nu = \frac{\partial L'}{\partial u_\nu} = mu_\nu + \frac{q}{c}A_\nu, \tag{6–58}$$

---

* It will be noticed that L' in (6–56) is a constant, equal to $-\frac{m}{2}c^2$. This is of no consequence, since all that is needed of L' is the correct *functional* dependence on u_ν so as to yield the proper equations. But it does mean that the expression given in (6–56) is by no means unique, indeed L' can have the form $mf(u_\nu u_\nu)$, where $f(x)$ is any function such that if f' denote $\partial f/\partial x$ then

$$f'(-c^2) = +\frac{1}{2}.$$

In Eq. (6–56) we have used $f(u_\nu u_\nu) = \frac{1}{2}u_\nu u_\nu$. Another possible choice is

$$f(u_\nu u_\nu) = -c\sqrt{-u_\nu u_\nu},$$

which is obtained directly if the variable of integration in (6–48) is changed from t to τ.

so that the Lagrange equations (6–54) appear as

$$\frac{d}{d\tau}\left(mu_\nu + \frac{q}{c}A_\nu\right) - \frac{\partial}{\partial x_\nu}\left(\frac{q}{c}u_\lambda A_\lambda\right) = 0$$

or

$$\frac{d}{d\tau}mu_\nu = \frac{\partial}{\partial x_\nu}\left(\frac{q}{c}u_\lambda A_\lambda\right) - \frac{q}{c}\frac{dA_\nu}{d\tau}.$$

But this is exactly the form of the generalized Newton's equation of motion (6–30), with the Minkowski force given by

$$K_\nu = \frac{\partial}{\partial x_\nu}\left(\frac{q}{c}u_\lambda A_\lambda\right) - \frac{q}{c}\frac{dA_\nu}{d\tau},$$

in agreement with the previously derived relation, Eq. (6–33).

The canonical four-momentum Eq. (6–58) again differs from the usual kinetic momentum by the added term involving the electromagnetic potential. It will be noticed that here p_4, instead of being simply iT/c as in the force-free case (cf. Eq. (6–43)) now becomes

$$p_4 = \frac{iT}{c} + \frac{iq\phi}{c} = \frac{i}{c}E,$$

where E is the total energy $T + q\phi$. Thus the momentum canonical to the *time* coordinate is proportional to the *total energy*. A similar connection between these two quantities will recur later in nonrelativistic theory. The relation between the spatial components of the momentum and the kinetic energy T can be derived here by the same route that led to Eq. (6–44). The spatial components of Eq. (6–58) can be written

$$mu_i = p_i - \frac{qA_i}{c},$$

so that

$$m^2u_\nu u_\nu = -m^2c^2 = \left(\mathbf{p} - \frac{q\mathbf{A}}{c}\right)^2 - \frac{T^2}{c^2},$$

yielding the desired connection:

$$T^2 = \left(\mathbf{p} - q\frac{\mathbf{A}}{c}\right)^2 c^2 + m^2c^4. \tag{6–59}$$

It must not be thought from these considerations that all of the aspects of nonrelativistic mechanics have unequivocal correspondences in a relativistic theory. We have already mentioned the difficulties caused by gravitational and other "action-at-a-distance" forces. The Lorentz transformation is concerned only with uniformly moving coordinate systems and transformations to accelerating systems, such as rotating axes, can be fitted into the framework of special relativity only with difficulty. The notion of constraints is also troublesome, for any constraint must be expressed in Lorentz-invariant fashion. In particular rigid body constraints do not fulfill this requirement as they involve only the space parts of the position four-vector. Hence the entire field of rigid-body dynamics is without a relativistic analog.

SUGGESTED REFERENCES

P. Bergmann, *An Introduction to the Theory of Relativity*. Undoubtedly the best treatment in English of the techniques of relativity. The tensor formulation is developed with an eye to general relativity (which occupies the second half of the book) resulting, unfortunately, in a notation that is often unduly complicated for the needs of special relativity.

R. B. Lindsay and H. Margenau, *Foundations of Physics*. Chapter 7 deals mainly with special relativity, discussing particularly the events leading up to the founding of the theory, and the physical interpretations of some of its immediate consequences. The four-dimensional representation is not developed in any detail.

Albert Einstein, *The Meaning of Relativity*. This is *not* a treatment designed for popular audiences. Little more than a third of this brief book is concerned with special relativity, but it contains a great deal of information. A considerable background in electrodynamics is assumed.

R. Becker, *Theorie der Elektrizität*, Vol. II. Many of the German treatises on electrodynamics include extensive discussions of special relativity. Perhaps the best of these (and certainly the most readily available) is in this second volume of the famous *Abraham and Becker*. The style is fluent and surprisingly easy to read for a book written in German. While the emphasis here is on electromagnetic phenomena, the discussion of mechanics is quite complete. In the more than one hundred pages devoted to special relativity both the physical and mathematical background are covered quite thoroughly.

EXERCISES

1. Verify the Einstein addition law (Eq. (6–20)) for the addition of two parallel velocities. The proof is performed most simply by considering the two Lorentz transformations as successive rotations in the x_3x_4 plane.

2. Obtain the Lorentz transformation in which the velocity is at an infinitesimal angle $d\theta$ counterclockwise from the z-axis, by means of a similarity transformation applied to Eq. (6–15). Show directly that the resulting matrix is orthogonal and that the inverse matrix is obtained by substituting $-v$ for v.

3. The Einstein addition law can also be obtained by remembering that the second velocity is related directly to the space components of a four-velocity, which may then be transformed back to the initial system by a Lorentz transformation. If the second system is moving with a speed v' relative to the first in the direction of their z-axes, while a third system is moving relative to the second with an arbi-

trarily oriented velocity $\mathbf{v}''$, show by this procedure that the magnitude of the velocity $\mathbf{v}$ between the first and third system is given by

$$\sqrt{1-\beta^2} = \frac{\sqrt{1-\beta'^2}\sqrt{1-\beta''^2}}{1+\beta'\beta_z''},$$

and that the components of $\mathbf{v}$ are

$$\beta_x = \frac{\beta_x''\sqrt{1-\beta'^2}}{1+\beta'\beta_z''}, \qquad \beta_y = \frac{\beta_y''\sqrt{1-\beta'^2}}{1+\beta'\beta_z''},$$

$$\beta_z = \frac{\beta' + \beta_z''}{1+\beta'\beta_z''}.$$

Here $\beta_x'' = v_x''/c$, and so forth. Note that the equation for β_z correctly reduces to Eq. (6–20) when $\mathbf{v}''$ is along the z-axis.

4. Consider the situation described in Exercise 3 in the special case that $\mathbf{v}''$ lies in the zx plane and is of infinitesimal magnitude compared to v'. Show that the z- and z''-axes are not parallel (despite the fact that both are taken parallel to the z'-axis) by demonstrating that the relative velocity between the two systems will be observed to make different angles with the z- and z''-axes. In particular, show that the angle between the two axes will be

$$d\theta = \frac{-\beta_x''\beta'}{2}.$$

If the velocity $\mathbf{v}''$ actually represents a change in the velocity $\mathbf{v}'$ during an infinitesimal time as the result of an acceleration $\mathbf{a}$, then this conclusion may be interpreted as saying that the axes appear to rotate with an angular frequency

$$\omega_t = \frac{\mathbf{v}' \times \mathbf{a}}{2c^2},$$

a phenomenon important in atomic physics and known as the *Thomas precession.*

5. By expanding the equation of motion (6–46) show that the force is parallel to the acceleration only when the velocity is either parallel or perpendicular to the acceleration, and verify Eq. (6–47) for the coefficients of the acceleration in these two cases.

6. From the transformation properties of the world acceleration show that the components of the acceleration $\mathbf{a}$ are given in terms of the transformed acceleration $\mathbf{a}'$ in a system momentarily at rest with respect to the particle by the formulas:

$$a_x' = \frac{a_x}{1-\beta^2}, \qquad a_y' = \frac{a_y}{1-\beta^2}, \qquad a_z' = \frac{a_z}{(1-\beta^2)^{3/2}},$$

the z-axis being chosen in the direction of the relative velocity.

7. In the β-disintegration considered in Exercise 1, Chapter 1, the electron has a mass equivalent to a rest energy of 0.511 Mev, while the neutrino has no mass. What are the total energies carried away by the electron and neutrino? What fraction of the nuclear mass is converted into kinetic energy (including the electron rest energy)?

8. A meson of mass π comes to rest and disintegrates into a meson of mass μ and a neutrino of zero mass. Show that the kinetic energy of motion of the μ-meson (i.e., without the rest mass energy) is

$$T = \frac{(\pi - \mu)^2}{2\pi} c^2.$$

9. A photon may be described classically as a particle of zero mass possessing nevertheless a momentum $h/\lambda = h\nu/c$, and therefore a kinetic energy $h\nu$. If the photon collides with an electron of mass m at rest it will be scattered off at some angle θ with a new energy $h\nu'$. Show that the change in energy is related to the scattering angle by the formula

$$\lambda' - \lambda = 2\lambda_c \sin^2 \frac{\theta}{2},$$

where $\lambda_c = h/mc$, known as the Compton wave length. Show also that the kinetic energy of the recoil motion of the electron is

$$T = h\nu \frac{2\left(\frac{\lambda_c}{\lambda}\right)\sin^2 \theta/2}{1 + 2\left(\frac{\lambda_c}{\lambda}\right)\sin^2 \theta/2}.$$

10. The theory of rocket motion developed in Exercise 3, Chapter 1, no longer applies in the relativistic region, in part because there is no longer conservation of mass. Instead, all the conservation laws are combined into the conservation of the world momentum; the change in each component of the rocket's world momentum in an infinitesimal time dt must be matched by the value of the same component of p_ν for the gases ejected by the rocket in that time interval. Show that if there are no external forces acting on the rocket the differential equation for its velocity as a function of the mass is

$$m \frac{dv}{dm} + a\left(1 - \frac{v^2}{c^2}\right) = 0,$$

where a is the constant velocity of the exhaust gases *relative to the rocket*. Verify that the solution can be put in the form

$$\beta = \frac{1 - \left(\frac{m}{m_0}\right)^{\frac{2a}{c}}}{1 + \left(\frac{m}{m_0}\right)^{\frac{2a}{c}}},$$

m_0 being the initial mass of the rocket. Since mass is not conserved, what happens to the mass that is lost?

11. Show that it follows immediately from Eq. (6–40) that the speed of a charged particle in a static magnetic field is constant. As a consequence of this fact, show that a charged particle describes a helix in a uniform magnetic field, the radius of the helix for a given type of particle being inversely proportional to the magnitude of the magnetic intensity B, and directly proportional to the component of the mechanical momentum perpendicular to the magnetic field. The product of the field strength and the radius of curvature, $B\rho$, is thus a direct measure of the particle's momentum and is often referred to as the magnetic rigidity of the particle.

12. A particle of rest mass m, charge q, and initial velocity $\mathbf{v}_0$ enters a uniform electric field $\mathbf{E}$ perpendicular to $\mathbf{v}_0$. Find the subsequent trajectory of the particle and show that it reduces to a parabola as the limit c becomes infinite.

13. Show that the relativistic motion of a particle in an attractive inverse square law of force is a precessing ellipse. Compute the precession of the perihelion of Mercury resulting from this effect. (The answer, about 7″ per century, is much smaller than the actual precession of 40″ per century which can be accounted for correctly only by general relativity.)

14. Starting from the equation of motion (6–46), derive the relativistic analog of the virial theorem which states that for motions bounded in space and such that the velocities involved do not approach c indefinitely close, then

$$\overline{L_0} + \overline{T} = -\overline{\mathbf{F} \cdot \mathbf{r}},$$

where L_0 is the form the Lagrangian takes in the absence of external forces. Note that although neither L_0 nor T corresponds exactly to the kinetic energy in nonrelativistic mechanics, their sum, $L_0 + T$, plays the same role as twice the kinetic energy in the nonrelativistic virial theorem, Eq. (3–26).

CHAPTER 7

THE HAMILTON EQUATIONS OF MOTION

The Lagrangian formulation of mechanics was developed largely in the first two chapters, and most of the subsequent discussion has been in the nature of application, but still within the framework of the Lagrangian procedure. In this chapter we resume the formal development of mechanics, turning our attention to an alternative statement of the structure of the theory known as the Hamiltonian formulation. Nothing new is added to the physics involved; we simply gain another (and more powerful) method of working with the physical principles already established. We shall assume in the following chapters that the mechanical systems are holonomic and that the forces are either derived from a potential dependent on position only, or from velocity-dependent generalized potentials of the type discussed in Section 1–5.

7–1 Legendre transformations and the Hamilton equations of motion. A system with n degrees of freedom possesses n Lagrange equations of motion:

$$\frac{d}{dt}\left(\frac{\partial L}{\partial \dot{q}_i}\right) - \frac{\partial L}{\partial q_i} = 0. \tag{7–1}$$

Since the equations are second order, the motion of the system for all time is completely specified only if initial values are given for all the n q_i's *and* all the n $\dot{q}_i$'s. In this sense the coordinates q_i and the velocities $\dot{q}_i$ together form a complete set of $2n$ independent variables necessary for describing the system motion. The Lagrangian formulation (in the nonrelativistic region) can thus be considered as a description of mechanics in terms of the generalized *coordinates* and *velocities*, with time as a parameter. We now obtain a formulation in which the independent variables are the generalized *coordinates* and the generalized *momenta* p_i defined as (cf. Eq. (2–41)):

$$p_i = \frac{\partial L(q_j, \dot{q}_j, t)}{\partial \dot{q}_i}. \tag{7–2}$$

The change in basis from the $(q, \dot{q}, t)$ set to the (q, p, t) set can best be accomplished by a mathematical procedure, known as the *Legendre transformation,* which is tailored for just this type of change of variable.

Consider a function of only two variables $f(x, y)$, so that a differential of f has the form

$$df = u\,dx + v\,dy, \tag{7–3}$$

where

$$u = \frac{\partial f}{\partial x}, \qquad v = \frac{\partial f}{\partial y}. \tag{7–4}$$

We wish now to change the basis of description from x, y to the independent variables u, y, so that differential quantities are expressed in terms of the differentials du and dy. Let g be a function of u and y defined by the equation

$$g = f - ux. \tag{7–5}$$

A differential of g is then given as

$$dg = df - u\,dx - x\,du,$$

or, by (7–3), as

$$dg = v\,dy - x\,du,$$

which is exactly in the form desired. The quantities x and v are now functions of the variables u and y given by the relations

$$x = -\frac{\partial g}{\partial u}, \qquad v = \frac{\partial g}{\partial y}, \tag{7–6}$$

which are in effect the converse of Eqs. (7–4).

The Legendre transformation so defined is used frequently in thermodynamics. For example, the enthalpy X is a function of the entropy S and the pressure P with the properties that

$$\frac{\partial X}{\partial S} = T, \qquad \frac{\partial X}{\partial P} = V,$$

so that

$$dX = T\,dS + V\,dP,$$

where T and V are temperature and volume respectively. The enthalpy is useful in considering isentropic and isobaric processes, but often one has to deal rather with isothermal and isobaric processes. In such case one wants a thermodynamic function of T and P alone. The Legendre transformation shows that the desired function may be defined as

$$G = X - TS$$

with

$$dG = -S\,dT + V\,dP. \tag{7–7}$$

G is the well-known Gibbs function, or free energy, whose properties are correctly given by Eq. (7–7).

The transformation from $(q, \dot{q}, t)$ to (q, p, t) differs from the type considered in Eqs. (7–3 to 7–5) only in that more than one variable is to be transformed. In place of the Lagrangian one deals with a function defined in analogy to Eq. (7–5), except for a minus sign:

$$H(p, q, t) = \sum_i \dot{q}_i p_i - L(q, \dot{q}, t). \tag{7–8}$$

H is known as the *Hamiltonian* and will be recognized as the same function H discussed previously in Section 2–6. Considered as a function of p, q, and t only, the differential of H is given by

$$dH = \sum_i \frac{\partial H}{\partial q_i} dq_i + \sum_i \frac{\partial H}{\partial p_i} dp_i + \frac{\partial H}{\partial t} dt, \tag{7–9}$$

but from the defining equation (7–8) we can also write:

$$dH = \sum_i \dot{q}_i \, dp_i + \sum_i p_i \, d\dot{q}_i - \sum_i \frac{\partial L}{\partial \dot{q}_i} d\dot{q}_i - \sum_i \frac{\partial L}{\partial q_i} dq_i - \frac{\partial L}{\partial t} dt. \tag{7–10}$$

The terms in $d\dot{q}_i$ in Eq. (7–10) cancel in consequence of the definition of generalized momentum:

$$\sum_i p_i \, d\dot{q}_i - \sum_i \frac{\partial L}{\partial \dot{q}_i} d\dot{q}_i = 0$$

and from Lagrange's equation it follows that

$$\frac{\partial L}{\partial q_i} = \dot{p}_i.$$

Eq. (7–10) therefore reduces to the simple form

$$dH = \sum_i \dot{q}_i \, dp_i - \sum_i \dot{p}_i \, dq_i - \frac{\partial L}{\partial t} dt. \tag{7–11}$$

Comparison with (7–9) furnishes the following set of $2n + 1$ relations, in analogy with Eqs. (7–6):

$$\dot{q}_i = \frac{\partial H}{\partial p_i}, \qquad -\dot{p}_i = \frac{\partial H}{\partial q_i}, \tag{7–12}$$

$$-\frac{\partial L}{\partial t} = \frac{\partial H}{\partial t}. \tag{7–13}$$

Eqs. (7–12) are known as the *canonical equations of Hamilton;* they constitute a set of $2n$ first order equations of motion replacing the Lagrange equations. In principle, the first step in solving mechanics problems in this canonical formulation is to set up the Lagrangian L as $L(q, \dot{q}, t)$. Using Eq. (7–2) the canonical momenta are then obtained, and with their aid the Hamiltonian H is formed as defined by Eq. (7–8). The equations of motion, which are now first order, then follow by substituting H in Eqs. (7–12).

7–2 Cyclic coordinates and Routh's procedure. It will be noted that the Hamiltonian procedure is especially adapted to the treatment of problems involving cyclic coordinates. According to the definition given in Section 2–6 a cyclic coordinate q_j is one which does not appear in the Lagrangian; by virtue of Lagrange's equations its conjugate momentum p_j is then a constant. But from Eqs. (7–12), if $\dot{p}_j$ vanishes, it follows that $\frac{\partial H}{\partial q_j}$ is likewise zero. A coordinate that is cyclic will thus also be absent from the Hamiltonian.* Conversely, if a generalized coordinate does not occur in H the conjugate momentum is conserved. So far the situation is exactly as in the Lagrangian formulation. However, when some coordinate, say q_n, is cyclic the Lagrangian as a function of q and $\dot{q}$ can be written

$$L = L(q_1 \ldots q_{n-1}, \dot{q}_1 \ldots \dot{q}_n, t).$$

All the generalized velocities still occur in the Lagrangian and in general will be functions of the time. We still have to solve a problem of n degrees of freedom even though one degree of freedom corresponds to a cyclic coordinate. A cyclic coordinate in the Hamiltonian formulation, on the other hand, truly deserves its alternative description as "ignorable," for in the same situation p_n is some constant α, and H has the form

$$H = H(q_1 \ldots q_{n-1}, p_1 \ldots p_{n-1}, \alpha, t).$$

In effect the Hamiltonian now describes a problem involving only $n - 1$ coordinates, which may be solved completely ignoring the cyclic coordinate except as it is manifested in the constant of integration α, to be determined from the initial conditions. The behavior of the cyclic coordinate itself with time is then found by integrating the equation of motion

$$\dot{q}_n = \frac{\partial H}{\partial \alpha}.$$

---

* This conclusion follows also from the definition Eq. (7–8), for H differs from $-L$ only by $\sum_i p_i \dot{q}_i$, which does not involve q_i explicitly.

The advantages of the Hamiltonian formulation in handling cyclic coordinates may be combined with the Lagrangian procedure by a method devised by Routh. Essentially, one carries out a transformation from the q, $\dot{q}$ basis to the q, p basis only for those coordinates which are cyclic, obtaining their equations of motion in the Hamiltonian form, while the remaining coordinates are governed by Lagrange equations. If the cyclic coordinates are labeled $q_1 \ldots q_s$, then a new function R, known as the Routhian,* may be introduced, defined by the relation:

$$R(q_1 \ldots q_n, p_1 \ldots p_s, \dot{q}_{s+1} \ldots \dot{q}_n, t) = \sum_{i=1}^{s} p_i \dot{q}_i - L. \tag{7-14}$$

A differential of R is therefore given by

$$dR = \sum_{i=1}^{s} \dot{q}_i \, dp_i - \sum_{i=s+1}^{n} \frac{\partial L}{\partial \dot{q}_i} d\dot{q}_i - \sum_{i=1}^{n} \frac{\partial L}{\partial q_i} dq_i - \frac{\partial L}{\partial t},$$

from which it follows that

$$\frac{\partial R}{\partial p_i} = \dot{q}_i, \qquad \frac{\partial R}{\partial q_i} = -\dot{p}_i \qquad i = 1 \ldots s \tag{7-15}$$

and

$$\frac{\partial R}{\partial \dot{q}_i} = -\frac{\partial L}{\partial \dot{q}_i}, \qquad \frac{\partial R}{\partial q_i} = -\frac{\partial L}{\partial q_i} \qquad i = s + 1 \ldots n. \tag{7-16}$$

Eqs. (7-15) for the coordinates $q_1 \ldots q_s$ are in the form of Hamilton's equations of motion with R as the Hamiltonian, while Eqs. (7-16) show that the remaining coordinates can be found from the Lagrange equations

$$\frac{d}{dt}\left(\frac{\partial R}{\partial \dot{q}_i}\right) - \frac{\partial R}{\partial q_i} = 0, \qquad i = s + 1 \ldots n, \tag{7-17}$$

with R as the Lagrangian! Up to this point no explicit use has been made of the cyclic nature of the coordinates q_1 to q_s. A coordinate absent from L will likewise not occur in the Routhian. The momenta p_1 to p_s conjugate to the cyclic coordinates are constants and may be replaced in the Routhian by a set of constants $\alpha_1 \ldots \alpha_s$ to be determined by the initial conditions. With these modifications the only variables in the Routhian are the $n - s$ noncyclic coordinates and their generalized velocities:

$$R(q_{s+1} \ldots q_n, \dot{q}_{s+1} \ldots \dot{q}_n, \alpha_1 \ldots \alpha_s, t).$$

* The function R as defined here is the negative of the form usually quoted, in order to bring it in line with the definition of H, Eq. (7-8).

The Lagrange equations for the noncyclic coordinates, Eqs. (7–17), can now be solved without any regard for the behavior of the cyclic coordinates, exactly as in the Hamiltonian formulation. One might describe the Routhian formulation as having a foot in both the Lagrangian and Hamiltonian camps. It would seem, however, that if one is going part of the way towards a Hamiltonian description, one might as well go the whole route.

7–3 Conservation theorems and the physical significance of the Hamiltonian. The conclusion that a coordinate cyclic in L is also cyclic in H means that the momentum conservation theorems of Section 2–6 can be transferred to the Hamiltonian formulation with no more than a substitution of H for L. In particular the connection between the symmetry properties of the physical system and the constants of the motion can also be derived in terms of the Hamiltonian. For example, if a system is symmetrical about a given axis so that H is invariant under rotation about this axis, then H obviously cannot involve the angle of rotation. The angle is therefore a cyclic coordinate and the corresponding angular momentum is conserved.

The physical significance of H has already been indicated in the discussion of Section 2–6. It was shown there that if L (and in consequence of Eq. (7–13), also H) is not an explicit function of t then H is a constant of the motion. This can also be seen directly from the equations of motion (7–12), by writing the total time derivative of the Hamiltonian as

$$\frac{dH}{dt} = \sum_i \left(\frac{\partial H}{\partial q_i} \dot{q}_i + \frac{\partial H}{\partial p_i} \dot{p}_i \right) + \frac{\partial H}{\partial t}. \tag{7–18}$$

Substituting $\dot{q}_i$ and $\dot{p}_i$ from Eqs. (7–12) this may be rearranged as

$$\frac{dH}{dt} = \sum_i \left(\frac{\partial H}{\partial q_i} \frac{\partial H}{\partial p_i} - \frac{\partial H}{\partial p_i} \frac{\partial H}{\partial q_i} \right) + \frac{\partial H}{\partial t},$$

and it therefore follows that

$$\frac{dH}{dt} = \frac{\partial H}{\partial t} = - \frac{\partial L}{\partial t}. \tag{7–19}$$

Further, it was proved in Section 2–6 that if the equations of transformation which define the generalized coordinates (1–36):

$$\mathbf{r}_m = \mathbf{r}_m(q_1 \ldots q_n, t),$$

do not depend explicitly on the time, and if the potential is velocity-independent then H is the total energy, $T + V$, a theorem which is valid even in the relativistic limit (cf. Section 6–5). The identification of H as a constant of the motion and as the total energy are two separate matters, and the conditions sufficient for the one are not enough for the other. It can happen that the Eqs. (1–36) do involve time explicitly but that H does not. In this case H is a constant of the motion but it is *not* the total energy.

In many problems the expressions for the canonical momenta are obvious from physical considerations. If it also happens that the Hamiltonian is the total energy then much of the formal procedure for obtaining the Hamilton's equations of motion can be sidestepped. Thus, to take a simple problem, consider the motion of a particle in a central force field. H is then the total energy,

$$H = T + V(r),$$

where

$$T = \frac{1}{2} mv^2 = \frac{1}{2} m(\dot{r}^2 + r^2\dot{\theta}^2).$$

To be usable in Hamilton's equations, H must be expressed in terms of the momenta conjugate to r and θ, which are respectively the linear momentum along $\mathbf{r}$, and the angular momentum of the particle:

$$p_r = mv_r = m\dot{r},$$
$$p_\theta = mrv_\theta = mr^2\dot{\theta}.$$

In terms of the momenta the generalized velocities are:

$$\dot{r} = \frac{p_r}{m}, \qquad \dot{\theta} = \frac{p_\theta}{mr^2}.$$

The Hamiltonian, therefore, follows immediately as

$$H = \frac{p_r^2}{2m} + \frac{p_\theta^2}{2mr^2} + V(r),$$

without the need for first writing the Lagrangian. There are four Hamilton's equations; the two equations for $\dot{q}$:

$$\dot{r} = \frac{\partial H}{\partial p_r} = \frac{p_r}{m}, \qquad \dot{\theta} = \frac{\partial H}{\partial p_\theta} = \frac{p_\theta}{mr^2},$$

which provide no new information, and the two for $\dot{p}$:

$$\dot{p}_r = -\frac{\partial H}{\partial r} = \frac{p_\theta^2}{mr^3} - \frac{\partial V}{\partial r}, \qquad \dot{p}_\theta = 0,$$

which agree with Eqs. (3–7) and (3–10) respectively.

As another example of this same procedure the relativistic Hamiltonian of a particle acted on by velocity-independent potentials has also been shown (Section 6–5) to be the total energy:

$$H = T + V.$$

The kinetic energy T in the Hamiltonian formulation must be expressed in terms of the momentum p, which is easily done by using Eq. (6–44):

$$T^2 = p^2c^2 + m^2c^4,$$

so that

$$H = \sqrt{p^2c^2 + m^2c^4} + V. \tag{7–20}$$

Introduction of velocity-dependent potentials into the Hamiltonian formulation poses no formal difficulty, but it is no longer clear that H represents the total energy. However, for the particular case of electromagnetic forces, the direct application of the definition of H, Eq. (7–8), proves that the Hamiltonian can still be written as a sum of kinetic and potential energies. The Lagrangian (nonrelativistic) for a single particle is

$$L = \frac{mv^2}{2} - q\phi + \frac{q}{c}\mathbf{A}\cdot\mathbf{v},$$

with canonical momenta

$$p_i = mv_i + \frac{q}{c}A_i. \tag{7–21}$$

From the definition, H is given by

$$\begin{aligned} H &= \sum_i p_i v_i - L \\ &= mv^2 + \frac{q}{c}\mathbf{A}\cdot\mathbf{v} - L, \end{aligned}$$

or finally

$$H = \frac{mv^2}{2} + q\phi = T + q\phi,$$

the total energy of the particle. In terms of the momenta, Eq. (7–21), the Hamiltonian appears as

$$H = \frac{1}{2m}\left(\mathbf{p} - \frac{q}{c}\mathbf{A}\right)^2 + q\phi. \tag{7–22}$$

A similar result is obtained for the relativistic Hamiltonian with electromagnetic forces. There too the canonical momenta (Eq. (6–52)) have the additional expression qA_i/c which is just such as to eliminate the term in the Hamiltonian involving the vector potential. The Hamiltonian is again the total energy

$$H = T + q\phi,$$

and it will be noticed that the fourth component of the world momentum, Eq. (6–58), is equal to iH/c. The relativistic kinetic energy is expressed in terms of the momenta by Eq. (6–59), so that H has the final form

$$H = \sqrt{\left(\mathbf{p} - \frac{q\mathbf{A}}{c}\right)^2 c^2 + m^2c^4} + q\phi. \tag{7–23}$$

It is interesting to compare the Hamiltonians (7–22) and (7–23) for a particle in an electromagnetic field with the corresponding Hamiltonians for velocity-independent potentials. For a particle with potential energy V, H is given by

$$H = \frac{p^2}{2m} + V,$$

and Eq. (7–22) can be obtained from this form merely by replacing V with $q\phi$, and the canonical momentum $\mathbf{p}$ with $\mathbf{p} - \frac{q}{c}\mathbf{A}$. The same recipe suffices to convert the relativistic Hamiltonian, Eq. (7–20), into (7–23), and is used frequently in quantum mechanics.

The Hamiltonians (7–20), (7–23) have been described as relativistic only in that they lead to the correct relativistic equations of motion; they are not themselves covariant. However, a truly covariant Hamiltonian may be derived by a Legendre transformation from the covariant Lagrangian L', discussed in the last chapter. The invariant parameter τ is used instead of t, and for a single particle there are four momentum components instead of the spatial three. In relativistic notation the invariant Hamiltonian for a single particle is defined as

$$H' = p_\lambda u_\lambda - L' \tag{7–24}$$

with the corresponding eight equations of motion:

$$\frac{\partial H'}{\partial p_\lambda} = \frac{dx_\lambda}{d\tau}, \qquad \frac{\partial H'}{\partial x_\lambda} = -\frac{dp_\lambda}{d\tau}. \tag{7–25}$$

For the particular case of electromagnetic forces L' has been given in Eq. (6–57) as

$$L' = \frac{1}{2} m u_\lambda u_\lambda + \frac{q}{c} A_\lambda u_\lambda$$

with canonical momenta (cf. Eq. 6–58)

$$p_\lambda = m u_\lambda + \frac{q}{c} A_\lambda.$$

The invariant H', by Eq. (7–24), is then

$$H' = mu_\lambda u_\lambda + \frac{q}{c} A_\lambda u_\lambda - \frac{1}{2} mu_\lambda u_\lambda - \frac{q}{c} A_\lambda u_\lambda$$
$$= \frac{1}{2} mu_\lambda u_\lambda,$$

or, in terms of the canonical momenta,

$$H' = \frac{1}{2m} \sum_\lambda \left(p_\lambda - \frac{q}{c} A_\lambda\right)^2. \tag{7–26}$$

With this covariant Hamiltonian the space part of the Eqs. (7–25) must obviously lead to the spatial equations of motion. In addition there are the two equations obtained if the index λ is set equal to 4. One of them simply states that p_4 is proportional to the total energy:

$$\frac{\partial H'}{\partial p_4} = \frac{p_4 - \frac{q}{c} A_4}{m} = u_4$$

or

$$p_4 = \frac{i}{c}(T + q\phi) = \frac{iE}{c},$$

a fact which has already been noted. The other can be written as

$$\frac{1}{\sqrt{1-\beta^2}} \frac{dp_4}{dt} = -\frac{1}{ic} \frac{\partial H'}{\partial t}$$

or

$$\frac{dH}{dt} = \sqrt{1-\beta^2} \frac{\partial H'}{\partial t}.$$

Comparison of the form of H' (7–26) with that of H (7–23) shows that

$$\frac{\partial H'}{\partial t} = \frac{T}{mc^2} \frac{\partial H}{\partial t},$$

so that the equation of motion reduces to

$$\frac{dH}{dt} = \frac{\partial H}{\partial t},$$

which is the result already expressed in Eq. (7–19).

As in the case of the Lagrangian, a covariant Hamiltonian can only be constructed providing the potentials involved can be expressed covariantly. It was emphasized in the preceding chapter that this is not often possible.

In fact, the example of electromagnetic forces is the only simple one for which a covariant formulation is possible at present.

Relativistic mechanics may thus in principle, at least, be fitted into the Hamiltonian procedure. For convenience, however, most of the subsequent discussion will deal only with the nonrelativistic formulation.

7–4 Derivation of Hamilton's equations from a variational principle. Lagrange's equations have been shown to be the consequence of a variational principle, namely, the Hamilton's principle of Section 2–1. Indeed, the variational method has often proved to be the preferable method of deriving Lagrange's equations, for it is applicable to types of systems not usually comprised within the scope of mechanics. It would be similarly advantageous if a variational principle could be found which leads directly to the Hamilton's equations of motion. Hamilton's principle,

$$\delta I \equiv \delta \int_{t_1}^{t_2} L \, dt = 0, \tag{7–27}$$

lends itself to this purpose if L is expressed in terms of the Hamiltonian by Eq. (7–8), resulting in:

$$\delta I \equiv \delta \int_{t_1}^{t_2} \Big(\sum_i p_i \dot{q}_i - H(q, p, t)\Big) \, dt = 0 \tag{7–28}$$

or

$$\delta \sum_i \int_{q_1}^{q_2} p_i \, dq_i - \delta \int_{t_1}^{t_2} H \, dt = 0. \tag{7–28'}$$

Eq. (7–28) is sometimes referred to as the *modified Hamilton's principle.* Although it will be used most frequently in connection with transformation theory (see Chapter 8) the main interest in it here is to show that the principle leads to Hamilton's canonical equations of motion.

It is well to review at this time the meaning given to the δ-variation process, as defined in Section 2–1. In a configuration space formed from the n generalized coordinates the initial and final configurations of the system, at times t_1 and t_2 respectively, are each represented by a point. The term "variation of the integral" refers to the variation in the value of the integral as we change the path traversed by the system point between the two end points (cf. Fig. 2–1). A further condition is imposed: the variation is described as occurring for constant time. The end-point times are therefore held fixed, so that the "travel time" between the two configurations is the same for all paths. In addition the varied paths are considered as built up by a succession of virtual displacements, involving no change in t, from the points of the actual path traversed. That the variation of the integral about the actual path is zero is equivalent to saying that the integral has an extremum for this path.

One can express the δ-variation process, it will be remembered, in terms of usual differentiation procedures by labeling each of the possible paths in configuration space with a parameter α. The coordinates of a point in the configuration space, q_i, then become functions of both the time t and α, the latter indicating which path of integration is used. The integral I can therefore be considered as a function of α, and the δ-variation is identified with the more familiar differential resulting from a change $d\alpha$ about the value of α corresponding to the correct path:

$$\delta \rightarrow d\alpha \frac{\partial}{\partial \alpha}. \tag{7–29}$$

This, in bare outline, was the method used to derive Lagrange's equations from Hamilton's principle, and it is in the main the procedure to be used here. The one difference will be to treat the variations of q and p as independent, for in the Hamiltonian formulation the generalized coordinates and momenta have equal status as coordinates necessary to specify the motion of the system.

In terms of the parameter α, the modified Hamilton's principle can be written as

$$\delta I = \frac{\partial I}{\partial \alpha} d\alpha = d\alpha \frac{\partial}{\partial \alpha} \int_{t_1}^{t_2} \Big(\sum_i p_i \dot{q}_i - H(q, p, t) \Big) \, dt = 0.$$

Since the end-point times are not varied, and hence are not functions of α, the differentiation may be taken inside the integral sign, yielding

$$d\alpha \int_{t_1}^{t_2} \sum_i \left(\frac{\partial p_i}{\partial \alpha} \dot{q}_i + p_i \frac{\partial \dot{q}_i}{\partial \alpha} - \frac{\partial H}{\partial q_i} \frac{\partial q_i}{\partial \alpha} - \frac{\partial H}{\partial p_i} \frac{\partial p_i}{\partial \alpha} \right) dt = 0. \tag{7–30}$$

The differentiations with respect to α and t in the second term in parentheses may be interchanged, as the variation is at constant time. The integrals involving this type of term may then be transformed by a parts integration:

$$\int_{t_1}^{t_2} p_i \frac{\partial \dot{q}_i}{\partial \alpha} \, dt = \int_{t_1}^{t_2} p_i \frac{d}{dt} \frac{\partial q_i}{\partial \alpha} \, dt = p_i \frac{\partial q_i}{\partial \alpha} \Bigg|_{t_1}^{t_2} - \int_{t_1}^{t_2} \dot{p}_i \frac{\partial q_i}{\partial \alpha} \, dt.$$

Since all varied paths have the same end points, $\dfrac{\partial q_i}{\partial \alpha}$ vanishes for t_1 and t_2 so that the integrated term is zero. In collecting terms in Eq. (7–30) we can use the correspondence given in Eq. (7-29) to recognize that

$$d\alpha \frac{\partial q_i}{\partial \alpha} = \delta q_i \quad \text{and} \quad d\alpha \frac{\partial p_i}{\partial \alpha} = \delta p_i.$$

After incorporating the parts integration, the final form of Eq. (7-30) becomes

$$\int_{t_1}^{t_2} \sum_i \left\{ \delta p_i \left(\dot{q}_i - \frac{\partial H}{\partial p_i} \right) - \delta q_i \left(\dot{p}_i + \frac{\partial H}{\partial q_i} \right) \right\} dt = 0.$$

As the variations δq_i and δp_i are independent the integral can vanish only if the coefficients separately vanish, resulting in the conditions that

$$\dot{q}_i = \frac{\partial H}{\partial p_i}, \qquad \dot{p}_i = -\frac{\partial H}{\partial q_i},$$

which are the desired Hamilton's equations.

The requirement of independent variation of q and p, so essential for the above derivation, highlights the fundamental difference between the Lagrangian and Hamiltonian formulations. It is true that in the Lagrangian procedure both the generalized coordinates q_i and generalized velocities $\dot{q}_i$ must be initially specified to completely determine the motion of the system. But $\dot{q}_i$ was always considered a dependent variable, closely tied to q_i by its definition as the time derivative of q_i. In the derivation of Lagrange's equations the variation $\delta\dot{q}_i$ had to be expressed in terms of the independent variations δq_i by means of a parts integration, which led to a second derivative of L and consequently to second order equations of motion. We obtained first order equations from the modified Hamilton's principle only because the variation δp_i, unlike $\delta\dot{q}_i$, was considered as independent of δq_i. The momenta therefore had to be elevated to the status of independent variables, on an equal basis with the coordinates and connected with them and the time *only through the medium of the equations of motion themselves* and not by any a priori defining relationship. Neither the coordinates q_i nor the momenta p_i are to be considered as the more fundamental set of variables; both are equally independent. Only by broadening the field of independent variables from n to $2n$ quantities are we enabled to obtain equations of motion that are of first order. In a sense the names "coordinates" and "momenta" are unfortunate, for they bring to mind pictures of spatial coordinates and linear, or at most, angular momenta. A wider meaning must now be given to the terms. The division into coordinates and momenta corresponds to no more than a separation of the independent variables describing the motion into two groups having an almost symmetrical relationship to each other through Hamilton's equations.

7–5 The principle of Least Action. Another variational principle associated with the Hamiltonian formulation is known as the principle of *Least Action*. In mechanics, action is a quantity defined most generally as the integral

$$A = \int_{t_1}^{t_2} \sum_i p_i \dot{q}_i \, dt.$$

The principle of Least Action states that in a system for which H is conserved

$$\Delta \int_{t_1}^{t_2} \sum_i p_i \dot{q}_i \, dt = 0, \tag{7–31}$$

where Δ represents a new type of variation of path, requiring detailed explanation.

The δ-variation has been shown to correspond to virtual displacements in which the time is held fixed and the coordinates varied subject to the constraints imposed on the system. Such a virtual displacement does not always coincide with a possible actual displacement occurring in the course of the motion. This will be the case, for example, when the constraints depend on time. Consequently the varied path in the δ-process need not correspond to a possible path of motion for the system; in particular, the Hamiltonian may not be conserved in the variation. In contrast the Δ-process deals with displacements which do involve a change, dt, in the time, so that the varied path is obtained by a succession of displacements each including a differential change dt. It is then possible to require further that the varied path shall be consistent with the physical motion, e.g., if H is conserved on the actual path, it must also be conserved on the varied path. The time of transit along the varied paths need no longer be the same; the system point may have to speed up or slow down in order to keep the Hamiltonian constant. As a result the Δ-process includes a variation of t *even at the end points*, where the variation of the q_i still remains zero.

It should be pointed out, by way of illustration of these ideas, that the actual varied path in configuration space may be the same in both the δ- and Δ-variation except that the system point travels along the path with different velocities in the two cases. In the former variation the speed is such that the time of transit is the same in the actual and varied path; the second process is such that H is constant.

The Δ-process can be represented by tagging the varied curves by a parameter α, as was done for the δ-process. Here, however, the variation includes the time associated with each point on the path, and therefore t must also be considered as a function of α. The variation in $q_i(t, \alpha)$ will involve not only the explicit dependence of q_i on α, but also the implicit dependence through t. Thus we must write Δq as

$$\Delta q \rightarrow d\alpha \left(\frac{dq}{d\alpha}\right) = d\alpha \left(\frac{\partial q}{\partial \alpha} + \dot{q}\,\frac{dt}{d\alpha}\right). \tag{7-32}$$

It will be recognized that the first term corresponds exactly to δq:

$$\delta q \rightarrow d\alpha\,\frac{\partial q}{\partial \alpha},$$

while the coefficient of $\dot{q}$ represents the change in t occurring as a result of the Δ-variation, and can therefore be designated as Δt. Consequently (7–32) has the form

$$\Delta q = \delta q + \dot{q}\,\Delta t. \tag{7-33}$$

(Note that the Δ-operation and time differentiation can no longer be interchanged, as was possible for the δ-variation.) The relation expressed by Eq. (7–33) between the Δ- and δ-variations can now be shown to hold for any function $f(q, t)$:

$$\Delta f = \delta f + \dot{f}\,\Delta t. \tag{7-34}$$

In general Δf will be given by the expression

$$\Delta f = \sum_i \frac{\partial f}{\partial q_i}\,\Delta q_i + \frac{\partial f}{\partial t}\,\Delta t,$$

which, by Eq. (7–33), may be written as

$$\Delta f = \sum_i \frac{\partial f}{\partial q_i}\,\delta q_i + \left(\sum_i \frac{\partial f}{\partial q_i}\,\dot{q}_i + \frac{\partial f}{\partial t}\right)\Delta t.$$

This expression reduces to Eq. (7–34), since the first term on the right is equal to δf, while the terms in the parentheses combine to form the total derivative of f, proving the desired lemma. Eq. (7–34) is the only important property of the Δ-variation needed to prove the principle of least action and we may dispense with the cumbersome parametric notation from here on.

Now, the action A can be written as

$$A = \int_{t_1}^{t_2} \sum_i p_i \dot{q}_i\,dt = \int_{t_1}^{t_2} (L + H)\,dt$$

$$= \int_{t_1}^{t_2} L\,dt + H(t_2 - t_1), \tag{7-35}$$

the last step following from the conservation of H. The Δ-variation of A therefore has the following appearance:

$$\Delta A = \Delta \int_{t_1}^{t_2} L\,dt + H(\Delta t_2 - \Delta t_1).$$

In applying the relation (7–34) to the integral appearing in this expression it must be remembered that the limits of the integral are also subject to the variation. If the integral is written as $I(t_2) - I(t_1)$, the variation takes the form

$$\Delta \int_{t_1}^{t_2} L\,dt = \Delta I(t_2) - \Delta I(t_1),$$

which, by Eq. (7–34), can also be written as:

$$\Delta \int_{t_1}^{t_2} L\,dt = \delta I(t_2) - \delta I(t_1) + \dot{I}(t_2)\,\Delta t_2 - \dot{I}(t_1)\,\Delta t_1,$$

or finally

$$\Delta \int_{t_1}^{t_2} L\,dt = \delta \int_{t_1}^{t_2} L\,dt + L\,\Delta t\Big|_{t_1}^{t_2}. \tag{7–36}$$

There may be a temptation to say that the first term on the right vanishes in consequence of Hamilton's principle, but this is not so. Hamilton's principle requires that the variation δq_i be zero at the initial and final configuration, whereas in the variation process involved here Δq_i instead becomes zero at the end points. The actual value of this term can be evaluated without much difficulty, however. From the nature of the δ-variation we have

$$\delta \int_{t_1}^{t_2} L\,dt = \int_{t_1}^{t_2} \sum_i \left(\frac{\partial L}{\partial q_i}\,\delta q_i + \frac{\partial L}{\partial \dot{q}_i}\,\delta \dot{q}_i\right) dt,$$

which by Lagrange's equations can be written

$$\begin{aligned}\delta \int_{t_1}^{t_2} L\,dt &= \sum_i \int \left(\frac{d}{dt}\left(\frac{\partial L}{\partial \dot{q}_i}\right) \delta q_i + \frac{\partial L}{\partial \dot{q}_i}\frac{d}{dt}\,\delta q_i\right) dt \\ &= \sum_i \int \frac{d}{dt}\left(\frac{\partial L}{\partial \dot{q}_i}\,\delta q_i\right) dt.\end{aligned}$$

Applying Eq. (7–33) this becomes:

$$\begin{aligned}\delta \int_{t_1}^{t_2} L\,dt &= \sum_i \int_{t_1}^{t_2} \frac{d}{dt}\left(\frac{\partial L}{\partial \dot{q}_i}\,\Delta q_i - \frac{\partial L}{\partial \dot{q}_i}\,\dot{q}_i\,\Delta t\right) dt \\ &= \sum_i \left(\frac{\partial L}{\partial \dot{q}_i}\,\Delta q_i - \frac{\partial L}{\partial \dot{q}_i}\,\dot{q}_i\,\Delta t\right)\Big|_{t_1}^{t_2}\end{aligned}$$

At the end points Δq_i vanishes, but since the time of transit is not constant, Δt does not. Consequently, the last expression can be written as

$$\delta \int_{t_1}^{t_2} L\, dt = -\sum_i p_i \dot{q}_i \,\Delta t \Big|_{t_1}^{t_2}. \tag{7-37}$$

Combining these various terms, the total variation of the action is

$$\Delta A = \left(-\sum_i p_i \dot{q}_i + L + H\right) \Delta t \Big|_{t_1}^{t_2},$$

which is identically zero, by virtue of the definition of H. This completes the proof of the principle of least action.*

The least action principle can be exhibited in a variety of forms. In nonrelativistic mechanics, if the defining transformation equations do not involve the time explicitly, it has been shown that

$$\sum_i p_i \dot{q}_i = 2T,$$

(cf. Eq. (2–56)) and in these circumstances the principle can be written as

$$\Delta \int T\, dt = 0. \tag{7-38}$$

If, further, there are no external forces on the system, as, for example, a rigid body with no net applied forces, then T is conserved along with the total energy H. The least action principle then takes the special form:

$$\Delta(t_2 - t_1) = 0. \tag{7-39}$$

Eq. (7–39) states that of all paths possible between two points, consistent with conservation of energy, the system moves along that particular path for which the time of transit is the least (more strictly, an extremum). In this form the principle of least action recalls Fermat's principle in geometrical optics that a light ray travels between two points along such a path that the time taken is the least. We shall have occasion to return to these considerations in Chapter 9 when we discuss the connection between the Hamiltonian formulation and geometrical optics.

If the system consists of only one particle the kinetic energy is given by

$$T = \frac{1}{2} mv^2 = \frac{1}{2} m \left|\frac{d\mathbf{r}}{dt}\right|^2$$

---

* The principle of least action is commonly associated with the name of Maupertuis. However, the original statement of the principle by Maupertuis (1747) was vaguely theological and could hardly pass muster today. The objective statement of the principle we owe to Euler and Lagrange.

or

$$dt = \sqrt{\frac{m\,|d\mathbf{r}|^2}{2T}}.$$

which presents dt in terms of the arc length of the particle trajectory. In such case the least action principle as expressed in Eq. (7–39) can be written

$$\Delta \int 2T\,dt = \Delta \int \sqrt{2mT}\,ds = \Delta \int \sqrt{2m(H - V)}\,ds = 0, \tag{7–40}$$

where $(ds)^2$ is written for $|d\mathbf{r}|^2$. This form of the least action principle can also be extended to systems of more than one particle and described by any set of generalized coordinates. The device required is a more general expression for dt in terms of an arc length. For the stipulated conditions that the equations defining q_i do not contain the time, the kinetic energy is always a homogeneous quadratic function of the velocities (cf. Eq. (1–62)):

$$T = \frac{1}{2} \sum_{i,k} m_{ik}\dot{q}_i\dot{q}_k,$$

(where $m_{ik} = 2a_{ik}$) which relation could also be written as

$$T = \frac{1}{2} \sum_{i,k} m_{ik} \frac{dq_i\,dq_k}{dt^2}. \tag{7–41}$$

We define now a new differential $d\rho$ by the equation

$$(d\rho)^2 = \sum_{i,k} m_{ik}\,dq_i\,dq_k, \tag{7–42}$$

so that the kinetic energy has the form

$$T = \frac{1}{2}\left(\frac{d\rho}{dt}\right)^2, \tag{7–43}$$

or equivalently

$$dt = \frac{d\rho}{\sqrt{2T}}.$$

With this relation for dt the least action principle becomes

$$\Delta \int_{t_1}^{t_2} T\,dt = \Delta \int \sqrt{T}\,d\rho = 0$$

or, finally,

$$\Delta \int \sqrt{H - V(q)}\,d\rho = 0, \tag{7–44}$$

an equivalent expression of the principle of least action which is formally similar to Eq. (7–40) for a single particle. Eq. (7–44) is often referred to as *Jacobi's form of the least action principle.* The procedure used in defining $d\rho$ appears quite abstract, but is susceptible of a very elegant interpretation. The differential $d\rho$ introduced in Eq. (7–42) is well known from differential geometry as the most general form for the path length in an n-dimensional space whose coordinates are $q_1 \ldots q_n$. In this application the coefficients m_{ik} are the elements of the so-called *metric tensor.* In a cartesian coordinate system the metric tensor is quite simple; the elements are

$$m_{ik} = \delta_{ik},$$

as can be seen by comparing the expression

$$(ds)^2 = (dx)^2 + (dy)^2 + (dz)^2$$

with Eq. (7–42). The metric tensor is diagonal for any general orthogonal curvilinear coordinate system, but except for cartesian systems, the diagonal elements are not all equal. For example, in cylindrical coordinates the element of arc length is

$$(ds)^2 = (dr)^2 + r^2(d\theta)^2 + (dz)^2,$$

so that the only nonvanishing elements of the corresponding metric tensor are

$$m_{rr} = 1, \qquad m_{\theta\theta} = r^2, \qquad m_{zz} = 1.$$

In general the nondiagonal elements will not be zero for an arbitrary curvilinear nonorthogonal set of coordinates.

The differential $d\rho$ defined by Eq. (7–42) is thus the element of path length in the configuration space of coordinates $q_1 \ldots q_n$ which is not in general cartesian but is curvilinear; in fact, so curved that the elements of the metric tensor are given by m_{ik} of Eq. (7–41). The square root of $2T$ represents the velocity of the system point along the path in configuration space. If there are no forces acting on the body then T is constant and the motion is such that the system point travels along the shortest path length in configuration space — i.e., travels along one of the *geodesics* of the space.

It should be emphasized that the Jacobi form of the principle of least action is concerned with the *path* of the system point rather than with its motion in *time.* Eq. (7–44) is a statement about the element of path length $d\rho$; the time nowhere appears, since H is a constant and V depends on q_i only. Indeed, the Jacobi form of the principle can be made to furnish differential equations for the path. This may best be accomplished by introducing some parameter marking, for example, distance along the path. One can then write Eq. (7–44) as

$$\Delta \int_{\theta_1}^{\theta_2} \sqrt{H - V} \sqrt{\sum_{i,k} m_{ik} \frac{dq_i}{d\theta} \frac{dq_k}{d\theta}} \, d\theta = 0, \tag{7–45}$$

where the parameter in question is designated by θ. The parameter must not be confused with the time; it is to be a geometrical property of the path, and can be so chosen as to be constant during the displacements which characterize the Δ-variation. With respect to θ, the Δ-variation is therefore identical with the δ-variation, and the differential equations derived from Eq. (7–45) have the familiar form of the Euler-Lagrange equations. If derivatives with respect to θ are designated by a prime then the integrand of Eq. (7–45) may be represented as a function $F(q_i, q_i', \theta, H)$. The Euler-Lagrange equations for the path of the system point are then

$$\frac{d}{d\theta}\left(\frac{\partial F}{\partial q_i'}\right) - \frac{\partial F}{\partial q_i} = 0. \tag{7–46}$$

When dealing with a system containing many particles the path determined by Jacobi's form of the principle is the path of the system point in configuration space. But if the system contains only one particle, and the coordinates q_i are position coordinates, then the path given by Eqs. (7–46) also represents the actual orbit of the particle in space. The coordinates q_i need not be cartesian coordinates, nor need the problem be confined to unconstrained motion in three dimensions. If, for example, the particle be constrained to move on a given surface, with two coordinates q_1 and q_2 giving the position of the particle on the surface, then the quantity $d\rho$ is obviously proportional to an element of the actual path length on the surface. Eqs. (7–46) can then be used to find the orbit described by the particle on the constraining surface. In particular, if there are no external forces acting on the particle, then the actual path on the surface, as well as the path in configuration space, is one of the geodesics of the surface. Thus a particle moving on the surface of a sphere, with no external forces acting on it, must travel along a great circle, a geodesic of the sphere.

A host of similar variational principles for classical mechanics can be derived in bewildering variety. To give one example out of many, the principle of least action leads immediately to *Hertz' principle of least curvature*, which states that a particle not under the influence of external forces travels along the path of least curvature. By Jacobi's principle such a path must be a geodesic, and the geometrical property of minimum curvature is one of the well-known characteristics of a geodesic. It has been pointed out that variational principles in themselves contain no new physical content, and they rarely simplify the practical solution of a given mechanical problem. Their value lies chiefly as starting points for new formulations of the theoretical structure of classical mechanics. For this purpose Hamilton's principle is especially fruitful, and to a lesser extent, so also is the principle of least action. The others have proved to be of little use, except as they have led to fruitless teleological speculations, and further discussion of them here seems pointless.

SUGGESTED REFERENCES

P. S. Epstein, *Textbook of Thermodynamics.* Sections 33 and 34 of this text illustrate the applications of the Legendre transformation in defining new thermodynamical functions.

E. T. Whittaker, *Analytical Dynamics.* The subject of the variational principles encountered in classical mechanics can become quite involved and possesses many far-flung roots in apparently unrelated fields. For example, there is a close connection between Hamilton's principle and the general theory of second order partial differential equations. Some of these topics we will discuss in the following chapters, but many are inappropriate for the treatment intended here. Similarly, such questions as whether the extremum in Hamilton's principle is a minimum or maximum cannot be taken up here, nor will it be feasible (or desirable) to consider the many subspecies of variational principles. The student interested in problems of this nature will find an abundant literature; indeed, there is an "embarras des richesses." Only a small fraction of the available references can be mentioned in this list, and of them Whittaker is one of the chief sources. Chapter IX and the first two sections of Chapter X are the portions pertinent to this chapter.

A. G. Webster, *Dynamics of Particles.* An extensive and often rambling discussion of the variational principles and their consequences will be found in Chapter IV, with many of the points illustrated by detailed examples. From the treatment one can obtain an excellent view of the distribution of emphasis in classical mechanics as it was taught at the turn of the century.

L. Nordheim, *Die Prinzipe der Dynamik,* Vol. V of the *Handbuch der Physik.* This article gives a quite complete discussion of the variety of integral and differential principles which can be used as the formulational basis of mechanics. Sections 1 and 2 of the following article, by Nordheim and Fues, form a readable introduction to Hamilton's equations.

C. Schaefer, *Die Prinzipe der Dynamik.* This little book of some 76 pages is an exhaustively thorough discussion of the various principles and of the types of variation processes encountered. The prospective reader should be strongly cautioned, however, that the author uses the most unorthodox notation of H for Lagrangian and R for Hamiltonian.

R. L. LINDSAY AND H. MARGENAU, *Foundations of Physics*. The statements of Hamilton's principle and the principle of least action appear to endow the mechanical system with conscious knowledge of the final state towards which the motion is directed. Such an appearance is of course illusory; the motion of the system is determined only by the initial conditions. But the view has given rise in the past to much philosophical speculation. Chapter 3 of this text presents an adequate discussion of this and similar points, and furnishes references for further reading for those inclined.

EXERCISES

1. Set up the problem of central force motion of two mass points in the Hamiltonian formulation, eliminating the cyclic variables and reducing the problem to quadratures.

2. Find the Hamiltonian for the system described in Exercise 5 of Chapter 5 and obtain Hamilton's equations of motion for the system.

3. Obtain the Hamiltonian of a heavy symmetrical top with one point fixed, and from it the Hamilton's equations of motion. Relate these to the equations of motion discussed in Section 5–7 and, in particular, show how the solution may be reduced to quadratures.

4. Consider a system consisting of one particle acted upon by conservative forces which are independent of the azimuth of the particle about the z-axis of an inertial system. Obtain the Hamiltonian of the particle in terms of its cartesian coordinates with respect to a system of axes rotating uniformly about the given z-axis with an angular velocity ω. What is the physical significance of the Hamiltonian in this case? Is it a constant of the motion?

5. In Exercise 4 of Chapter 1 there is given the velocity-dependent potential assumed in Weber's electrodynamics. What is the Hamiltonian for a single particle moving under the influence of such a potential?

6. It was pointed out in Chapter 6 that the term in the covariant relativistic Lagrangian corresponding to the kinetic energy is to some extent arbitrary, and Eq. (6–57) represents only one of the possible forms for the Lagrangian. An alternative form is obtained if we directly transform the parametric variable t in Hamilton's principle, (6–48), to the Lorentz-invariant world time τ, and use the new integrand as L'. Derive the covariant Hamiltonian for a particle in an electromagnetic field resulting from such an alternative Lagrangian. Show that the *value* of the Hamiltonian is identically zero. In obtaining the equations of motion the value of the Hamiltonian is, of course, of no physical significance; we are interested only in its *functional dependence* on the coordinates and momenta.

7. When the covariant Lagrangian is obtained in the manner specified in the preceding exercise, show that the Hamilton's equation of motion for the total derivative of p_4 with respect to τ reduces identically to Eq. (7–19), without the need of invoking properties of the specific form of H'.

CHAPTER 8

CANONICAL TRANSFORMATIONS

When applied in a straightforward manner the Hamiltonian formulation usually does not materially decrease the difficulty of solving any given problem in mechanics. One winds up with practically the same differential equations to be solved as are provided by the Lagrangian procedure. The advantages of the Hamiltonian formulation lie not in its use as a calculational tool, but rather in the deeper insight it affords into the formal structure of mechanics. The equal status accorded to coordinates and momenta as independent variables encourages a greater freedom in selecting the physical quantities to be designated as "coordinates" and "momenta." As a result we are led to newer, more abstract ways of presenting the physical content of mechanics. While often of considerable help in practical applications to mechanical problems, these more abstract formulations are primarily of interest to us today because of their essential role in constructing the more modern theories of matter. Thus, one or another of these formulations of classical mechanics serves as a point of departure for both statistical mechanics and quantum theory. It is to such formulations, arising as outgrowths of the Hamiltonian procedure, that this and the next chapter are devoted.

8–1 The equations of canonical transformation. There is one type of problem for which the solution of the Hamilton's equations is trivial. Consider a situation in which the Hamiltonian is a constant of the motion, and where *all* coordinates q_i are cyclic. Under these conditions the conjugate momenta p_i are all constant:

$$p_i = \alpha_i,$$

and since the Hamiltonian cannot be an explicit function of either the time or the cyclic coordinates, it may be written as

$$H = H(\alpha_1 \ldots \alpha_n).$$

Consequently, the Hamilton's equations for $\dot{q}_i$ are simply

$$\dot{q}_i = \frac{\partial H}{\partial \alpha_i} = \omega_i, \tag{8–1}$$

where the ω_i's are functions of the α_i's only and therefore are also constant in time. Eqs. (8–1) have the immediate solutions

$$q_i = \omega_i t + \beta_i, \tag{8–2}$$

where the β_i's are constants of integration, determined by the initial conditions.

It would seem that the solution to this type of problem, easy as it is, can only be of academic interest, for it rarely occurs in practice that one finds that all the generalized coordinates are cyclic. But a given system can be described by more than one set of generalized coordinates. Thus, to discuss motion of a particle in a plane one may use as generalized coordinates either the cartesian coordinates

$$q_1 = x, \qquad q_2 = y,$$

or the plane polar coordinates

$$q_1 = r, \qquad q_2 = \theta.$$

Both choices are equally valid, but one or the other set may be more convenient for the problem under consideration. It will be noticed that for central forces neither x nor y is cyclic, while the second set does contain a cyclic coordinate in the angle θ. The number of cyclic coordinates thus depends on the choice of generalized coordinates, and for each problem there may be one particular choice for which all coordinates are cyclic. If we can find this set the remainder of the job is trivial. Since the obvious generalized coordinates suggested by the problem will not normally be cyclic, we must first derive a specific procedure for *transforming* from one set of variables to some other set which may be more suitable.

The transformations considered in the previous chapters have involved going from one set of coordinates q_i to a new set Q_i by transformation equations of the form

$$Q_i = Q_i(q, t). \tag{8-3}$$

For example, the equations of an orthogonal transformation, or of the change from cartesian to plane polar coordinates, have the general form of Eqs. (8-3). We shall characterize such transformations as *point transformations.* But in the Hamiltonian formulation the momenta are also independent variables on the same level as the generalized coordinates. The concept of transformation of coordinates must therefore be widened to include the simultaneous transformation of the independent *coordinates* and *momenta,* q_i, p_i, to a new set Q_i, P_i, with equations of transformation:

$$\begin{aligned} Q_i &= Q_i(q, p, t), \\ P_i &= P_i(q, p, t). \end{aligned} \tag{8-4}$$

Thus the new coordinates will be defined not only in terms of the old coordinates but also in terms of the old momenta.

In developing Hamiltonian mechanics, only those transformations can be of interest for which the new Q, P are canonical coordinates. This requirement will be satisfied provided there exists some function $K(Q, P, t)$ such that the equations of motion in the new set are in the Hamiltonian form

$$\dot{Q}_i = \frac{\partial K}{\partial P_i}, \qquad \dot{P}_i = -\frac{\partial K}{\partial Q_i}. \tag{8-5}$$

Transformations for which Eqs. (8-5) are valid are said to be *canonical.** The function K plays the role of the Hamiltonian in the new coordinate set.

If the Q_i and P_i are to be canonical coordinates they must satisfy a modified Hamilton's principle of the form:

$$\delta \int_{t_1}^{t_2} \left(\sum P_i \dot{Q}_i - K(Q, P, t) \right) dt = 0. \tag{8-6}$$

At the same time the old coordinates, of course, satisfy a similar principle:

$$\delta \int_{t_1}^{t_2} \left(\sum p_i \dot{q}_i - H(q, p, t) \right) dt = 0. \tag{8-7}$$

The simultaneous validity of Eqs. (8-6) and (8-7) does not mean that the integrands of the two integrals are equal, but that they can differ at most by a total time derivative of an arbitrary function F. The integral between the two end points of such a difference term is then

$$\int_{t_1}^{t_2} \frac{dF}{dt}\, dt = F(2) - F(1),$$

and the variation of this integral is automatically zero for any function F since the variation vanishes at the end points. The arbitrary function F

---

* The designation *contact transformation* will also be used. A distinction can be made between a contact and a canonical transformation, but the practice in the literature is not uniform. Some authors (e.g., A. Sommerfeld in *Atomic Structure and Spectral Lines*) confine contact transformation to those in which the equations of transformation, Eqs. (8-4), do not involve time explicitly. Others (probably more correctly) consider contact transformations to be those which provide for a transformation of the time coordinate in addition to the coordinates and momenta, as would be required in a covariant relativistic theory. Physicists, however, tend to treat the two terms as synonymous and this usage will be followed here. The significance of contact transformations in projective geometry (where, as might be guessed, the name originated) is briefly described in Sommerfeld, loc. cit., and in Carathéodory, *Variationsrechnung.*

is called the *generating function* of the transformation for, as we shall see, once F is given the transformation equations (8–4) are completely specified.

In order to effect the transformation between the two sets of canonical variables, F must be a function of both the old and new variables. Besides the time t the generating function may thus be a function of $4n$ variables in all. But only $2n$ of these are independent, because the two sets of coordinates are connected by the $2n$ transformation equations (8–4). The generating function can therefore be written as a function of independent variables in one of only four forms:

$$F_1(q, Q, t), \qquad F_2(q, P, t), \qquad F_3(p, Q, t), \qquad F_4(p, P, t).$$

The circumstances of the problem will dictate which form is to be chosen. For example, if we are dealing with a point transformation as defined by Eq. (8–3), then q and Q are not independent variables, and generating functions of the form F_1 must be excluded, while any of the others may be used.

If the first form F_1 is a suitable choice then the integrands of Eqs. (8–6) and (8–7) can be connected by the relation:

$$\sum p_i\dot{q}_i - H = \sum P_i\dot{Q}_i - K + \frac{d}{dt}F_1(q, Q, t). \tag{8–8}$$

The total time derivative of F can be expanded as:

$$\frac{dF}{dt} = \sum_i \frac{\partial F}{\partial q_i}\dot{q}_i + \sum_i \frac{\partial F}{\partial Q_i}\dot{Q}_i + \frac{\partial F}{\partial t}.$$

Since the old and new coordinates, q_i and Q_i, are considered here as independent, Eq. (8–8) can hold identically only if the coefficients of the $\dot{q}_i$ and $\dot{Q}_i$ separately vanish:

$$p_i = \frac{\partial F_1}{\partial q_i}, \tag{8–9a}$$

$$P_i = -\frac{\partial F_1}{\partial Q_i}, \tag{8–9b}$$

finally leaving:

$$K = H + \frac{\partial F_1}{\partial t}. \tag{8–9c}$$

The n equations (8–9a) are n relations involving only p_i, q_i, Q_i, and t. These can be solved for the n Q_i's in terms of the p_i, q_i, t; thus yielding the first half of the transformation Eqs. (8–4). Once the relations between Q_i and (q_i, p_i) have been established, Eqs. (8–9b) furnish the remaining half of the transformation equations, giving P_i in terms of (q_i, p_i). To complete the story, Eq. (8–9c) provides the connection between the new Hamiltonian K and the old one.

If the independent arguments of F are to be q_i and P_i then the generating function is of the type F_2. It will be noticed that the transition from q, Q as independent variables to q, P can be accomplished by a Legendre transformation, since by Eq. (8–9b):

$$\frac{\partial F_1}{\partial Q_i} = -P_i.$$

This suggests that the generating function F_2 can therefore be suitably defined in terms of F_1 according to the relation

$$F_2(q, P, t) = F_1(q, Q, t) + \sum_i P_i Q_i. \tag{8–10}$$

On solving Eq. (8–10) for F_1 and substituting, Eq. (8–8) becomes:

$$\begin{aligned}\sum_i p_i \dot{q}_i - H &= \sum_i P_i \dot{Q}_i - K + \frac{d}{dt}\left(F_2(q, P, t) - \sum_i Q_i P_i\right)\\ &= -\sum_i Q_i \dot{P}_i - K + \frac{d}{dt} F_2(q, P, t).\end{aligned}$$

Repeating the procedure followed for F_1 by differentiating F_2 and collecting coefficients of $\dot{q}_i$, $\dot{P}_i$, we obtain the transformation equations:

$$p_i = \frac{\partial F_2}{\partial q_i}, \tag{8–11a}$$

$$Q_i = \frac{\partial F_2}{\partial P_i}, \tag{8–11b}$$

with

$$K = H + \frac{\partial F_2}{\partial t}. \tag{8–11c}$$

Equations (8–11a) can be solved for P_i as functions of q_i, p_i, t, and therefore correspond to the second half of Eqs. (8–4). The remaining half of the transformation equations is then provided by Eqs. (8–11b).

The third type of generating function $F_3(p, Q, t)$ can again be connected with F_1 by a Legendre transformation, by virtue of the form of $\frac{\partial F}{\partial q_i}$ as given by Eqs. (8–9a). The new generating function can therefore be defined by:

$$F_1(q, Q, t) = \sum q_i p_i + F_3(Q, p, t). \tag{8–12}$$

Eq. (8–8) now appears in the form:

$$-\sum q_i \dot{p}_i - H = \sum P_i \dot{Q}_i - K + \frac{d}{dt} F_3(p, Q, t), \tag{8–13}$$

and the same process of equating coefficients yields the transformation equations:

$$q_i = -\frac{\partial F_3}{\partial p_i}, \tag{8–14a}$$

$$P_i = -\frac{\partial F_3}{\partial Q_i}, \tag{8–14b}$$

and

$$K = H + \frac{\partial F_3}{\partial t}. \tag{8–14c}$$

As in the two previous cases, Eqs. (8–14a) give Q_i as functions of q, p, t; Eqs. (8–14b) then furnish the new momenta P_i in terms of the old variables.

Finally, when p and P are to be taken as the independent variables the corresponding generating function F_4 can be related to F_1 by a double Legendre transformation

$$F_4(p, P, t) = F_1(q, Q, t) + \sum_i P_i Q_i - \sum_i p_i q_i, \tag{8–15}$$

and Eq. (8–8) reduces to:

$$-\sum q_i \dot{p}_i - H = -\sum Q_i \dot{P}_i - K + \frac{d}{dt} F_4(p, P, t). \tag{8–16}$$

The last set of transformation equations is therefore:

$$q_i = -\frac{\partial F_4}{\partial p_i}, \tag{8–17a}$$

$$Q_i = \frac{\partial F_4}{\partial P_i}, \tag{8–17b}$$

$$K = H + \frac{\partial F_4}{\partial t}. \tag{8–17c}$$

In these considerations the time has been treated as an invariant parameter, and no provision has been made for a transformation of the time scales along with the coordinates and momenta. Of course, a transformation of time occurs automatically in the four-dimensional relativistic Hamiltonian formulation, for then the invariant parameter of the system is the world time τ, and the ordinary time t appears as one of the coordinates of the particle. It is of interest, however, to see how a change of the time scales, other than that produced by a Lorentz transformation, can be incorporated into the ordinary scheme of canonical transformations.

Since the time is no longer an invariant parameter of the motion of the system, it is necessary to introduce some other parameter which can take its place. In the covariant relativistic formulation this parameter is supplied by the particle proper time, but in general any invariant quantity marking the progress of the system along its path in configuration space will be satisfactory. If such a parameter is designated by θ, the modified Hamilton's principle may be written as

$$\delta \int_{\theta_1}^{\theta_2} \left(\sum_i^n p_i \dot{q}_i \frac{dt}{d\theta} - H \frac{dt}{d\theta} \right) d\theta = 0,$$

or as

$$\delta \int_{\theta_1}^{\theta_2} \left(\sum_i^n p_i \frac{dq_i}{d\theta} - H \frac{dt}{d\theta} \right) d\theta = 0.$$

The form of this expression suggests that t may be considered as the $n + 1$ generalized coordinate, with $-H$ as its conjugate momentum.* With this designation the principle takes on the form

$$\delta \int_{\theta_1}^{\theta_2} \sum_{i=1}^{n+1} p_i q_i' \, d\theta = 0,$$

where the prime denotes a derivative with respect to θ. Upon canonical transformation of the coordinates (including t) the modified Hamilton's principle will appear as

$$\delta \int_{\theta_1}^{\theta_2} \sum_{i=1}^{n+1} P_i Q_i' \, d\theta = 0.$$

Here Q_{n+1} is the transformed time variable, and $-P_{n+1}$ is equal to the new Hamiltonian K. Following the same procedure as used previously, we may define a generating function $G(q_i, P_i)$ in terms of which the transformation equations are given by the $2n + 2$ relations

$$p_i = \frac{\partial G}{\partial q_i},$$

$$Q_i = \frac{\partial G}{\partial P_i}.$$

While it is thus formally possible to perform canonical transformations involving changes in the time scale, it will be assumed in the following sections that the time is an invariant parameter of the motion.

---

* This choice will be reminiscent of the relativistic result that iH/c is conjugate to $x_4 = ict$. Here, however, the identification is purely formal, and does not imply or require any of the physical content of special relativity.

8–2 Examples of canonical transformations. The nature of canonical transformations and the role played by the generating function can best be illustrated by some simple yet important examples. Consider, first, a generating function of the second type with the particular form:

$$F_2 = \sum_i q_i P_i. \tag{8–18}$$

From Eqs. (8–11) the transformation equations are:

$$p_i = \frac{\partial F_2}{\partial q_i} = P_i,$$

$$Q_i = \frac{\partial F_2}{\partial P_i} = q_i,$$

$$K = H.$$

The new and old coordinates are the same; hence F_2 merely generates the *identity transformation.*

A more general type of transformation is described by the generating function:

$$F_2 = \sum_i f_i(q_1 \ldots q_n, t) P_i, \tag{8–19}$$

where the f_i may be any desired set of functions. By Eqs. (8–11b) the new coordinates Q_i are given by

$$Q_i = \frac{\partial F_2}{\partial P_i} = f_i(q, t). \tag{8–20}$$

Thus, with this generating function the new coordinates depend only on the old coordinates and the time, and do not involve the old momenta. Such a transformation is therefore an example of the class of point transformations defined by Eqs. (8–3). Since the functions f_i are completely arbitrary we may conclude that *all point transformations are canonical.* Eq. (8–11c) furnishes the new Hamiltonian in terms of the old and of the time derivatives of the f_i functions.

Orthogonal transformations, which have been discussed in great detail in Chapters 4 and 6, are special cases of point transformations with the functions f_i given as

$$f_i = Q_i = \sum_k a_{ik} q_k.$$

The generating function then has the form

$$F_2 = \sum_{i,k} a_{ik} q_k P_i,$$

and the new momenta are to be found from Eqs. (8-11a),

$$p_k = \frac{\partial F_2}{\partial q_k} = \sum_i a_{ik} P_i. \tag{8-21}$$

These equations may be solved for P_i by multiplying by a_{jk} and summing over k:

$$\sum_k a_{jk} p_k = \sum_{i,k} a_{jk} a_{ik} P_i = \sum_i \delta_{ij} P_i,$$

the last step following from the orthogonality conditions. The final summation gives:

$$P_i = \sum_k a_{ik} p_k, \tag{8-22}$$

so that the momenta also transform orthogonally, as would be expected a priori.

An instructive transformation is provided by the generating function of the first kind, $F_1(q, Q, t)$, of the form:

$$F_1 = \sum_k q_k Q_k.$$

The corresponding transformation equations, from (8-9a, b) are:

$$p_i = \frac{\partial F_1}{\partial q_i} = Q_i,$$

$$P_i = -\frac{\partial F_1}{\partial Q_i} = -q_i.$$

In effect the transformation interchanges the momenta and coordinates; the new coordinates are the old momenta and the new momenta are essentially the old coordinates. This simple example should emphasize the independent status of generalized coordinates and momenta. They are both needed to describe the motion of the system in the Hamiltonian formulation, and the distinction between them is practically one of nomenclature. One can shift the names around with at most no more than a change in sign. There is no longer present in the theory any lingering remnant of the concept of q_i as a spatial coordinate and p_i as a mass times a velocity. Incidentally, one may see directly from Hamilton's equations:

$$\dot{p}_i = -\frac{\partial H}{\partial q_i}, \qquad \dot{q}_i = \frac{\partial H}{\partial p_i},$$

that this exchange transformation is canonical. If q_i is substituted for p_i the equations remain in the canonical form only if $-p_i$ is substituted for q_i.

As a final example, consider the generating function

$$F_1 = \frac{m}{2}\omega q^2 \cot Q, \tag{8-23}$$

where m and ω are constants whose significance will be discussed later. With this generating function Eqs. (8-9a, b) reduce to

$$p = \frac{\partial F_1}{\partial q} = m\omega q \cot Q, \tag{8-24}$$

and

$$P = -\frac{\partial F_1}{\partial Q} = \frac{m\omega q^2}{2\sin^2 Q}. \tag{8-25}$$

These two equations may be solved for Q and P in terms of q and p, but it is more desirable for our purposes to obtain the old variables as functions of the new. From Eq. (8-25) q is given by

$$q = \sqrt{\frac{2P}{m\omega}}\sin Q, \tag{8-26}$$

which, on substitution in Eq. (8-24), yields p as

$$p = \sqrt{2m\omega P}\cos Q. \tag{8-27}$$

As the generating function does not involve the time explicitly, the value of the Hamiltonian is not affected by the transformation and it is only necessary to express H in terms of the new Q and P by means of Eqs. (8-26, 27). Now, the constants in F_1 have been selected with foresight for application to the problem of the linear harmonic oscillator. If the force constant of the linear restoring force acting on the particle is k, then the potential energy is:

$$V = \frac{kq^2}{2},$$

and the Hamiltonian has the form

$$H = \frac{m\dot{q}^2}{2} + \frac{kq^2}{2} = \frac{p^2}{2m} + \frac{kq^2}{2}.$$

Designating the ratio k/m by ω^2, H can also be written as

$$H = \frac{p^2}{2m} + \frac{m\omega^2}{2}q^2. \tag{8-28}$$

Substitution of the transformation equations (8-26) and (8-27) in (8-28) furnishes the form of the Hamiltonian when expressed in the new coordinates:

$$H = \omega P\cos^2 Q + \omega P\sin^2 Q = \omega P. \tag{8-29}$$

The Hamiltonian is thus cyclic in Q and the conjugate momentum P is therefore a constant. It is seen from Eq. (8–29) that P is in fact equal to the constant energy divided by ω:

$$P = \frac{E}{\omega}.$$

The equation of motion for Q reduces to the simple form

$$\dot{Q} = \frac{\partial H}{\partial P} = \omega,$$

with the immediate solution:

$$Q = \omega t + \alpha,$$

where α is a constant of integration fixed by the initial conditions. From Eq. (8–26) the solution for q is:

$$q = \sqrt{\frac{2E}{m\omega^2}} \sin(\omega t + \alpha), \tag{8–30}$$

which is the customary solution for a harmonic oscillator.

It would seem that the use of contact transformations to solve the harmonic oscillator problem is similar to "cracking a peanut with a sledge hammer." We have here, however, a simple example of how the Hamiltonian can be reduced to a form cyclic in all coördinates by means of canonical transformations. Discussion of general schemes for the solution of mechanical problems by this technique will be reserved for the next chapter. For the present we shall continue to examine the formal properties of canonical transformations.

8–3 The integral invariants of Poincaré. The canonical transformations are defined as having the property of preserving the form of Hamilton's equations of motion under the transformation. The question arises whether there are other expressions invariant under canonical transformations. One such set are the integral invariants found by Poincaré. In analogy to configuration space in Lagrangian formulation, we define a $2n$-dimensional cartesian space formed of coordinates $q_1 \ldots q_n, p_1 \ldots p_n$, and known as *phase space*. The complete dynamical specification of a mechanical system will be given by a point in such a space. The theorem of Poincaré then states that the integral

$$J_1 = \iint_S \sum_i dq_i\, dp_i, \tag{8–31}$$

is invariant under canonical transformation, where S indicates that the integrals are to be evaluated over any arbitrary two-dimensional surface in phase space.

The proof of invariance starts with the observation that the position of a point on any two-dimensional surface is specified completely by not more than two parameters. Let u and v be such parameters appropriate to the surface S, so that on this surface $q_i = q_i(u, v)$ and $p_i = p_i(u, v)$. As is well known, the element of area $dq_i\, dp_i$ transforms to the area element $du\, dv$ by means of the Jacobian determinant, written symbolically as:

$$\frac{\partial(q_i, p_i)}{\partial(u, v)} = \begin{vmatrix} \frac{\partial q_i}{\partial u} & \frac{\partial p_i}{\partial u} \\ \frac{\partial q_i}{\partial v} & \frac{\partial p_i}{\partial v} \end{vmatrix}, \tag{8-32}$$

according to the relation

$$dq_i\, dp_i = \frac{\partial(q_i, p_i)}{\partial(u, v)}\, du\, dv. \tag{8-33}$$

Thus, the statement that J_1 has the same value for all canonical coordinates:

$$\iint_S \sum_i dq_i\, dp_i = \iint_S \sum_k dQ_k\, dP_k,$$

can be written also as:

$$\iint_S \sum_i \frac{\partial(q_i, p_i)}{\partial(u, v)}\, du\, dv = \iint_S \sum_k \frac{\partial(Q_k, P_k)}{\partial(u, v)}\, du\, dv.$$

Since the region of integration is arbitrary, the integrals can be equal only if the integrands are identical:

$$\sum_i \frac{\partial(q_i, p_i)}{\partial(u, v)} = \sum_k \frac{\partial(Q_k, P_k)}{\partial(u, v)}. \tag{8-34}$$

The proof of the invariance of J_1 has thus been reduced to showing that the sum of the Jacobians is invariant.

It is convenient to consider the canonical transformation from q, p to Q, P as being obtained from a generating function of the type $F_2(q, P, t)$.* The columns of the determinants on the left in Eq. (8-34) can then be expressed in terms of the new set of variables by means of the generating function. From Eqs. (8-11a) we have that

$$\frac{\partial p_i}{\partial u} = \frac{\partial}{\partial u}\left(\frac{\partial F_2}{\partial q_i}\right).$$

* This is not a necessary restriction. A proof can be carried through using any one of the other generating functions, as can be readily verified by the reader.

The quantity in the parentheses is a function of u only through its arguments q_k and P_k, and the expression can therefore be expanded as:

$$\frac{\partial p_i}{\partial u} = \sum_k \frac{\partial^2 F_2}{\partial q_i\, \partial P_k} \frac{\partial P_k}{\partial u} + \sum_k \frac{\partial^2 F_2}{\partial q_i\, \partial q_k} \frac{\partial q_k}{\partial u}, \tag{8-35}$$

with a similar expression for the partial derivative of p_i with respect to v. Substituting these expansions in the sum of determinants appearing in (8-34), the following relation is obtained:

$$\sum_i \frac{\partial(q_i, p_i)}{\partial(u, v)} = \sum_i \begin{vmatrix} \dfrac{\partial q_i}{\partial u} & \displaystyle\sum_k \frac{\partial^2 F_2}{\partial q_i\, \partial P_k} \frac{\partial P_k}{\partial u} + \sum_k \frac{\partial^2 F_2}{\partial q_i\, \partial q_k} \frac{\partial q_k}{\partial u} \\ \dfrac{\partial q_i}{\partial v} & \displaystyle\sum_k \frac{\partial^2 F_2}{\partial q_i\, \partial P_k} \frac{\partial P_k}{\partial v} + \sum_k \frac{\partial^2 F_2}{\partial q_i\, \partial q_k} \frac{\partial q_k}{\partial v} \end{vmatrix}.$$

By the rules of determinant manipulation one may take the summation signs outside the determinant, and likewise remove any factors common to all terms in a column. With this authority one may write the determinant sum as

$$\sum_i \frac{\partial(q_i, p_i)}{\partial(u, v)} = \sum_{i,k} \frac{\partial^2 F_2}{\partial q_i\, \partial q_k} \begin{vmatrix} \dfrac{\partial q_i}{\partial u} & \dfrac{\partial q_k}{\partial u} \\ \dfrac{\partial q_i}{\partial v} & \dfrac{\partial q_k}{\partial v} \end{vmatrix} + \sum_{i,k} \frac{\partial^2 F_2}{\partial q_i\, \partial P_k} \begin{vmatrix} \dfrac{\partial q_i}{\partial u} & \dfrac{\partial P_k}{\partial u} \\ \dfrac{\partial q_i}{\partial v} & \dfrac{\partial P_k}{\partial v} \end{vmatrix}.$$

The terms of the first series are antisymmetric under exchange of indices i and k, since the two columns of the determinant are thereby interchanged. On the other hand the value of the summation cannot be affected by an exchange of dummy indices; hence the series must vanish identically. We are therefore at liberty to put in its place a similarly constructed series which also sums to zero:

$$\sum_i \frac{\partial(q_i, p_i)}{\partial(u, v)} = \sum_{i,k} \frac{\partial^2 F_2}{\partial P_i\, \partial P_k} \begin{vmatrix} \dfrac{\partial P_i}{\partial u} & \dfrac{\partial P_k}{\partial u} \\ \dfrac{\partial P_i}{\partial v} & \dfrac{\partial P_k}{\partial v} \end{vmatrix} + \sum_{i,k} \frac{\partial^2 F_2}{\partial q_i\, \partial P_k} \begin{vmatrix} \dfrac{\partial q_i}{\partial u} & \dfrac{\partial P_k}{\partial u} \\ \dfrac{\partial q_i}{\partial v} & \dfrac{\partial P_k}{\partial v} \end{vmatrix}.$$

The operation of expanding the determinant sum can now be reversed, except that it is the sum over i, and not over k, which is to be readmitted into the determinant. A typical element of the first column in the determinant occurring in the resulting series is

$$\sum_i \frac{\partial^2 F_2}{\partial P_i\, \partial P_k} \frac{\partial P_i}{\partial u} + \sum_i \frac{\partial^2 F_2}{\partial q_i\, \partial P_k} \frac{\partial q_i}{\partial u} = \frac{\partial}{\partial u} \frac{\partial F_2}{\partial P_k},$$

in analogy with Eq. (8–35). But the transformation equations (8–11b) state that

$$\frac{\partial F_2}{\partial P_k} = Q_k,$$

and the determinant sum must therefore reduce to

$$\sum_i \frac{\partial(q_i, p_i)}{\partial(u, v)} = \sum_k \begin{vmatrix} \frac{\partial Q_k}{\partial u} & \frac{\partial P_k}{\partial u} \\ \frac{\partial Q_k}{\partial v} & \frac{\partial P_k}{\partial v} \end{vmatrix} = \sum_k \frac{\partial(Q_k, P_k)}{\partial(u, v)},$$

which establishes the desired theorem.

In similar fashion, though the proof is more complicated, one can show that

$$J_2 = \iiiint\limits_S \sum dq_i\, dp_i\, dq_k\, dp_k \tag{8–36}$$

is an invariant under a canonical transformation, where S here is an arbitrary 4-dimensional surface of the $2n$-dimensional phase space. This chain of integral invariants can be extended further, reaching eventually to the invariant:

$$J_n = \int \cdots \int dq_1 \ldots dq_n dp_1 \ldots dp_n, \tag{8–37}$$

where now the integral can be extended over any arbitrary region in phase space. The invariance of J_n is equivalent to the statement that volume in phase space is invariant under canonical transformation. As we shall show later, it follows then that the volume in phase space is constant in time.

8–4 Lagrange and Poisson brackets as canonical invariants. The condition for the invariance of the sum of the Jacobian determinants, Eq. (8–34), can also be written as

$$\sum_i \left(\frac{\partial q_i}{\partial u}\frac{\partial p_i}{\partial v} - \frac{\partial p_i}{\partial u}\frac{\partial q_i}{\partial v}\right) = \sum_i \left(\frac{\partial Q_i}{\partial u}\frac{\partial P_i}{\partial v} - \frac{\partial P_i}{\partial u}\frac{\partial Q_i}{\partial v}\right). \tag{8–38}$$

Each side of this equation is in the form of a *Lagrange bracket* of u and v, defined as

$$\{u, v\}_{q,p} = \sum_i \left(\frac{\partial q_i}{\partial u}\frac{\partial p_i}{\partial v} - \frac{\partial p_i}{\partial u}\frac{\partial q_i}{\partial v}\right). \tag{8–39}$$

Eq. (8–38) states that the Lagrange brackets are invariant under contact transformations. It is therefore immaterial which set of canonical coordinates is used to express the Lagrange brackets, and the subscripts q, p will be omitted from now on. Note as a general relation that

$$\{u, v\} = -\{v, u\}. \tag{8–40}$$

It will be remembered that u, v are coordinates of a two-dimensional region of phase space, and we can take the q_i, q_j plane to be this region. In evaluating the corresponding Lagrange bracket, $\{q_i, q_j\}$, any set of canonical coordinates may be used, and it is obvious that the q, p set is the most convenient. The Lagrange bracket may therefore be expanded as:

$$\{q_i, q_j\} = \sum_k \left(\frac{\partial q_k}{\partial q_i}\frac{\partial p_k}{\partial q_j} - \frac{\partial q_k}{\partial q_j}\frac{\partial p_k}{\partial q_i}\right).$$

Since the q's and p's are independent coordinates, it follows that

$$\frac{\partial p_k}{\partial q_i} = 0 = \frac{\partial p_k}{\partial q_j},$$

and hence the Lagrange bracket vanishes:

$$\{q_i, q_j\} = 0. \tag{8–41a}$$

A parallel proof shows that

$$\{p_i, p_j\} = 0. \tag{8–41b}$$

However, if $u = q_i$ and $v = p_j$ the Lagrange bracket is

$$\{q_i, p_j\} = \sum_k \left(\frac{\partial q_k}{\partial q_i}\frac{\partial p_k}{\partial p_j} - \frac{\partial q_k}{\partial p_j}\frac{\partial p_k}{\partial q_i}\right),$$

and while the second term in parentheses vanishes by the same type of reasoning, the first term is no longer zero. The partial derivatives occurring in this term have the values

$$\frac{\partial p_k}{\partial p_j} = \delta_{kj}, \quad \frac{\partial q_k}{\partial q_i} = \delta_{ki},$$

and the Lagrange bracket consequently reduces to

$$\{q_i, p_j\} = \sum_k \delta_{jk}\,\delta_{ki} = \delta_{ij}. \tag{8–41c}$$

Equations (8–41) obviously are valid for all sets of canonical variables and are often referred to as the *fundamental Lagrange brackets.*

Of greater usefulness are the so-called *Poisson brackets*, which are defined as:

$$[u, v]_{q,p} = \sum_k \left(\frac{\partial u}{\partial q_k} \frac{\partial v}{\partial p_k} - \frac{\partial u}{\partial p_k} \frac{\partial v}{\partial q_k} \right), \tag{8–42}$$

with the identity:

$$[u, v] = -[v, u]. \tag{8–43}$$

There exists an intimate relation between the Lagrange and Poisson brackets. Considering them solely as mathematical expressions, without regard to any physical meaning, one can prove the following theorem: if u_l, $l = 1 \ldots 2n$, form a set of $2n$ independent functions, such that each u is a function of the $2n$ coordinates $q_1 \ldots q_n, p_1 \ldots p_n$, then

$$\sum_{l=1}^{2n} \{u_l, u_i\}[u_l, u_j] = \delta_{ij}.^* \tag{8–44}$$

The proof is straightforward, although rather clumsy. By Eqs. (8–39) and (8–42) the sum can be written

$$\sum_l^{2n} \{u_l, u_i\}[u_l, u_j] = \sum_l^{2n} \sum_k^n \sum_m^n \left(\frac{\partial q_k}{\partial u_l} \frac{\partial p_k}{\partial u_i} - \frac{\partial q_k}{\partial u_i} \frac{\partial p_k}{\partial u_l} \right) \left(\frac{\partial u_l}{\partial q_m} \frac{\partial u_j}{\partial p_m} - \frac{\partial u_j}{\partial q_m} \frac{\partial u_l}{\partial p_m} \right).$$

There will be four terms in the expansion, the first of which is

$$\sum_{k,m}^n \frac{\partial p_k}{\partial u_i} \frac{\partial u_j}{\partial p_m} \sum_l^{2n} \frac{\partial q_k}{\partial u_l} \frac{\partial u_l}{\partial q_m} = \sum_{k,m} \frac{\partial p_k}{\partial u_i} \frac{\partial u_j}{\partial p_m} \frac{\partial q_k}{\partial q_m} = \sum_{k,m} \frac{\partial p_k}{\partial u_i} \frac{\partial u_j}{\partial p_m} \delta_{km},$$

or finally,

$$= \sum_k^n \frac{\partial u_j}{\partial p_k} \frac{\partial p_k}{\partial u_i}. \tag{8–45}$$

The last of the four terms has exactly the same form as Eq. (8–45), except that the roles of q_k and p_k are interchanged. Consequently the two terms together form

$$\sum_k^n \left(\frac{\partial u_j}{\partial p_k} \frac{\partial p_k}{\partial u_i} + \frac{\partial u_j}{\partial q_k} \frac{\partial q_k}{\partial u_i} \right) = \frac{\partial u_j}{\partial u_i}. \tag{8–46}$$

---

* The Poisson brackets, as defined in (8–42), vaguely resemble reciprocals of the Lagrange brackets. Eq. (8–44) gives a precise meaning to this intuitive feeling. If $\{u_l, u_i\}$ be considered as a matrix element L_{il}, and $-[u_l, u_j]$ as the matrix element P_{lj}, where the matrices are $2n$ square, then Eq. (8–44) reads

$$\mathsf{LP} = 1,$$

or

$$\mathsf{P} = \mathsf{L}^{-1}.$$

As can easily be seen, the remaining two terms of the expansion do not contribute to the sum. One of them contains a factor involving u_l, given by

$$\sum_l \frac{\partial q_k}{\partial u_l}\frac{\partial u_l}{\partial p_m} = \frac{\partial q_k}{\partial p_m},$$

which must always vanish, while the other merely has p and q interchanged. Hence Eq. (8-46) is the entire sum:

$$\sum_l^{2n}\{u_l, u_i\}[u_l, u_j] = \frac{\partial u_j}{\partial u_i} = \delta_{ij},$$

which completes the proof.

Note that the particular coordinate system q, p involved in this derivation was entirely immaterial; any set of $2n$ independent variables Q, P would have served equally well. Thus Eq. (8-44) is invariant under *all* transformations of the coordinates, without restriction even to canonical coordinates. The theorem can therefore be used to evaluate certain Poisson brackets without stipulating a particular coordinate set. For the $2n$ independent functions u_l let us choose the set $q_1 \ldots q_n$, $p_1 \ldots p_n$, and further take u_i to be q_i and u_j to be p_j. As u_i is a coordinate and u_j a momentum, the two functions can never be identical for any value of i or j, and Eq. (8-44) becomes

$$\sum_l^n \{p_l, q_i\}[p_l, p_j] + \sum_l^n \{q_l, q_i\}[q_l, p_j] = 0.$$

Since

$$\{p_l, q_i\} = -\delta_{il} \quad \text{and} \quad \{q_l, q_i\} = 0$$

for all *canonical* coordinates, it follows that

$$[p_i, p_j] = 0, \tag{8-47a}$$

which must likewise hold for all canonical coordinates. The Poisson bracket of q_i and q_j similarly vanishes:

$$[q_i, q_j] = 0. \tag{8-47b}$$

Finally, if $u_i = q_i$ and $u_j = q_j$ the summation (8-44) takes the form

$$\sum_l^n \{q_l, q_i\}[q_l, q_j] + \sum_l^n \{p_l, q_i\}[p_l, q_j] = \delta_{ij},$$

which reduces to

$$-\sum_l \delta_{il}[p_l, q_j] = \delta_{ij},$$

or

$$[q_i, p_j] = \delta_{ij}. \tag{8-47c}$$

Eqs. (8–47) constitute the *fundamental Poisson bracket* relations in analogy to the corresponding equations for the Lagrange brackets, Eqs. (8–41). It might be objected that these relations could have been derived with less labor by direct evaluation, as was done for the Lagrange brackets. But the whole point of the calculation has been to evaluate the fundamental Poisson brackets *without involving any particular set of canonical variables.* The advantage of the proof is that it demonstrates that Eqs. (8–47) are canonical *invariants.*

The fundamental Poisson brackets are basic building stones; with their aid it can be shown that the values of all Poisson brackets are independent of the set of canonical coordinates they are expressed in. If F and G are two arbitrary functions, their Poisson bracket with respect to the q, p set is

$$[F, G]_{q,p} = \sum_j \left(\frac{\partial F}{\partial q_j} \frac{\partial G}{\partial p_j} - \frac{\partial F}{\partial p_j} \frac{\partial G}{\partial q_j} \right). \tag{8–48}$$

Considering q_j and p_j to be functions of the transformed set of variables Q_k, P_k, Eq. (8–48) can be written as

$$[F, G]_{q,p} = \sum_{j,k} \left[\frac{\partial F}{\partial q_j} \left(\frac{\partial G}{\partial Q_k} \frac{\partial Q_k}{\partial p_j} + \frac{\partial G}{\partial P_k} \frac{\partial P_k}{\partial p_j} \right) - \frac{\partial F}{\partial p_j} \left(\frac{\partial G}{\partial Q_k} \frac{\partial Q_k}{\partial q_j} + \frac{\partial G}{\partial P_k} \frac{\partial P_k}{\partial q_j} \right) \right],$$

which, after some rearrangement of terms, reduces to

$$[F, G]_{q,p} = \sum_k \left(\frac{\partial G}{\partial Q_k} [F, Q_k]_{q,p} + \frac{\partial G}{\partial P_k} [F, P_k]_{q,p} \right). \tag{8–49}$$

Eq. (8–49) can itself be used to evaluate the Poisson brackets which occur within the parentheses. If we substitute Q_k for F, and replace G by the symbol F, then Eq. (8–49) becomes:

$$[Q_k, F]_{q,p} = \sum_j \frac{\partial F}{\partial Q_j} [Q_k, Q_j] + \sum_j \frac{\partial F}{\partial P_j} [Q_k, P_j]. \tag{8–50}$$

The subscripts have been omitted from the Poisson brackets on the right since these are the fundamental Poisson brackets, which have already been shown to be invariant. In consequence of (8–47), Eq. (8–50) reduces to

$$[Q_k, F]_{q,p} = \sum_j \frac{\partial F}{\partial P_j} \delta_{jk},$$

or

$$[F, Q_k] = - \frac{\partial F}{\partial P_k}, \tag{8–51}$$

which is itself a useful, canonically invariant result. In a similar fashion the other Poisson bracket is found from the equation

$$[P_k, F]_{q,p} = \sum_j \frac{\partial F}{\partial Q_j}[P_k, Q_j] + \sum_j \frac{\partial F}{\partial P_j}[P_k, P_j],$$

resulting in

$$[F, P_k] = \frac{\partial F}{\partial Q_k}. \tag{8-52}$$

Substituting the relations (8-51) and (8-52), Eq. (8-49) becomes

$$[F, G]_{q,p} = \sum_k \left(\frac{\partial F}{\partial Q_k}\frac{\partial G}{\partial P_k} - \frac{\partial F}{\partial P_k}\frac{\partial G}{\partial Q_k}\right) = [F, G]_{Q,P}.$$

Accordingly, the subscripts on all Poisson brackets will be omitted from here on.

Some simple algebraic properties of Poisson brackets may be listed here for future reference. It is obvious from the definition (8-42) that the Poisson bracket of a function with itself is identically zero:

$$[u, u] = 0. \tag{8-53}$$

Likewise, the Poisson bracket of u with a quantity c not depending on either q or p must vanish:

$$[u, c] = 0. \tag{8-54}$$

From the elementary properties of differentiation it follows that

$$[u + v, w] = [u, w] + [v, w], \tag{8-55}$$

and

$$[u, vw] = [u, v]w + v[u, w].^* \tag{8-56}$$

8-5 The equations of motion in Poisson bracket notation. If the function F in Eqs. (8-51) and (8-52) be chosen as the Hamiltonian then the equations read

$$[q_i, H] = \frac{\partial H}{\partial p_i} = \dot{q}_i, \tag{8-57a}$$

---

* Overstepping the bounds of classical physics for a moment, it may be stated that the Poisson bracket of classical mechanics corresponds in quantum mechanics to $\frac{2\pi}{ih}$ times the *commutator* of the two quantities:

$$[u, v] \rightarrow \frac{2\pi}{ih}(uv - vu),$$

where h is Planck's constant. It will be readily verified that the algebraic relations (8-53 to 8-56) hold equally well in terms of commutators.

and

$$[p_i, H] = -\frac{\partial H}{\partial q_i} = \dot{p}_i. \tag{8-57b}$$

These two relations constitute the canonical equations of motion as written in terms of Poisson brackets; they are special cases of a general formula for the total time derivative of a function $u(q, p, t)$. Mathematically the total time derivative is defined by

$$\frac{du}{dt} = \sum_i \left(\frac{\partial u}{\partial q_i} \dot{q}_i + \frac{\partial u}{\partial p_i} \dot{p}_i \right) + \frac{\partial u}{\partial t}.$$

Expressing $\dot{q}_i$ and $\dot{p}_i$ in terms of the Hamiltonian by means of the equations of motion, we can write

$$\frac{du}{dt} = \sum_i \left(\frac{\partial u}{\partial q_i} \frac{\partial H}{\partial p_i} - \frac{\partial u}{\partial p_i} \frac{\partial H}{\partial q_i} \right) + \frac{\partial u}{\partial t},$$

or

$$\frac{du}{dt} = [u, H] + \frac{\partial u}{\partial t}. \tag{8-58}$$

Eqs. (8–57) obviously follow from this relation if u is taken as q_i and p_i respectively. Another familiar result is obtained if u is the Hamiltonian itself, for from (8–53) we then have

$$\frac{dH}{dt} = \frac{\partial H}{\partial t},$$

which is identical with Eq. (7–19), obtained previously.

For systems in which t does not occur explicitly in the quantities of interest (and we shall confine future discussions to only such systems) the total time derivative is simply the Poisson bracket with H. All functions whose Poisson brackets with the Hamiltonian vanish will therefore be constants of the motion, and conversely the Poisson brackets of all constants of the motion with H must be zero. We thus have a general test for seeking and identifying the constants of the system.

Once any two constants of the motion are known it is possible to construct other constants by means of *Jacobi's identity*, which states that if u, v, and w are three functions of q and p then

$$[u, [v, w]] + [v, [w, u]] + [w, [u, v]] = 0. \tag{8-59}$$

The proof of this relationship is based on an examination of the form of the first two terms on the left in Eq. (8–59):

$$[u, [v, w]] - [v, [u, w]]. \tag{8-60}$$

It will be shown that these terms together contain no second derivative of w. The Poisson bracket $[v, w]$ can be considered as a linear differential operator D_v acting on w, where D_v has the general form

$$D_v = \sum_{k=1}^{n}\left(\frac{\partial v}{\partial q_k}\frac{\partial}{\partial p_k} - \frac{\partial v}{\partial p_k}\frac{\partial}{\partial q_k}\right),$$

which can also be represented as

$$D_v = \sum_{i=1}^{2n}\alpha_i\frac{\partial}{\partial \xi_i}.$$

In terms of the differential operators so defined the expression (8-60) can be written as

$$D_uD_vw - D_vD_uw = \sum_{i,k}\beta_i\frac{\partial}{\partial \eta_i}\left(\alpha_k\frac{\partial w}{\partial \xi_k}\right) - \alpha_k\frac{\partial}{\partial \xi_k}\left(\beta_i\frac{\partial w}{\partial \eta_i}\right).$$

Now, the only terms in this expression involving the second derivatives of w are

$$\sum_{i,k}\left(\beta_i\alpha_k\frac{\partial^2 w}{\partial \eta_i\,\partial \xi_k} - \alpha_k\beta_i\frac{\partial^2 w}{\partial \xi_k\,\partial \eta_i}\right),$$

which vanish identically. Hence Eq. (8-60) can involve only first derivatives of w, and must have the form

$$[u, [v, w]] - [v, [u, w]] = \sum_k\left(A_k\frac{\partial w}{\partial p_k} + B_k\frac{\partial w}{\partial q_k}\right), \tag{8-61}$$

where A_k and B_k are functions of u and v yet to be determined, but which do not involve w. If we consider the special case where $w = p_i$ (which does not change A or B) then (8-61) reduces to

$$[u, [v, p_i]] - [v, [u, p_i]] = A_i,$$

or, from (8-52):

$$\left[u, \frac{\partial v}{\partial q_i}\right] + \left[\frac{\partial u}{\partial q_i}, v\right] = A_i,$$

so that finally

$$A_i = \frac{\partial}{\partial q_i}[u, v].$$

Similarly, by taking $w = q_i$ one finds

$$B_i = -\frac{\partial[u, v]}{\partial p_i},$$

and Eq. (8-61) can be written as

$$\begin{aligned}[u, [v, w]] + [v, [w, u]] &= \sum_k\left(\frac{\partial[u, v]}{\partial q_k}\frac{\partial w}{\partial p_k} - \frac{\partial[u, v]}{\partial p_k}\frac{\partial w}{\partial q_k}\right),\\ &= [[u, v], w],\end{aligned}$$

which is identical with Jacobi's identity, Eq. (8–59). If u and v are two constants of the motion and w is taken to be the Hamiltonian the first two terms in the identity vanish, and the relation reduces to

$$[H, [u, v]] = 0.$$

Hence the Poisson bracket of two constants of the motion is itself a constant of the motion.* It is sometimes possible to construct an entire sequence of constants of motion in this manner. All too often, however, the constants of motion so obtained are found to be trivial functions of the old, and are therefore of little help.

8–6 Infinitesimal contact transformations, constants of the motion, and symmetry properties. Further properties of the Poisson bracket are revealed by introducing the concept of *infinitesimal contact transformations.* As in the case of infinitesimal rotations such a transformation is one in which the new coordinates differ from the old only by infinitesimals. Only first-order terms in these infinitesimals are to be retained in all calculations. The transformation equations can then be written in the form,

$$Q_i = q_i + \delta q_i, \tag{8–62a}$$
$$P_i = p_i + \delta p_i, \tag{8–62b}$$

where δq_i and δp_i do *not* represent virtual displacements but are simply the infinitesimal changes in coordinates and momenta. Clearly the generating function will differ only by an infinitesimal amount from that for the identity transformation, which is given by Eq. (8–18). One can therefore write the generating function as

$$F_2 = \sum_i q_i P_i + \epsilon G(q, P), \tag{8–63}$$

where ϵ is some infinitesimal parameter of the transformation. The transformation equations for the new momenta P_i are found from Eqs. (8–11a):

$$\frac{\partial F_2}{\partial q_i} = p_i = P_i + \epsilon \frac{\partial G}{\partial q_i},$$

or

$$P_i - p_i = \delta p_i = -\epsilon \frac{\partial G}{\partial q_i}. \tag{8–64a}$$

* This result is sometimes referred to as Poisson's theorem.

Similarly, the transformation equations for Q_i are obtained from Eqs. (8-11b):

$$Q_i = \frac{\partial F_2}{\partial P_i} = q_i + \epsilon \frac{\partial G}{\partial P_i}.$$

Since the second term is already linear in ϵ, and P differs from p only by an infinitesimal, it will be correct to first order to replace P_i by p_i in the derivative. G is then considered as a function of q, p only. Consequently, the transformation equation for Q_i becomes

$$\delta q_i = \epsilon \frac{\partial G}{\partial p_i}. \tag{8-64b}$$

Although strictly speaking the term "generating function" is restricted to the quantity F, it is also customary to designate G in this manner, a practice which we shall follow.

An interesting application of these results is obtained if we consider an infinitesimal canonical transformation in which $G = H(q, p)$ and ϵ is a small time interval dt. The corresponding infinitesimal changes in the coordinates and momenta are then

$$\delta q_i = dt \frac{\partial H}{\partial p_i} = \dot{q}_i \, dt = dq_i, \tag{8-65a}$$

and

$$\delta p_i = -dt \frac{\partial H}{\partial q_i} = \dot{p}_i \, dt = dp_i. \tag{8-65b}$$

These equations state that the transformation changes the coordinates and momenta at the time t to the values they have at the time $t + dt$. Thus the motion of the system in a time interval dt can be described by an infinitesimal contact transformation generated by the Hamiltonian. Correspondingly, the system motion in a finite time interval from t_0 to t is represented by a succession of infinitesimal contact transformations. Since the result of two canonical transformations applied one after the other is equivalent to a single canonical transformation, the values of q and p at any time t can be obtained from their initial values by a canonical transformation which is a continuous function of time. According to this view the motion of a mechanical system corresponds to the continuous evolution or unfolding of a canonical transformation. In a very literal sense, the *Hamiltonian is the generator of the system motion with time.*

Conversely, there must exist a canonical transformation from the values of the coordinates and momenta at any time t to their constant initial values. Obtaining such a transformation is obviously equivalent to solving the problem of the system motion. At the beginning of the chapter it was pointed out that a mechanical problem could be reduced to finding the canonical transformation for which all momenta are constants of the motion. The present considerations indicate the possibility of an alternative solution by means of the canonical transformation for which *both* the momenta and coordinates are constants of the motion. These two suggestions will be elaborated in the next chapter in order to show how formal solutions may be obtained for any mechanical problem.

An important connection between infinitesimal contact transformations and Poisson brackets is revealed by considering the change in some function $u(q, p)$ as a result of the transformation. It is necessary to explain carefully what is meant by the "change" in the function. In the previous discussions a quantity $u(q, p)$ has been "transformed" to new coordinates by considering the arguments q and p as functions of the new Q and P, $q(Q, P)$ and $p(Q, P)$, substituting these in u to obtain its dependence on the new variables. Following the change in variables the *functional* dependence of u upon Q and P will not in general be the same as the previous dependence upon q and p. On the other hand, the numerical value for a given system configuration is unaffected by the transformation. The function $u(q, p)$ is a "point" function in phase space, and its value at a given point of phase space is clearly unchanged by any transformation of the coordinate axes of the space. The "change" to be discussed now is of quite a different nature. Here we mean the numerical change in u as the result of substituting Q for q and P for p everywhere in the function. In such a change the dependence of u upon the new and old variables remains the same. In effect the transformation has shifted the point in phase space at which the function is to be evaluated. Thus, if the infinitesimal canonical transformation is generated by the Hamiltonian the result of substituting the new variables for the old is to change u from its value at time t to the value it has at a time dt later.

In accordance with this meaning the change in a function u as the result of an infinitesimal canonical transformation is

$$\delta u = u(q_i + \delta q_i, p_i + \delta p_i) - u(q_i, p_i).$$

To first order of the infinitesimal parameters, a Taylor's series expansion shows the difference to be

$$\delta u = \sum_i \left(\frac{\partial u}{\partial q_i} \delta q_i + \frac{\partial u}{\partial p_i} \delta p_i \right).$$

From the equations of transformation (8–64) δu can also be written as

$$\delta u = \epsilon \sum_i \left(\frac{\partial u}{\partial q_i} \frac{\partial G}{\partial p_i} - \frac{\partial u}{\partial p_i} \frac{\partial G}{\partial q_i} \right),$$

or finally as

$$\delta u = \epsilon[u, G]. \tag{8–66}$$

Consequently the change in the Hamiltonian under an infinitesimal canonical transformation is

$$\delta H = \epsilon[H, G]. \tag{8–67}$$

It has already been shown that if a function $G(q, p)$ is a constant of the motion its Poisson bracket with H vanishes. Eq. (8–67) therefore states that such a constant generates an infinitesimal canonical transformation which does not change the value of the Hamiltonian. Equivalently, *the constants of the motion are the generating functions of those infinitesimal canonical transformations which leave the Hamiltonian invariant.* Now, the symmetry properties of the system dictate which are transformations that do not change the value of H. Clearly, if the physical system is symmetrical under a given operation the Hamiltonian must remain unaffected under the corresponding transformation. One can therefore determine all the constants of the motion (which is tantamount to solving the problem) by an examination of the symmetry properties of the Hamiltonian! This is not the first instance of a connection between constants of the motion and symmetry characteristics. We encountered it previously (Section 2–6) in dealing with the conservation of generalized momenta. Here, however, the statement is truly more elegant, and also more complete, for it embraces all constants of motion and not merely the conserved generalized momenta.

The momentum conservation theorems appear now as a special case of the general statement. If a coordinate q_i is cyclic the Hamiltonian will be independent of q_i, and will certainly be invariant under an infinitesimal contact transformation which involves a displacement in q_i alone. The equations of transformation would then be of the form

$$\begin{aligned} \delta q_j &= \epsilon\, \delta_{ij}, \\ \delta p_j &= 0, \end{aligned} \tag{8–68}$$

where ϵ is the infinitesimal displacement of q_i. By Eqs. (8–64) it is seen that the only generating function which produces such a transformation is

$$G = p_i, \tag{8–69}$$

namely, the momentum conjugate to q_i. We readily recognize this as the familiar momentum conservation theorem: if a coordinate is cyclic its conjugate momentum is a constant of the motion.

As a specific illustration of these concepts consider the infinitesimal contact transformation of the dynamical variables which produces a rotation of the system as a whole by an angle $d\theta$. The physical significance of the corresponding generating function cannot depend upon the choice of initial canonical coordinates, and it is convenient to use for this purpose the cartesian coordinates of all particles in the system. Nor will there be any loss in generality if the axes are so oriented that the infinitesimal rotation is along the z-axis. The change in the particle coordinates as each particle is rotated by an angle $d\theta$ is equivalent to the change produced if the system were held stationary and the coordinate axes rotated by an angle $-d\theta$. To first order in $d\theta$ the resultant new coordinates (cf. Eq. 4–90) are:

$$\begin{aligned} X_i &= x_i - y_i\, d\theta, \\ Y_i &= y_i + x_i\, d\theta, \\ Z_i &= z_i. \end{aligned}$$

The infinitesimal change in the coordinates is then

$$\delta x_i = -y_i\, d\theta, \qquad \delta y_i = x_i\, d\theta, \qquad \delta z_i = 0. \tag{8–70}$$

There are similar equations for the momentum components, since they transform under rotations in the same manner as do the position components. By comparison of Eqs. (8–70) with the transformation equations (8–64) the corresponding generating function is

$$G = \sum_i (x_i p_{iy} - y_i p_{ix}), \tag{8–71}$$

with $d\theta$ as the infinitesimal parameter ϵ. For a direct check note that

$$\begin{aligned} \delta x_i &= d\theta \frac{\partial G}{\partial p_{ix}} = -y_i\, d\theta, &\qquad \delta p_{ix} &= -d\theta \frac{\partial G}{\partial x_i} = -p_{iy}\, d\theta, \\ \delta y_i &= d\theta \frac{\partial G}{\partial p_{iy}} = x_i\, d\theta, &\qquad \delta p_{iy} &= -d\theta \frac{\partial G}{\partial y_i} = p_{ix}\, d\theta, \end{aligned}$$

agreeing with Eqs. (8–70). The generating function (8–71) in addition has the physical significance of being the z-component of the total angular momentum:

$$G = L_z.$$

Since the z-axis was arbitrarily chosen, one can state that the generating function corresponding to an infinitesimal rotation about an axis denoted by the unit vector $\mathbf{n}$ is

$$G = \mathbf{L} \cdot \mathbf{n}. \tag{8–72}$$

Just as the Hamiltonian is the generator of a displacement of the system in time so the angular momentum is the generator of the rotational motion of the system.

The same observation, of course, follows directly and perhaps more concisely from the general result given in Eq. (8–69). If one of the canonical coordinates is chosen as an angle denoting rotation of the system as a whole, then the corresponding canonical momentum has been shown (Section 2–6) to be the component of the angular momentum along the axis of rotation. Thus Eq. (8–72) becomes a special case of Eq. (8–69).

8–7 The angular momentum Poisson bracket relations. The identification of angular momentum as the rotation generator leads to a number of very interesting and important Poisson bracket relations. According to Eq. (8–66) the change of a vector function $\mathbf{F}(q, p)$ under an infinitesimal rotation of the system is given by

$$\delta\mathbf{F} = d\theta[\mathbf{F}, \mathbf{L} \cdot \mathbf{n}]. \tag{8–73}$$

The special meaning used here for "change of a function" must be constantly borne in mind. Eq. (8–73), of course, stands for three scalar equations. For example, if $A(q, p)$ is the x-component of $\mathbf{F}$ then Eq. (8–73) states that

$$\delta A = A(Q, P) - A(q, p) = d\theta[A, \mathbf{L} \cdot \mathbf{n}], \tag{8–74}$$

with similar expressions for the y- and z-components which may be denoted as $B(q, p)$ and $C(q, p)$ respectively. It was noted in the previous section that the transformation of a scalar function $u(q, p)$ *at a point in phase space* under a coordinate transformation corresponds to an entirely different process. The value of the function remains the same but its functional dependence on the coordinates will in general be changed. In dealing with a vector function under a transformation corresponding to rotation there is a still greater difference. Not only does the functional dependence of the component functions change because of the transformation of their arguments, but the value of the components themselves changes in accordance with the rotation properties of vectors. Thus for an infinitesimal rotation about the z-axis the initial components of $\mathbf{F}$ are related to the new components by the equations

$$\begin{aligned} A(q, p) &= A'(Q, P) + B'(Q, P)\, d\theta, \\ B(q, p) &= B'(Q, P) - A'(Q, P)\, d\theta, \\ C(q, p) &= C'(Q, P). \end{aligned} \tag{8–75}$$

Here the primes on the transformed functions indicate that they are different functions of the new arguments.

We have gone into some detail in explaining these distinctions so that it can be clearly understood under what conditions a connection does exist between the ordinary transformation of a vector under rotation and the "change" involved in Eq. (8–74). It may happen that the vector function **F** has the property that the functional dependence of the old and new components on their respective coordinates is exactly the same, i.e., $A'(Q, P)$ can be written $A(Q, P)$, and similarly for the other components. In such case the arguments and the components transform together so that the functional dependence remains unchanged. For example, the x-component of the angular momentum is

$$L_x = \sum_i (y_i p_{iz} - z_i p_{iy}),$$

and after rotation, the X-component of **L** is

$$L_X = \sum_i (Y_i P_{iZ} - Z_i P_{iY}),$$

which is the same function of Q and P that L_x was of q and p. With vector functions possessing this property, *and only for such vectors,* the transformation equations (8–75) become

$$\begin{aligned} A(q, p) &= A(Q, P) + B(Q, P)\, d\theta, \\ B(q, p) &= B(Q, P) - A(Q, P)\, d\theta, \\ C(q, p) &= C(Q, P). \end{aligned}$$

Further, the term $B(Q, P)\, d\theta$ can be written to first order infinitesimals as $B(q, p)\, d\theta$. Hence the change (as defined in (8–74)) in the vector components under a small rotation about the z-axis would be

$$\begin{aligned} A(Q, P) - A(q, p) &= \delta A = -B\, d\theta, \\ B(Q, P) - B(q, p) &= \delta B = A\, d\theta, \end{aligned}$$

and

$$C(Q, P) - C(q, p) = \delta C = 0.$$

These equations agree exactly with the change in the components of a fixed vector under a z rotation of the coordinate axes by an angle $-d\theta$, Eq. (4–94), which in this case has the form

$$d\mathbf{F} = \mathbf{k}\, d\theta \times \mathbf{F} = \delta\mathbf{F}.$$

For a general infinitesimal rotation about an arbitrary axis the change $\delta\mathbf{F}$ is therefore

$$\delta\mathbf{F} = \mathbf{n}\, d\theta \times \mathbf{F}, \tag{8–76}$$

and by (8–73) it follows that

$$[\mathbf{F}, \mathbf{L} \cdot \mathbf{n}] = \mathbf{n} \times \mathbf{F}. \tag{8–77}$$

Although we have been careful to restrict Eq. (8–77) to a particular class of vector functions it should be noted that most vectors encountered in mechanical problems satisfy the necessary conditions. Any vector function of **r** and **p** which does not involve a fixed vector independent of the system will meet the requirements. By using dyadic notation, Eq. (8–77) can be stated in a more general form as

$$[\mathbf{F}, \mathbf{L}] = -\mathbf{1} \times \mathbf{F}, \tag{8–78}$$

where $\mathbf{1}$ is the unit dyadic $\mathbf{ii} + \mathbf{jj} + \mathbf{kk}$. Eq. (8–78) readily reduces to (8–77) by taking the dot product of both sides with **n**. The best known application of this relationship is obtained by setting **F** equal to **L** itself:

$$[\mathbf{L}, \mathbf{L} \cdot \mathbf{n}] = \mathbf{n} \times \mathbf{L},$$

or

$$[\mathbf{L}, \mathbf{L}] = -\mathbf{1} \times \mathbf{L}. \tag{8–79}$$

A special case of Eq. (8–79) is the Poisson bracket of L_x with L_y, given by

$$[L_x, L_y] = (\mathbf{j} \times \mathbf{L})_x = L_z.$$

The nonvanishing scalar components of Eq. (8–79) can therefore be written as

$$[L_i, L_j] = L_k, \qquad i, j, k \text{ in cyclic order.} \tag{8–80}$$

A number of interesting consequences follow from Eqs. (8–79) and (8–80). If L_x and L_y are constants of the motion, so that their Poisson brackets with H vanish, it follows then from Poisson's theorem that $L_z = [L_x, L_y]$ is also a constant of the motion. If any two components of the angular momentum are constant, the total angular momentum vector is conserved. Of greater consequence is the relation:

$$[L^2, \mathbf{L} \cdot \mathbf{n}] = 0. \tag{8–81}$$

To prove this lemma note that the left-hand side can be written as

$$[\mathbf{L} \cdot \mathbf{L}, \mathbf{L} \cdot \mathbf{n}] = 2\mathbf{L} \cdot [\mathbf{L}, \mathbf{L} \cdot \mathbf{n}],$$

which, by Eq. (8–79) is

$$2\mathbf{L} \cdot (\mathbf{n} \times \mathbf{L}) = 0.$$

Indeed the same argument holds for any vector which satisfies Eq. (8–77) and we can state as a general theorem that

$$[F^2, \mathbf{L} \cdot \mathbf{n}] = 0. \tag{8–82}$$

It will be remembered from Eqs. (8–41b) that the Poisson bracket of any two canonical momenta must always be zero. But, from (8–80), L_i

does not have a vanishing Poisson bracket with any of the other components of **L**. Thus, if one of the components of the angular momentum along a fixed direction is taken as a canonical momentum, the two perpendicular components cannot simultaneously be canonical momenta! In contrast, from Eq. (8–81), the magnitude of **L**, and any of its components, can be simultaneous canonical momenta.*

8–8 Liouville's theorem. As a final application of Poisson brackets we shall briefly discuss a fundamental theorem of statistical mechanics known as Liouville's theorem. While the exact motion of any system is completely determined in classical mechanics by the initial conditions, it is often impracticable to calculate an exact solution for complex systems. It would be obviously hopeless, for example, to calculate completely the motion of some 10^{23} molecules in a volume of gas. In addition, the initial conditions are often only incompletely known. We may be able to state that at t_0 a given mass of gas has a certain energy, but we cannot determine the initial coordinates and velocities of each molecule. Statistical mechanics therefore makes no attempt to obtain a complete solution for systems containing many particles. Its aim, instead, is to make predictions about certain average properties by examining the motion of a large number of identical systems. The values of the desired quantities are then computed by forming averages over all the systems in this *ensemble*. All the members of the ensemble are as like the actual system as our imperfect knowledge permits, but they may have any of the initial conditions that are consistent with this incomplete information. Since each system is

---

* It has been remarked previously that the correspondence between quantum and classical mechanics is such that the Poisson bracket goes over essentially into the quantum-mechanical commutator. Much of the formal structure of quantum mechanics appears as a close transcript of the Poisson bracket formulation of classical mechanics, reading the symbol [ ] everywhere as "commutator" (except for a multiplicative constant). All the results of this section have close quantum analogs. For example, the fact that two components of **L** cannot be simultaneous canonical momenta appears as the well-known statement that L_i and L_j cannot have simultaneous eigenvalues. But L^2 and any L_i can be quantized together. Indeed, most of these relations are known far better in their quantum form than as classical theorems. Thus one of the earliest references to the classical Poisson brackets for angular momentum appears to be the 1930 treatise by Born and Jordan on *Elementare Quantenmechanik*. Again, while the general change of a vector function under rotation, Eq. (8–78), has long been used in quantum mechanics (cf. Condon and Shortley, *The Theory of Atomic Spectra*, p. 59), so far as I know it was first proved for classical mechanics in recent years, by Professor J. Schwinger.

represented by a single point in phase space the ensemble of systems corresponds to a swarm of points in phase space. Liouville's theorem states that the density of systems in the neighborhood of some given system in phase space remains constant in time.

The density, D, as defined above can vary with time through two separate mechanisms. Since it is the density in the neighborhood of a given system point, there will be an *implicit* dependence as the coordinates of the system (q_i, p_i) vary with time, and the system point wanders through phase space. There may also be an explicit dependence on time. The density may still vary with time even when evaluated at a fixed point in phase space. By Eq. (8–58) the total time derivative of D, due to both types of variation with time, can be written as

$$\frac{dD}{dt} = [D, H] + \frac{\partial D}{\partial t}, \tag{8–83}$$

where the Poisson bracket arises from the implicit dependence, and the last term from the explicit dependence.

Consider an infinitesimal volume in phase space surrounding the given system point, with the boundary of the volume formed by some surface of neighboring system points at the time $t = 0$. In the course of time the system points defining the volume move about in phase space and the volume will take on different shapes as time progresses. The dotted curve in Fig. 8–1 schematically indicates the evolution of the infinitesimal volume with time. It is clear that the number of systems within the volume remains constant, for a system initially inside can never get out. If some system point were to cross the border it would occupy at some time the same position in phase space as one of the system points defining the boundary surface. Since the subsequent motion of a system is uniquely determined by its location in phase space at a particular time, the two systems would travel together from there on. Hence the system can never leave the volume. By the same token, a system initially outside can never enter the volume.

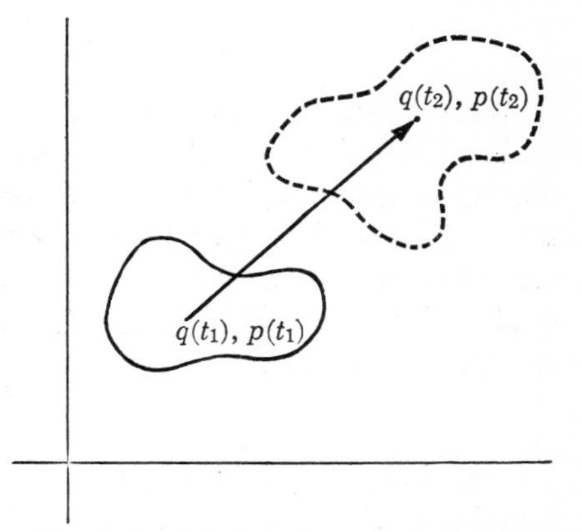

FIG. 8–1. Motion of a volume in phase space.

It has been shown that the motion of a system point in time is simply a particular contact transformation of the canonical coordinates in phase space. Consequently the change with time of the infinitesimal region under discussion can be represented by a contact transformation. Now, one of Poincaré's integral invariants asserts that the volume of a region of phase space remains invariant under contact transformations. The size of the volume therefore cannot vary with time.

Thus, both the number of systems in the infinitesimal region, dN, and the volume, dV, are constants, and consequently the density

$$D = \frac{dN}{dV}$$

must also be constant in time, i.e.,

$$\frac{dD}{dt} = 0,$$

which proves Liouville's theorem. An alternative statement of the theorem follows from Eq. (8–83) as

$$\frac{\partial D}{\partial t} = -[D, H]. \tag{8–84}$$

When the ensemble of systems is in statistical equilibrium, the number of systems in a given state must be constant in time, which is to say that the density of system points at a given spot in phase space does not change with time. The variation of D with time at a fixed point corresponds to the partial derivative with respect to t, which therefore must vanish in statistical equilibrium. By Eq. (8–84) it follows that the equilibrium condition can be expressed as

$$[D, H] = 0.$$

We can insure equilibrium, therefore, by choosing the density D to be a function of the constants of the motion of the system, for then the Poisson bracket with H must vanish. Thus, for conservative systems D can be any function of the energy, and the equilibrium condition is automatically satisfied. The characteristics of the ensemble will be determined by the choice of function for D. As an example, one well-known ensemble, the *microcanonical* ensemble, occurs if D is constant for systems having a given energy, zero otherwise.

These considerations have been presented here to illustrate the usefulness of the Poisson bracket formulation in statistical mechanics. Further discussion of these points would carry us far outside our field.

SUGGESTED REFERENCES

L. Nordheim and E. Fues, *Die Hamilton-Jacobische Theorie der Dynamik*, in Vol. V of the *Handbuch der Physik*. While no single reference covers all the material of this chapter, the article of Nordheim and Fues comes the closest of any. The article deals chiefly with canonical transformations and the Poisson bracket formulation and it is undoubtedly the best available exposition of these subjects. Despite the title, Hamilton-Jacobi theory (cf. Chapter 9) per se enters only in the last few sections.

E. T. Whittaker, *Analytical Dynamics*. Much of the material given in the Nordheim and Fues article is also discussed by Whittaker in Chapters IX and X, more from the viewpoint of the mathematician, and it is interesting to contrast the two treatments. It should be noted that only those transformations are discussed for which the generating function does not involve time explicitly.

M. Born, *The Mechanics of the Atom*. The subject of the canonical transformations of classical mechanics played an important role in the first formulations of both the older Bohr quantum theory and the newer quantum mechanics. Hence many treatises ostensibly devoted to one or the other of these forms of quantum mechanics often contain detailed expositions of the needed branches of classical mechanics. Outstanding among them is this 1924 volume of Born, written before the days of wave mechanics. The first chapter succinctly discusses canonical transformations and gives many interesting physical illustrations. There is no mention of Poisson brackets for they become of special interest to the modern physicist only with the advent of Heisenberg's and Dirac's formulation of quantum mechanics.

M. Born and P. Jordan, *Elementare Quantenmechanik*. In the preface to his 1924 book Born admitted the shortcomings of the then existent quantum theory, and stated that the difficulties could probably be overcome only by radical revision of fundamental principles (a situation very similar to the present day feeling of frustration in the theory of nuclear forces). To this revision of quantum mechanics, when and if it came, he promised to devote a future volume. Born's prediction of subsequent developments proved remarkably accurate, and in 1929 he and Jordan were able to publish the "future volume." As in the previous treatise it was found necessary to devote some space to classical mechanics, but this time it was the Poisson bracket formulation that formed the most interesting aspect of mechanics. Appendix III is concerned with this topic and contains, among other items, the Poisson bracket relations for angular momentum.

A. Sommerfeld, *Atomic Structure and Spectral Lines*. This classic treatise on the older quantum mechanics contains much of interest on Hamilton's equations and canonical transformations. The material will be found scattered in the chapter on the hydrogen atom and in some of the appendices.

R. C. Tolman, *The Principles of Statistical Mechanics*. A veritable encyclopedia of theoretical physics, Chapter II of this bulky volume gives a brief but clear discussion of canonical transformations and similar topics in classical mechanics. The properties of Poisson brackets are included in the treatment. Section 19, Chapter III, is concerned with Liouville's theorem.

C. CARATHÉODORY, *Variationsrechnung*. It is not possible here to go into the wider mathematical implications of canonical transformations, which are especially important in the theory of partial differential equations. The interested reader will find Carathéodory's volume an admirable introduction to this aspect of the subject, with a wealth of material on canonical and contact transformations, and on the various types of bracket expressions. A somewhat shorter discussion will be found in the chapter on calculus of variations by the same author in Vol. I of Frank and Von Mises', *Die Differential- und Integralgleichungen der Mechanik und Physik.*

P. A. M. DIRAC, *The Principles of Quantum Mechanics.* The standard reference for the application of Poisson brackets to quantum mechanics is still Dirac's treatise. Unfortunately the book has acquired an almost legendary reputation for being difficult to understand. This is not so for the later editions, and a student with some previous acquaintance with the physical basis of quantum mechanics should be able to master the material. The pertinent data for this chapter will be found in Sections 25 to 30.

EXERCISES

1. Show directly that the transformation

$$Q = \log\left(\frac{1}{q}\sin p\right), \qquad P = q\cot p$$

is canonical.

2. In deriving the transformation equations in terms of the generating functions F_2, F_3, and F_4, Legendre transformations were used merely to suggest the connections between the generating functions. Show that Eqs. (8–11, 14, and 17) may be derived directly from Eqs. (8–9) in consequence of the properties of the Legendre transformation as given in Chapter 7.

3. The transformation equations between two sets of coordinates are

$$Q = \log(1 + q^{\frac{1}{2}}\cos p),$$
$$P = 2(1 + q^{\frac{1}{2}}\cos p)q^{\frac{1}{2}}\sin p.$$

(a) Show directly from these transformation equations that Q, P are canonical variables if q and p are. (b) Show that the function which generates this transformation is

$$F_3 = -(e^Q - 1)^2\tan p.$$

4. For what values of α and β do the equations

$$Q = q^\alpha\cos\beta p, \qquad P = q^\alpha\sin\beta p$$

represent a canonical transformation? What is the form of the generating function F_3 for this case?

5. A particle of mass m moves in a potential field that is cylindrically symmetric about the z-axis. Obtain the generating function for a canonical transformation to coordinates rotating about the z-axis with a constant frequency ω. What is the physical significance of the new Hamiltonian? Compare with the result obtained in Exercise 4, Chapter 7. Derive the new canonical equations of motion and identify physically each of the terms occurring in the equations.

6. Show that the equations of transformation in which t is considered a canonical variable reduce to the ordinary equations (8–11) if the transformation does not affect the time scale.

7. If the canonical variables are not all independent, but are connected by auxiliary conditions of the form

$$\psi_k(q_i, p_i, t) = 0,$$

show that the canonical equations of motion can be written

$$\frac{\partial H}{\partial p_i} + \sum_k \lambda_k \frac{\partial \psi_k}{\partial p_i} = \dot{q}_i, \qquad \frac{\partial H}{\partial q_i} + \sum_k \lambda_k \frac{\partial \psi_k}{\partial q_i} = -\dot{p}_i,$$

where the λ_k are the undetermined Lagrange multipliers. The formulation of the Hamiltonian equations in which t is a canonical variable is a case in point, since a relation exists between p_{n+1} and the other canonical variables:

$$H(q_1 \cdots q_{n+1}, p_1 \cdots p_n) + p_{n+1} = 0.$$

Further, it will be seen from the appearance of the modified Hamilton's principle that the "Hamiltonian" of the $2n + 2$ variables is always zero (cf. Exercise 6, Chapter 7). Show that as a result of these circumstances the $2n + 2$ Hamilton's equations of this formulation can be reduced to the $2n$ ordinary Hamilton's equations plus Eq. (7–19) and the relation

$$\lambda = \frac{dt}{d\theta}.$$

Note that while these results are reminiscent of the relativistic covariant Hamiltonian formulation, they have been arrived at entirely within the framework of nonrelativistic mechanics.

8. Show directly by substitution in Eq. (8–8) that $\sum_i q_i Q_i$ generates the exchange transformation. Show that $F_4 = \sum_i p_i P_i$ also generates an exchange of coordinates and momenta, and that $F_3 = -\sum_i Q_i p_i$ generates the identity transformation.

9. Show that it is possible to rearrange the elements of the $2n \times 2n$ functional determinant of a canonical transformation

$$D = \frac{\partial(Q_k, P_k)}{\partial(q_i, p_i)}$$

in such a manner that the elements of the determinant D^2 are the fundamental Lagrange brackets, and prove from this result that $D^2 = 1$. (In fact it can be seen from the Poincaré integral invariant J_n that D is always $+1$.)

10. From the equations of transformation (8–9, 11, 14, 17) prove the following relations for a canonical transformation:

$$\frac{\partial q_i}{\partial Q_k} = \frac{\partial P_k}{\partial p_i}, \qquad \frac{\partial q_i}{\partial P_k} = -\frac{\partial Q_k}{\partial p_i}, \qquad \frac{\partial p_i}{\partial Q_k} = -\frac{\partial P_k}{\partial q_i}, \qquad \frac{\partial p_i}{\partial P_k} = \frac{\partial Q_k}{\partial q_i}.$$

Hence show, for example, that

$$[q_i, p_j] = \{q_j, p_i\},$$

which could be used for a separate proof of the invariance of the fundamental Poisson brackets. As another consequence show that the determinant of the inverse transformation $(Q, P) \longrightarrow (q, p)$, denoted by D^{-1}, is equal to D, which constitutes another proof that $D^2 = 1$.

11. A class of operations is said to have the *group-property* if (1) it contains the identity operator, (2) the inverse to each operator is also a member of the class, and (3) the product of any two operators also belongs to the class. Show that the canonical transformations for a system of n degrees of freedom has the group-property.

12. In the text the invariance of the fundamental Poisson brackets was shown to be a necessary condition that the transformation be canonical. It can also be shown to be a sufficient condition for demonstrating the canonical nature of a transformation. For the particularly simple case that the transformation equations do not involve the time explicitly demonstrate directly that Q, P satisfy equations of the canonical form provided that q, p are canonical variables and that

$$[Q_i, Q_j]_{q,p} = 0 = [P_i, P_j]_{q,p}, \qquad [Q_i, P_j]_{q,p} = \delta_{ij}.$$

The easiest method of proof is to compute $\frac{dQ}{dt}$ and $\frac{dP}{dt}$ in terms of the old canonical variables.

13. Prove that the Poisson bracket of two constants of the motion is itself a constant of the motion even when the constants depend on time explicitly.

14. (a) Show that if the Hamiltonian and a quantity F are constants of the motion then $\partial F/\partial t$ must also be a constant.

(b) As an illustration of this result consider the uniform motion of a free particle of mass m. The Hamiltonian is certainly conserved, and there exists a constant of the motion

$$F = x - \frac{pt}{m}.$$

Show by direct computation that the constant of the motion $\frac{\partial F}{\partial t}$ agrees with $[H, F]$.

15. Set up the problem of the spherical pendulum in the Hamiltonian formulation, using spherical polar coordinates for the q_i. Evaluate directly in terms of these canonical variables the following Poisson brackets:

$$[L_x, L_y], \qquad [L_y, L_z], \qquad [L_z, L_x],$$

showing that they have the values predicted by Eq. (8–80). Why is it that p_θ and p_ψ can be used as canonical momenta, although they are perpendicular components of the angular momentum?

CHAPTER 9

HAMILTON–JACOBI THEORY

It has already been mentioned that canonical transformations may be used to provide a general procedure for solving mechanical problems. Two methods have been suggested. If the Hamiltonian is conserved then a solution could be obtained by transforming to new canonical coordinates which are all cyclic, since the integration of the new equations of motion becomes trivial. An alternative technique is to seek a canonical transformation from the coordinates and momenta, (q, p), at the time t, to a new set of constant quantities which may be the $2n$ initial values, (q_0, p_0), at $t = 0$. With such a transformation, the equations of transformation relating the old and new canonical variables are then exactly the desired solution of the mechanical problem:

$$q = q(q_0, p_0, t)$$
$$p = p(q_0, p_0, t),$$

for they give the coordinates and momenta as a function of their initial values and the time. This last procedure is the more general one, especially as it is applicable, in principle at least, even when the Hamiltonian involves the time. We shall therefore begin our discussion by considering how such a transformation may be found.

9–1 The Hamilton-Jacobi equation for Hamilton's principal function. We can automatically insure that the new variables are constant in time by requiring that the transformed Hamiltonian, K, shall be identically zero, for then the equations of motion are

$$\frac{\partial K}{\partial P_i} = \dot{Q}_i = 0$$
$$-\frac{\partial K}{\partial Q_i} = \dot{P}_i = 0. \tag{9–1}$$

K is related to the old Hamiltonian and to the generating function by the equation

$$K = H + \frac{\partial F}{\partial t},$$

and hence will be zero if F satisfies the equation

$$H(q, p, t) + \frac{\partial F}{\partial t} = 0. \tag{9–2}$$

It is convenient to take F as a function of the old coordinates q_i, the new constant momenta P_i and the time; in the notation of the previous chapter we would designate the generating function as $F_2(q, P, t)$. To write the Hamiltonian in Eq. (9–2) as a function of the same variables, use may be made of the equations of transformation (cf. Eq. (8–11a)):

$$p_i = \frac{\partial F_2}{\partial q_i},$$

so that Eq. (9–2) becomes

$$H\left(q_1, \ldots q_n, \frac{\partial F_2}{\partial q_1}, \ldots \frac{\partial F_2}{\partial q_n}, t\right) + \frac{\partial F_2}{\partial t} = 0. \tag{9–3}$$

Eq. (9–3), known as the *Hamilton-Jacobi equation*, constitutes a partial-differential equation in $(n + 1)$ variables, $q_1 \ldots q_n, t$, for the desired generating function. It is customary to denote the solution of Eq. (9–3) by S, and to call it *Hamilton's principal function.*

Of course, the integration of Eq. (9–3) only provides the dependence on the old coordinates and time; it would not appear to tell how the new momenta are contained in S. Indeed the new momenta have not yet been specified except that we know they must be constants. However, the nature of the solution indicates how the new P_i's are to be selected.

Mathematically Eq. (9–3) has the form of a first-order partial differential equation in $n + 1$ variables. Consequently a complete solution must involve $n + 1$ independent constants of integration: $\alpha_1 \ldots \alpha_n, \alpha_{n+1}$. It will be noted however that S itself does not appear in Eq. (9–3), only its partial derivatives with respect to q or t are involved. If S is some solution of the differential equation then $S + \alpha$, where α is anv constant, must also be a solution, for the additive constant does not affect the values of the partial derivatives. One of the $n + 1$ constants of integration must therefore be an additive constant tacked on to S. But by the same token this constant has no importance as far as the transformation is concerned, since only partial derivatives of S occur in the transformation equations. Hence a complete solution of Eq. (9–3) can be written in the form

$$S = S(q_1 \ldots q_n, \alpha_1 \ldots \alpha_n, t), \tag{9–4}$$

where none of the n constants is solely additive. It will be seen that this mathematical form tallies exactly with the previously established physical description of the generating function, inasmuch as Eq. (9–4) states that S is a function of n coordinates q_i, the time t, and n independent constants α_i. We are therefore at liberty to take the n constants of integration to be the new (constant) momenta:

$$P_i = \alpha_i. \tag{9–5}$$

Such a choice does not contradict the original assertion that the new momenta are connected with the initial values of q and p at the time t_0. The n transformation Eqs. (8–11a) can now be written as

$$p_i = \frac{\partial S(q_i, \alpha_i, t)}{\partial q_i}. \tag{9–6}$$

At the time t_0 these constitute n equations relating the n α's with the initial q and p values, thus enabling one to evaluate the constants of integration in terms of the specific initial conditions of the problem. The other half of the equations of transformation, which provide the new constant coordinates, appear as

$$Q_i = \beta_i = \frac{\partial S(q_i, \alpha_i, t)}{\partial \alpha_i}. \tag{9–7}$$

The constant β's can be similarly obtained from the initial conditions, simply by calculating the value of the right side of Eq. (9–7) at $t = t_0$ with the known initial values of q_i. Eqs. (9–7) can then be "turned inside out" to furnish q in terms of α_i, β_i, and t:

$$q = q(\alpha_i, \beta_i, t), \tag{9–8}$$

which solves the problem by giving the coordinates as functions of time and the initial conditions.*

Hamilton's principal function is thus the generator of a contact transformation to constant coordinates and momenta; *when solving the Hamilton-Jacobi equation we are at the same time obtaining a solution to the mechanical problem.* Mathematically speaking, we have established an equivalence between the $2n$ canonical equations of motion, which are first-order dif-

---

* As a mathematical point it may be questioned whether the process of "turning inside out" is feasible for Eqs. (9–6) and (9–7), i.e., whether they can be solved for α_i and q_i respectively. The question hinges on whether the equations in each set are independent, for otherwise they are obviously not sufficient to determine the n independent quantities α_i or q_i as the case may be. That the derivatives $\frac{\partial S}{\partial \alpha_i}$ in (9–7) form independent functions of the q's follows directly from the nature of the Hamilton-Jacobi equation, indeed this is what we mean by saying the n constants of integration are independent. Consequently the Jacobian of $\frac{\partial S}{\partial \alpha_i}$ with respect to q_j cannot vanish. Since the order of differentiation is immaterial this is equivalent to saying that the Jacobian of $\frac{\partial S}{\partial q_j}$ with respect to α_i cannot vanish, which proves the independence of Eqs. (9–6).

ferential equations, to the first-order partial differential Hamilton-Jacobi equation. This correspondence is not restricted to equations governed by the Hamiltonian, indeed the general theory of first-order partial differential equations is largely concerned with the properties of the equivalent set of first-order ordinary differential equations. Essentially, the connection can be traced to the fact that both the partial differential equation and its canonical equations stem from a common variational principle, in this case Hamilton's modified principle.

To a certain extent the choice of the α_i's as the new momenta is arbitrary. One could just as well choose any n quantities, γ_i, which are independent functions of the α_i constants of integration:

$$\gamma_i = \gamma_i(\alpha_1 \ldots \alpha_n). \tag{9–9}$$

By means of these defining relations Hamilton's principal function can be written as a function of q_i, γ_i, and t, and the rest of the derivation then goes through unchanged. It often proves convenient to take some particular set of γ_i's as the new momenta, rather than the constants of integration that appear naturally in integrating the Hamilton-Jacobi equation.

Further insight into the physical significance of S is furnished by an examination of its total time derivative, which can be computed from the formula

$$\frac{dS}{dt} = \sum_i \frac{\partial S}{\partial q_i}\dot{q}_i + \frac{\partial S}{\partial t},$$

since the P_i's are constant in time. By Eqs. (9–6) and (9–3) this relation can also be written

$$\frac{dS}{dt} = \sum_i p_i\dot{q}_i - H = L, \tag{9–10}$$

so that Hamilton's principal function differs at most from the indefinite time integral of the Lagrangian only by a constant:

$$S = \int L\,dt + \text{constant}. \tag{9–11}$$

Now, Hamilton's principle is a statement about the definite integral of L, and from it we obtained the solution of the problem via the Lagrange equations. Here the same integral, in an indefinite form, furnishes another way of solving the problem. In actual calculations the result expressed by Eq. (9–11) is of no help, because one cannot integrate the Lagrangian with respect to time until q_i and p_i are known as functions of time, i.e., until the problem is solved.*

9–2 The harmonic oscillator problem as an example of the Hamilton-Jacobi method. To illustrate the Hamilton-Jacobi technique for solving the motion of mechanical systems we shall work out in detail the simple problem of a one-dimensional harmonic oscillator. The Hamiltonian is

$$H = \frac{p^2}{2m} + \frac{kq^2}{2},$$

where k is the spring constant. One obtains the Hamilton-Jacobi equation for S by setting p equal to $\frac{\partial S}{\partial q}$ and substituting in the Hamiltonian; the requirement that the new Hamiltonian vanishes becomes

$$\frac{1}{2m}\left(\frac{\partial S}{\partial q}\right)^2 + \frac{kq^2}{2} + \frac{\partial S}{\partial t} = 0. \tag{9–12}$$

Since the explicit dependence of S on t is involved only in the last term, a solution for Eq. (9–12) can be found in the form

$$S(q, \alpha, t) = W(q, \alpha) - \alpha t, \tag{9–13}$$

where α is a constant of integration (to be designated later as the transformed momentum). With this choice of solution the time can be eliminated from Eq. (9–12), which reduces to

$$\frac{1}{2m}\left(\frac{\partial W}{\partial q}\right)^2 + \frac{kq^2}{2} = \alpha. \tag{9–14}$$

Eq. (9–14) can be integrated immediately to

$$W = \sqrt{mk}\int dq\,\sqrt{\frac{2\alpha}{k} - q^2},$$

so that

$$S = \sqrt{mk}\int dq\,\sqrt{\frac{2\alpha}{k} - q^2} - \alpha t. \tag{9–15}$$

---

* Historically the recognition by Hamilton that the time integral of L is a special solution of a partial differential equation came before it was seen how the Hamilton-Jacobi equation can furnish the solution to a mechanical problem. It was Jacobi who realized that the converse was true, that by the techniques of canonical transformations any complete solution of the Hamilton-Jacobi equation could be used to describe the motion of the system.

While the integration involved in Eq. (9–15) is not particularly difficult there is no reason to carry it out at this stage, for what is desired is not S but its partial derivatives. The solution for q arises out of the transformation equation (9–7).

$$\beta = \frac{\partial S}{\partial \alpha} \equiv \sqrt{\frac{m}{k}} \int \frac{dq}{\sqrt{\frac{2\alpha}{k} - q^2}} - t,$$

which can be integrated without trouble as

$$t + \beta = -\sqrt{\frac{m}{k}} \arccos q \sqrt{\frac{k}{2\alpha}}. \tag{9–16}$$

Writing ω for $\sqrt{\frac{k}{m}}$, Eq. (9–16) can be "turned inside out" to furnish q as a function of t and the two constants of integration, α and β:

$$q = \sqrt{\frac{2\alpha}{k}} \cos \omega(t + \beta), \qquad \omega = \sqrt{\frac{k}{m}} \tag{9–17}$$

which is the familiar solution for a harmonic oscillator. To complete the story, the constants α and β must be connected with the initial conditions. Suppose at time $t = 0$, the particle is initially stationary, $p_0 = 0$, but is displaced from the equilibrium position by an amount q_0. The straightforward procedure for finding α is to evaluate Eq. (9–12) at $t = 0$:

$$\left(\frac{\partial S}{\partial q}\right)_0 = p_0 = 0 = \sqrt{2m} \sqrt{\alpha - \frac{kq_0^2}{2}},$$

which yields

$$\alpha = \frac{kq_0^2}{2} = \frac{m\omega^2 q_0^2}{2}. \tag{9–18}$$

The constant α is therefore the initial total energy of the system. Since the forces are conservative the energy must at all times be equal to α. Indeed, the identity of α with the total energy could have been recognized directly from Eq. (9–13) and the relation

$$\frac{\partial S}{\partial t} + H = 0,$$

which gives immediately:

$$H = \alpha.$$

With Eq. (9–18) for α in terms of q_0, the solution for q, Eq. (9–17), reduces to

$$q = q_0 \cos \omega(t + \beta),$$

which shows that β is zero under the given initial conditions. Thus, Hamilton's principal function is the generator of a contact transformation to a canonical momentum identified as the total energy, and to a coordinate which (for the particular initial conditions assumed) vanishes identically.

In terms of Eq. (9–18) Hamilton's principal function can be written as

$$S = m\omega \int \sqrt{q_0^2 - q^2}\, dq - \frac{m\omega^2 q_0^2 t}{2},$$

or, upon substituting Eq. (9–17),* as:

$$= m\omega^2 q_0^2 \int \left(\sin^2 \omega t - \frac{1}{2}\right) dt.$$

Now, the Lagrangian is

$$\begin{aligned} L &= \frac{m\dot{q}^2}{2} - \frac{m\omega^2 q^2}{2} \\ &= \frac{m\omega^2 q_0^2}{2} (\sin^2 \omega t - \cos^2 \omega t) \\ &= m\omega^2 q_0^2 \left(\sin^2 \omega t - \frac{1}{2}\right), \end{aligned}$$

so that S is the time integral of the Lagrangian, in agreement with the general relation (9–11). Note that the identity could not be proved until *after* the solution to the problem had been obtained.

9–3 The Hamilton-Jacobi equation for Hamilton's characteristic function. It was possible to integrate the Hamilton-Jacobi equation for the simple harmonic oscillator primarily because S could be separated into two parts, one involving q only and the other only time. Such a separation of variables is always possible *whenever the old Hamiltonian does not involve time explicitly.*

If H is not an explicit function of t then the Hamilton-Jacobi equation for S becomes

$$\frac{\partial S}{\partial t} + H\left(q_i, \frac{\partial S}{\partial q_i}\right) = 0.$$

---

* In doing so one must take the negative square root of $q_0^2 - q^2$, for it will be recognized that $\sqrt{q_0^2 - q^2} = \frac{p}{m\omega} = \frac{\dot{q}}{\omega}$, which by Eq. (9–17) is $-q_0 \sin \omega t$.

The first term involves only the dependence on t, whereas the second is concerned only with the dependence of S on the q_i. The time variable can therefore be separated by assuming a solution for S of the form

$$S(q_i, \alpha_i, t) = W(q_i, \alpha_i) - \alpha_1 t. \tag{9–19}$$

Upon substituting this trial solution, the differential equation reduces to the expression

$$H\left(q_i, \frac{\partial W}{\partial q_i}\right) = \alpha_1, \tag{9–20}$$

which no longer involves the time. One of the constants of integration appearing in S, namely α_1, is thus equal to the constant value of H. (Normally H will be the energy, but it should be remembered that this need not always be the case, cf. Exercise 4, Chapter 7.)

The time-independent function W appears here merely as a part of the generating function S when H is constant. It can also be shown that W separately generates its own contact transformation with properties quite different from that generated by S. Consider a canonical transformation in which the new momenta are all constants of the motion α_i, and where α_1 in particular is the constant of motion H. If the generating function for this transformation be denoted by $W(q, P)$ then the equations of transformation are

$$p_i = \frac{\partial W}{\partial q_i}, \qquad Q_i = \frac{\partial W}{\partial P_i} = \frac{\partial W}{\partial \alpha_i}. \tag{9–21}$$

While these equations resemble Eqs. (9–6, 7) for Hamilton's principal function S, the condition now determining W is that H shall be equal to the new momentum α_1:

$$H(q_i, p_i) = \alpha_1.$$

Using Eqs. (9–21) this requirement becomes a partial differential equation for W:

$$H\left(q_i, \frac{\partial W}{\partial q_i}\right) = \alpha_1,$$

which is seen to be identical with Eq. (9–20). Since W does not involve the time, the new and old Hamiltonians are equal, and it follows that $K = \alpha_1$.

The function W, *known as Hamilton's characteristic function*, thus generates a canonical transformation in which all the new coordinates are cyclic. It was noted in the previous chapter that when H is a constant of the motion a transformation of this nature in effect solves the mechanical problem involved, for the integration of the new equations of motion is then trivial. The canonical equations for $\dot{P}_i$, in fact, merely repeat the statement that the momenta conjugate to the cyclic coordinates are all constant:

$$\dot{P}_i = -\frac{\partial K}{\partial Q_i} = 0, \qquad P_i = \alpha_i. \tag{9-22a}$$

As the new Hamiltonian depends on only one of the momenta α_i the equations of motion for $\dot{Q}_i$ are

$$\begin{aligned} \dot{Q}_i = \frac{\partial K}{\partial \alpha_i} &= 1 \qquad i = 1, \\ &= 0 \qquad i \neq 1, \end{aligned}$$

with the immediate solutions

$$\begin{aligned} Q_1 &= t + \beta_1 \equiv \frac{\partial W}{\partial \alpha_1}, \\ Q_i &= \quad\;\; \beta_i \equiv \frac{\partial W}{\partial \alpha_i} \qquad i \neq 1. \end{aligned} \tag{9-22b}$$

The only coordinate which is not simply a constant of the motion is Q_1, which is equal to the time plus a constant. We have here another instance of the conjugate relationship between the time as a coordinate and the Hamiltonian as its conjugate momentum.

The dependence of W on the old coordinates q_i is determined by the partial differential equation (9-20), which, like Eq. (9-3), is also referred to as the Hamilton-Jacobi equation. There will now be n constants of integration, but again one of them must be merely an additive constant. The $n - 1$ remaining independent constants, $\alpha_2 \ldots \alpha_n$, together with α_1 may then be taken as the new constant momenta. When evaluated at t_0 the first half of Eqs. (9-21) serve to relate the n constants α_i with the initial values of q_i and p_i. Finally, Eqs. (9-22b) can be solved for the q_i as a function of α_i, β_i, and the time t, thus completing the solution of the problem. It will be noted that $(n - 1)$ of the Eqs. (9-22b) do not involve the time at all. One of the q_i's can be chosen as an independent variable, and the remaining coordinates can then be expressed in terms of it by solving only these time-independent equations. We are thus led directly to the *orbit equations* of the motion. In central force motion, for example, this technique would furnish r as a function of θ, without the need for separately finding r and θ as functions of time.

It is not necessary always to take α_1 and the constants of integration in W as the new constant momenta. Occasionally it is desirable rather to use some particular set of n independent functions of the α_i's as the transformed momenta. Designating these constants by γ_i the characteristic function W can then be expressed in terms of q_i and γ_i as the independent variables. The Hamiltonian will in general depend upon more than one of the γ_i's and the equations of motion for $\dot{Q}_i$ become

$$\dot{Q}_i = \frac{\partial K}{\partial \gamma_i} = \nu_i,$$

where the ν_i's are functions of γ_i. In this case all the new coordinates are linear functions of the time:

$$Q_i = \nu_i t + \beta_i. \tag{9–22'}$$

The characteristic function W possesses a physical significance similar to that for S. As W does not involve time explicitly, its total time derivative is

$$\frac{dW}{dt} = \sum_i \frac{\partial W}{\partial q_i} \dot{q}_i = \sum_i p_i \dot{q}_i,$$

and hence

$$W = \int \sum_i p_i \dot{q}_i \, dt = \int \sum_i p_i \, dq_i.$$

The above integrals will be recognized as defining the action A, as used in Section 7–5. Again, this information is of little practical help; the form of W cannot be found a priori without obtaining a complete integral of the Hamilton-Jacobi equation. The procedures involved in solving a mechanical problem by either Hamilton's principal or characteristic function may now be summarized in the following tabular form:

The two methods of solution are applicable when the Hamiltonian

is any general function of q, p, t: $H(q, p, t)$.	is conserved: $H(q, p) = \text{constant}$.

We seek canonical transformations to new variables such that

all the coordinates and momenta Q_i, P_i are constants of the motion.	all the momenta P_i are constants.

To meet these requirements it is sufficient to demand that the new Hamiltonian

shall vanish identically: $K = 0$.	shall be cyclic in all the coordinates: $K = H(P_i) = \alpha_1$.

Under these conditions the new equations of motion become

$$\dot{Q}_i = \frac{\partial K}{\partial P_i} = 0, \qquad \dot{P}_i = -\frac{\partial K}{\partial Q_i} = 0,$$

$$\dot{Q}_i = \frac{\partial K}{\partial P_i} = \nu_i, \qquad \dot{P}_i = -\frac{\partial K}{\partial Q_i} = 0,$$

with the immediate solutions:

$$Q_i = \beta_i, \qquad P_i = \gamma_i,$$

$$Q_i = \nu_i t + \beta_i \qquad P_i = \gamma_i$$

which satisfy the stipulated requirements.

The generating function producing the desired transformation is Hamilton's

Principal Function:

$$S(q, P, t),$$

Characteristic Function:

$$W(q, P),$$

satisfying the Hamilton-Jacobi partial differential equation:

$$H\left(q, \frac{\partial S}{\partial q}, t\right) + \frac{\partial S}{\partial t} = 0.$$

$$H\left(q, \frac{\partial W}{\partial q}\right) - \alpha_1 = 0.$$

A complete solution to the equation contains

n nontrivial constants of integration $\alpha_1 \ldots \alpha_n$.

$n - 1$ nontrivial constants of integration which together with α_1 form a set of n independent constants $\alpha_1 \ldots \alpha_n$.

The new constant momenta, $P_i = \gamma_i$, can be chosen as any n independent functions of the n constants of integration:

$$P_i = \gamma_i(\alpha_1 \ldots \alpha_n), \qquad P_i = \gamma_i(\alpha_1 \ldots \alpha_n),$$

so that the complete solutions to the Hamilton-Jacobi equation may be considered as functions of the new momenta:

$$S = S(q_i, \gamma_i, t). \qquad W = W(q_i, \gamma_i).$$

In particular, the γ_i's may be chosen to be the α_i's themselves. One-half of the transformation equations:

$$p_i = \frac{\partial S}{\partial q_i}, \qquad p_i = \frac{\partial W}{\partial q_i},$$

are fulfilled automatically, since they have been used in constructing the Hamilton-Jacobi equation. The other half,

$$Q_i = \frac{\partial S}{\partial \gamma_i} = \beta_i, \qquad\qquad Q_i = \frac{\partial W}{\partial \gamma_i} = \nu_i(\gamma_j)t + \beta_i,$$

can be solved for q_i in terms of t and the $2n$ constants β_i, γ_i. The solution to the problem is then completed by evaluating these $2n$ constants in terms of the initial values, (q_{i0}, p_{i0}), of the coordinates and momenta.

When the Hamiltonian does not involve time explicitly, both methods are suitable, and the generating functions are then related to each other according to the formula

$$S(q, P, t) = W(q, P) - \alpha_1 t.$$

9–4 Separation of variables in the Hamilton-Jacobi equation. It might appear from the preceding section that little practical advantage has been gained through the introduction of the Hamilton-Jacobi procedure. Instead of solving the $2n$ ordinary differential equations which make up the canonical equations of motion, one now must solve the partial differential Hamilton-Jacobi equation, and partial differential equations are notoriously complicated to solve. Under certain conditions, however, it is possible to separate the variables in the Hamilton-Jacobi equation, and the solution can then always be reduced to quadratures. In practice the Hamilton-Jacobi technique becomes a useful computational tool only when such a separation can be effected.

The subsequent discussion will be restricted to systems for which the Hamiltonian is one of the constants of motion, though it may not necessarily be the total energy. It will therefore be sufficient to consider only the contact transformation generated by W and its corresponding Hamilton-Jacobi differential equation. The variables q_i occurring in this equation are said to be separable if a solution of the form

$$W = \sum_i W_i(q_i, \alpha_1 \ldots \alpha_n)$$

splits the Hamilton-Jacobi equation into n equations of the form

$$H_i\left(q_i, \frac{\partial W_i}{\partial q_i}, \alpha_1 \ldots \alpha_n\right) = \alpha_i. \tag{9–23}$$

Each of the Eqs. (9–23) involves only one of the coordinates q_i and the corresponding partial derivative of W_i with respect to q_i. They are therefore a set of ordinary differential equations of a particularly simple form. Since the equations are only of first order, it is always possible to reduce them to quadratures; one has only to solve for $\frac{\partial W_i}{\partial q_i}$ and then integrate with respect to q_i.

No simple criterion can be given to indicate when the Hamilton-Jacobi equation is separable.* For some problems it is not at all possible to perform the separation of variables, the famous three-body problem being one illustration. Fortunately, the systems of interest in current atomic physics are almost always separable. Certainly, those problems of classical mechanics which are soluble in closed form have separable Hamilton-Jacobi equations. It must be emphasized that the question of whether or not the Hamilton-Jacobi equation is separable depends upon the system of generalized coordinates employed. Thus, the one-body central force problem is separable in polar coordinates, but not in cartesian coordinates. In many cases there will be more than one set of coordinates for which separation is feasible.

A partial separation of variables has already been used in reducing the Hamilton-Jacobi equation for S when H is not an explicit function of time. In effect, a solution for S was sought in the form

$$S(q_i, \alpha_i, t) = W(q_i, \alpha_i) + S_2(t, \alpha_i).$$

With this trial solution the Hamilton-Jacobi equation becomes

$$H\left(q_i, \frac{\partial W}{\partial q_i}\right) + \frac{\partial S_2}{\partial t} = 0.$$

Since the first term involves only the q_i's while the second involves only the time, the equation can hold for all values of the variables only if the two terms are equal and opposite constants:

$$\frac{\partial S_2}{\partial t} = -\alpha_1, \tag{9–24a}$$

$$H\left(q_i, \frac{\partial W}{\partial q_i}\right) = \alpha_1. \tag{9–24b}$$

The first equation yields $S_2 = -\alpha_1 t$, as in Eq. (9–13), while the second is the Hamilton-Jacobi equation for W.

A similar separation of variables in Hamilton's characteristic function is feasible whenever all but one of the generalized coordinates are cyclic.

* More complete discussions of the separable types of Hamilton-Jacobi equations will be found in the Nordheim-Fues article in Vol. V of the *Handbuch der Physik*, in Chapter 2, Section 5, Vol. 2 of *Differential Gleichungen der Physik* by Frank and Von Mises, and in the references cited therein.

Suppose q_1 is the only noncyclic coordinate. We seek a solution for W of the form

$$W = \sum_i W_i(q_i, P_i).$$

Since the momenta conjugate to the cyclic coordinates are constants, the transformation equations for $i \neq 1$ can then be written as

$$\frac{\partial W_i}{\partial q_i} = p_i = \alpha_i \qquad i \neq 1. \tag{9–25a}$$

The Hamilton-Jacobi equation in these circumstances reduces to

$$H\left(q_1, \frac{\partial W_1}{\partial q_1}, \alpha_2 \ldots \alpha_n\right) = \alpha_1, \tag{9–25b}$$

which is an ordinary first-order differential equation for W_1 and hence is immediately soluble. Eqs. (9–25a) and (9–25b) together completely define Hamilton's characteristic function W. The integration of Eqs. (9–25a) is trivial and leads to the results

$$W_i = \alpha_i q_i \qquad i \neq 1,$$

so that W can be written simply as

$$W = W_1 + \sum_{i=2}^{n} \alpha_i q_i. \tag{9–26}$$

There is an obvious resemblance between Eq. (9–26) and the form S assumes when H is not an explicit function of time, Eq. (9–13). Indeed, both equations can be considered as arising under similar circumstances. We have seen that t may be considered as a generalized coordinate with $-H$ as its canonical momentum. If H is conserved, then t may be treated as a cyclic coordinate, and Eq. (9–24a) is then merely a particular one of the Eqs. (9–25a) valid for all cyclic coordinates.*

---

* The form of (9–26) may also be arrived at by the following considerations. It must be remembered that W is to be the generating function for a transformation to new coordinates which are all cyclic. But if $q_2 \ldots q_n$ are already cyclic, no further transformation is needed for them. As far as they are concerned, W can be the identity transformation. Since the α_i's are the new momenta, the summation in (9–26) can be written as

$$\sum_{i=2}^{n} P_i q_i$$

which will be recognized as the generator of the identity transformation (cf. Eq. (8–19)) for the coordinates $q_2 \ldots q_n$.

As an example of solving the Hamilton-Jacobi equation by separation of the variables we may consider the motion of a particle in a plane under the action of a central force. The Hamiltonian has the form

$$H = \frac{1}{2m}\left(p_r^2 + \frac{p_\phi^2}{r^2}\right) + V(r),$$

and is cyclic in ϕ. Consequently Hamilton's characteristic function appears as

$$W = W_1(r) + \alpha_\phi \phi, \tag{9-27}$$

where α_ϕ is the constant angular momentum p_ϕ conjugate to ϕ. The Hamilton-Jacobi equation then becomes

$$\frac{1}{2m}\left[\left(\frac{\partial W_1}{\partial r}\right)^2 + \frac{\alpha_\phi^2}{r^2}\right] + V(r) = \alpha_1, \tag{9-28}$$

where α_1 is the constant identified physically as the total energy of the system. Solving Eq. (9-28) for the partial derivative of W_1 we obtain

$$\frac{\partial W_1}{\partial r} = \sqrt{2m(\alpha_1 - V) - \frac{\alpha_\phi^2}{r^2}},$$

so that W is

$$W = \int dr \sqrt{2m(\alpha_1 - V) - \frac{\alpha_\phi^2}{r^2}} + \alpha_\phi \phi.$$

With this form for the characteristic function the transformation Eqs. (9-22b) appear as

$$t + \beta_1 = \frac{\partial W}{\partial \alpha_1} = \int \frac{m\, dr}{\sqrt{2m(\alpha_1 - V) - \frac{\alpha_\phi^2}{r^2}}}, \tag{9-29a}$$

and

$$\beta_2 = \frac{\partial W}{\partial \alpha_\phi} = -\int \frac{\alpha_\phi\, dr}{r^2\sqrt{2m(\alpha_1 - V) - \frac{\alpha_\phi^2}{r^2}}} + \phi. \tag{9-29b}$$

Eq. (9-29a) furnishes r as a function of t, and agrees with the corresponding solution, Eq. (3-18), found in Chapter 3, with α_1 and α_ϕ written explicitly as E and l respectively. It has been remarked previously that the remaining transformation equations for Q_i, here only Eq. (9-29b), should provide the orbit equation. If the variable of integration in Eq. (9-29b) is changed to $u = \frac{1}{r}$ the equation reduces to

$$\phi = \beta_2 - \int \frac{du}{\sqrt{\frac{2m}{\alpha_\phi^2}(\alpha_1 - V) - u^2}},$$

which agrees with Eq. (3–37) previously found for the orbit, identifying β_2 as ϕ_0.

In this simple example some of the power and elegance of the Hamilton-Jacobi method begins to be apparent. In a few short steps we have obtained the dependence of r on t and the orbit equation, results obtained earlier only with considerable labor. Separation of variables in the Hamilton-Jacobi equation is not restricted, of course, solely to situations where only one coordinate is noncyclic. For example, if the Hamiltonian for central force motion be written in *spherical* polar coordinates, only the azimuth angle ϕ is cyclic, yet the Hamilton-Jacobi equation is still separable, as will be shown in Section 9–7.

9–5 Action-angle variables. Of especial importance in many branches of physics are systems in which the motion is periodic. Very often we are interested not so much in the details of the orbit as in the frequencies of the motion. A very elegant and powerful method of handling such systems is provided by a variation of the Hamilton-Jacobi procedure. In this technique the integration constants α_i appearing directly in the solution of the Hamilton-Jacobi equation are not themselves chosen to be the new momenta. Instead we use suitably defined constants J_i, which form a set of n independent functions of the α_i's, and which are known as the *action variables.*

Before introducing these variables, it will be necessary to explain in detail what is meant by the term "periodic motion." Consider first a system with a single degree of freedom. For such a system phase space is a two-dimensional plane. Two types of periodic motion may be distinguished:

1. The first type occurs whenever both q and p are periodic functions of the time with the same frequency. This motion is characteristic of oscillatory systems, such as a one-dimensional linear harmonic oscillator. It is often designated by the astronomical name *libration.* Periodic motion of this nature will be found when the initial position of the system point lies between two zeros of the kinetic energy. Since q and p return to their original starting values in one period, the system point retraces its steps every period, and the orbit in phase space is *closed,* as illustrated in Fig. 9–1a.

2. In the second type of periodic motion q itself is not periodic, but is such that when q is increased by some value, say q_0, the configuration of the system remains essentially unchanged. The most familiar example is that of a rigid body constrained to rotate about a given axis, with q as the angle of rotation. Increasing q by 2π then produces no essential change

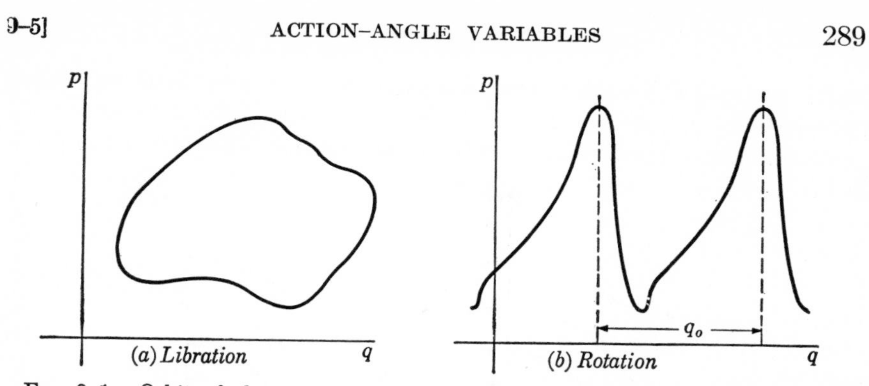

FIG. 9-1. Orbit of the system point in phase space for periodic motion of one-dimensional systems.

in the state of the system. Indeed, the position coordinate in this type of periodicity is invariably an angle of rotation, and the motion will be referred to simply as *rotation*, in contrast to libration. The values of q are no longer bounded, but can increase indefinitely. In phase space the system point does not travel in a closed bounded orbit but instead p will be some periodic function of q, with the period q_0, as illustrated in Fig. 9-1b.

It may serve to clarify these ideas to note that both types of periodicity may occur in the same physical system. The classic example is the simple pendulum where q is the angle of deflection θ. If the length of the pendulum is l the constant energy of the system is given by

$$E = \frac{p_\theta^2}{2ml^2} - mgl \cos \theta. \tag{9-30}$$

Solving Eq. (9-30) for p_θ, the equation of the path of the system point in phase space is

$$p_\theta = \sqrt{2ml^2(E + mgl \cos \theta)}. \tag{9-31}$$

If E is less than mgl then physical motion of the system can only occur for $|\theta|$ less than a bound, θ', defined by the equation

$$\cos \theta' = -\frac{E}{mgl}.$$

Under these conditions the pendulum oscillates between $-\theta'$ and $+\theta'$, which is a periodic motion of the libration type. The system point then traverses some such path in phase space as the curve 1 of Fig. 9-2. However, if $E > mgl$ all values of θ correspond to physical motion and θ can increase without limit to produce a periodic motion of the rotation type. What happens physically in this case is that the pendulum has so much energy that it can swing through the vertical position $\theta = \pi$, and therefore continues rotating. Curve 3 in Fig. 9–2 corresponds to the rotation motion of the pendulum. The limiting case when $E = mgl$ is illustrated by curve 2 in Fig. 9–2.

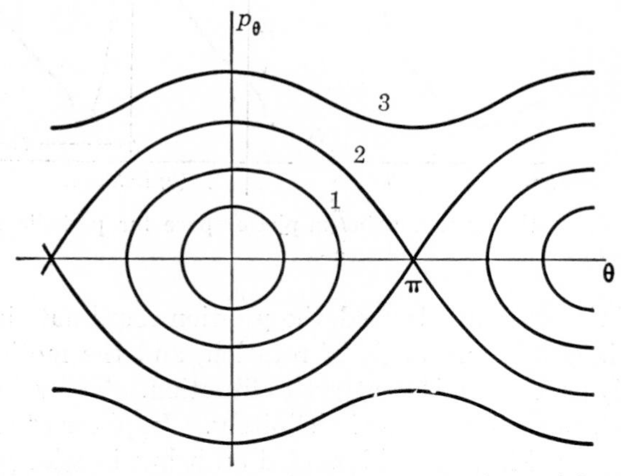

FIG. 9–2. Phase space orbits for the simple pendulum.

In dealing with systems of more than one degree of freedom we shall restrict our discussion to those problems for which the Hamilton-Jacobi equation for W is separable in at least one set of canonical variables. As the system moves the corresponding point will trace out a complicated path in the multidimensional phase space of the separation variables (q, p). The motion of the system is said to be periodic if the projection of the system point orbit on each (q_i, p_i) plane is simply periodic in the sense defined for motion of only one degree of freedom. Because the variables are separable, these projected motions are independent of each other, and their nature may be readily examined. The equations of the canonical transformation state that

$$p_i = \frac{\partial W_i(q_i, \alpha_1 \ldots \alpha_n)}{\partial q_i}, \tag{9–32}$$

which furnishes each p_i as a function of the corresponding q_i and the n constants α_j:

$$p_i = p_i(q_i, \alpha_1 \ldots \alpha_n). \tag{9–33}$$

It will be recognized that Eq. (9–33) is the projected orbit equation of the system point in the (q_i, p_i) plane. We will have periodic motion if Eq. (9–33) describes either a closed orbit or a periodic function of q_i.

It is not necessary that all q_i, p_i sets exhibit the same frequency of periodic motion. Thus, in a three-dimensional harmonic oscillator the force constants for motion along the three axes may all be different, but the system will still be called periodic. The complete motion of the particle in this example need not itself be simply periodic. Indeed, if the separate frequencies are not rational fractions of each other the particle will not traverse a closed curve in space, but will describe an open "Lissajous figure." Such motion will be described as *conditionally periodic.*

Having prescribed the type of motion under discussion, we can now introduce the action variables, $J_1 \ldots J_n$, which are to replace the constants $\alpha_1 \ldots \alpha_n$ as the transformed constant momenta P_i. The action variable J_i corresponding to the pair of separation variables (q_i, p_i) is defined as

$$J_i = \oint p_i \, dq_i, \tag{9-34}$$

where the integration is to be carried over a complete period of oscillation or rotation of q_i, as the case may be. The designation as action variable stems from the resemblance of (9-34) to the action A (cf. Section 7-5) defined as

$$A = \int \sum_i p_i \, dq_i = \int \sum_i p_i \dot{q}_i \, dt.$$

By Eq. (9-32) J_i can also be written as

$$J_i = \oint \frac{\partial W_i(q_i, \alpha_1 \ldots \alpha_n)}{\partial q_i} \, dq_i. \tag{9-35}$$

Since q_i is here merely a variable of integration, each action variable J_i is a function only of the n constants of integration appearing in the solution of the Hamilton-Jacobi equation. Further, it follows from the independence of the separate variable pairs (q_i, p_i) that the J_i's form n independent functions of the α_i's, and hence are suitable for use as a set of new constant momenta. Expressing the α_i's as functions of the action variables, the characteristic function W can be written in the form

$$W = W(q_1 \ldots q_n; J_1 \ldots J_n),$$

while the Hamiltonian appears as a function of the J_i's only:

$$H = \alpha_1 = H(J_1 \ldots J_n). \tag{9-36}$$

It will be noted from the definition, Eq. (9-34), that the dimensions of the action variables are always those of angular momentum.

If one of the separation coordinates is cyclic, its conjugate momentum is constant. The corresponding orbit in the q_i, p_i plane of phase space is then a horizontal straight line, which would not appear to be in the nature of a periodic motion. Actually the motion can be considered as a limiting case of the rotation type of periodicity, in which q_i may be assigned any arbitrary period. Since the coordinate in a rotation periodicity is invariably an angle, such a cyclic q_i always has a natural period of 2π. Accordingly, the integral in the definition of the action variable corresponding to a cyclic angle coordinate is to be evaluated from 0 to 2π, and hence

$$J_i = 2\pi p_i \tag{9–34$'$}$$

for all cyclic variables.

The generalized coordinates conjugate to J_i are known as the *angle variables* w_i, and are furnished by the transformation equations:

$$w_i = \frac{\partial W}{\partial J_i}. \tag{9–37}$$

Correspondingly, the equations of motion for the angle variables are

$$\dot{w}_i = \frac{\partial H(J_1 \ldots J_n)}{\partial J_i} = \nu_i(J_1 \ldots J_n), \tag{9–38}$$

where the ν_i's are a set of constant functions of the action variables. Eqs. (9–38) have the immediate solutions

$$w_i = \nu_i t + \beta_i, \tag{9–39}$$

so that the w_i's are linear functions of time, exactly as in Eqs. (9–22′).

Eqs. (9–37) and (9–39) can always be combined to furnish the q_i's as functions of t, ν_i, and β_i, as was done when the α_i's were taken as the new momenta. But when employed in this fashion the action-angle variables exhibit no great advantage over the α_i coordinates. Their particular merit arises from the physical interpretation which can be given to the ν_i's. Let us seek the change in an angle variable w_i when one of the coordinates q_j goes through a complete cycle of libration or rotation, all other coordinates being kept constant. This change may be represented as

$$\Delta w_i = \oint \delta w_i,$$

where δw_i is the infinitesimal change as a result of an increment in q_j alone:

$$\delta w_i = \frac{\partial w_i}{\partial q_j}\, dq_j. \tag{9–40}$$

Combining Eqs. (9–40) and (9–37), Δw_i can be written as

$$\Delta w_i = \oint \frac{\partial w_i}{\partial q_j}\, dq_j = \oint \frac{\partial^2 W}{\partial q_j\, \partial J_i}\, dq_j.$$

The derivative with respect to J_i can be taken outside the integral sign, reducing the expression to

$$\Delta w_i = \frac{\partial}{\partial J_i} \oint \frac{\partial W}{\partial q_j} dq_j = \frac{\partial}{\partial J_i} \oint p_j \, dq_j,$$

where use has been made of the transformation equations. It will be recognized that the integral involved here defines the action variable J_j, so that

$$\begin{aligned} \Delta w_i &= \frac{\partial J_j}{\partial J_i} \\ &= \delta_{ij}. \end{aligned} \tag{9-41}$$

Eq. (9–41) states that w_i changes by unity when q_i goes through a complete period, but is unaffected by a similar change in q_j, $j \neq i$. If τ_i is the period associated with q_i, then it follows from Eq. (9–39) that

$$\Delta w_i = 1 = \nu_i \tau_i.$$

Hence the constant ν_i can be identified as the reciprocal of the period,

$$\nu_i = \frac{1}{\tau_i}, \tag{9-42}$$

and therefore is *the frequency associated with the periodic motion of* q_i. The action-angle variables thus provide a powerful technique for obtaining the frequencies of periodic motion *without finding a complete solution for the motion of the system.* If it is known a priori that the system is periodic all that is necessary for finding the frequencies is to evaluate the action variables by the defining Eq. (9–34), and to express the energy in terms of the J_i's. The derivative of H with respect to J_i, by Eq. (9–38), then directly gives the frequency corresponding to q_i. The designation of w_i as an *angle* variable becomes obvious from the identification of ν_i in Eq. (9–39) as a frequency. The name is also consistent with the fact that J_i has the dimensions of an angular momentum, since the coordinate conjugate to an angular momentum is an angle.

As an illustration of the use of action-angle variables to find the frequencies, consider the familiar linear harmonic oscillator problem. There is only one action variable, given by

$$J = \oint p \, dq = \oint \frac{\partial W(q, \alpha)}{\partial q} dq = \sqrt{mk} \oint \sqrt{\frac{2\alpha}{k} - q^2} \, dq,$$

by Eq. (9–15). The substitution

$$q = \sqrt{\frac{2\alpha}{k}} \sin \theta$$

reduces the integral to

$$J = 2\alpha\sqrt{\frac{m}{k}}\int_0^{2\pi}\cos^2\theta\,d\theta, \tag{9–43}$$

where the limits are such as to correspond to a complete cycle in q. The integration is readily performed and results in

$$J = 2\pi\alpha\sqrt{\frac{m}{k}}. \tag{9–44}$$

Solving Eq. (9–44) for α we have

$$\alpha = H = \frac{J}{2\pi}\sqrt{\frac{k}{m}},$$

and the frequency of oscillation is therefore

$$\frac{\partial H}{\partial J} = \nu = \frac{1}{2\pi}\sqrt{\frac{k}{m}} = \frac{\omega}{2\pi},$$

which is the customary formula for the frequency of a linear harmonic oscillator.

9–6 Further properties of action-angle variables.

The results of the preceding section show that when the value of the angle coordinate changes by unity, the corresponding separation coordinate q_i goes through a complete cycle. For the libration type of periodicity a complete cycle of q_i means that it returns to its original value. Hence, in the case of libration q_i must be a periodic function of the angle coordinate w_i, with the fundamental period $\Delta w_i = 1$. It is therefore possible to represent a libration coordinate q_k as a Fourier series of the form

$$q_k = \sum_{j=-\infty}^{\infty} a_j e^{2\pi i j w_k}, \qquad \text{(libration)} \tag{9–45a}$$

or, by (9–39) as

$$q_k = \sum_{j=-\infty}^{\infty} a_j e^{2\pi i j(\nu_k t+\beta_k)}, \qquad \text{(libration)} \tag{9–46a}$$

Here the index j is an integer varying from $-\infty$ to $+\infty$. The Fourier coefficients a_j may be found in the standard manner from the equation

$$a_j = \int_0^1 q_k e^{-2\pi i j w_k}\,dw. \tag{9–47}$$

When the motion is in the nature of a rotation a change in w_k by unity does not return q_k to its original value, but instead increases q_k by the value of its period q_{0k}. Such a rotation coordinate is therefore not itself purely periodic, but it is seen that the function $q_k - w_k q_{0k}$ must be a periodic function of w_k with the period unity, and may be expanded in a Fourier series:

$$q_k - w_k q_{0k} = \sum_{j=-\infty}^{\infty} a_j e^{2\pi i j w_k}, \qquad \text{(rotation).} \quad (9\text{–}45\text{b})$$

The dependence of the function on time is similarly given by the expansion:

$$q_k - (\nu_k t + \beta_k)\, q_{0k} = \sum_{j=-\infty}^{\infty} a_j e^{2\pi i j(\nu_k t + \beta_k)}, \qquad \text{(rotation).} \quad (9\text{–}46\text{b})$$

It is thus always possible to express a periodic separable coordinate as a sum of simple harmonic motions involving the fundamental frequency ν_k and all its harmonics. If, however, we consider a function containing several of the q_k's, then in a Fourier expansion of this function there must appear the frequencies ν_k of each q_k coordinate involved. For example, the cartesian coordinates of the particles of the system, x_i, are often not separable coordinates. They may be expressed, nevertheless, in terms of the separable coordinates q_k, and any Fourier expansion of the x_i's will contain all possible linear combinations of the fundamental frequencies of the separation coordinates. We are thus led to an expansion of the form

$$x_i = \sum_{j_1=-\infty}^{\infty} \sum_{j_2=-\infty}^{\infty} \cdots \sum_{j_n=-\infty}^{\infty} a_{j_1 \ldots j_n} e^{2\pi i (j_1 w_1 + j_2 w_2 \ldots j_n w_n)}, \qquad (9\text{–}48)$$

where the j's are n integral indices running from $-\infty$ to $+\infty$. As a function of time Eq. (9–48) appears finally as

$$x_i = \sum_{j_1=-\infty}^{\infty} \cdots \sum_{j_n=-\infty}^{\infty} a_{j_1 \ldots j_n} e^{2\pi i (j_1 \nu_1 + \ldots j_n \nu_n) t + 2\pi i (j_1 \beta_1 + \ldots j_n \beta_n)}. \qquad (9\text{–}49)$$

Now, unless the ν_i's are rational fractions of each other, Eq. (9–49) does not represent a simply periodic function of time, as does, for example, (9–46). While the factor

$$e^{2\pi i j_1 \nu_1 t}$$

will return to its original value when t changes by $1/\nu_1$, the other exponential factors do not exhibit this same periodicity so long as the various frequencies are incommensurable. The function as a whole therefore is not simply periodic; rather, it is said to be *multiply* or *conditionally* periodic. The orbit of a harmonic oscillator of more than one dimension has already been cited as an example of conditionally periodic motion. Consider a two-dimensional oscillator with restoring forces acting along the x- and y-axes. These cartesian coordinates are therefore the separation variables, and each will exhibit simple harmonic motion with frequencies ν_x and ν_y

respectively. Suppose now that the coordinates be rotated 45° about the z-axis; the components of the motion along the new x', y' axes will be

$$\begin{aligned} x' &= \frac{1}{\sqrt{2}}\,[x_0 \cos 2\pi(\nu_x t + \beta_x) + y_0 \cos 2\pi(\nu_y t + \beta_y)], \\ y' &= \frac{1}{\sqrt{2}}\,[y_0 \cos 2\pi(\nu_y t + \beta_y) - x_0 \cos 2\pi(\nu_x t + \beta_x)]. \end{aligned} \tag{9–50}$$

If ν_x/ν_y is a rational number, these two expressions will be simply periodic, corresponding to a closed Lissajous figure. But if ν_x and ν_y are incommensurable, the Lissajous figure never exactly retraces its steps and Eqs. (9–50) provide simple examples of multiply periodic series expansions of the form (9–49).

Again barring commensurability of all the frequencies, a multiply periodic function can always be formed from the generating function W. The defining equation for J_i, Eq. (9–35), in effect states that when q_i goes through a complete cycle, i.e., when w_i changes by unity, the characteristic function increases by J_i. It follows that the function

$$W' = W - \sum_k w_k J_k \tag{9–51}$$

remains unchanged when *each* w_k is increased by unity, all the other angle variables remaining constant. Eq. (9–51) therefore represents a multiply periodic function which can be expanded in terms of the w_i by a series of the form of Eq. (9–48) or in terms of the frequencies ν_i as in Eq. (9–49). Since the transformation equations for the angle variables are

$$w_k = \frac{\partial W}{\partial J_k},$$

it will be recognized that Eq. (9–51) defines a Legendre transformation from the q, J basis to the q, w basis. Indeed, comparison with Eq. (8–10) shows that $W'(q, w)$ is the generating function of the form $F_1(q, Q)$ which performs the canonical transformation from the q, p variables to the w, J variables. While W' thus generates the same transformation as W it is of course *not* a solution of the Hamilton-Jacobi equation.*

* Action-angle variables have been defined here in terms of the simply periodic separable coordinates. It was then shown that the motion of the system as a whole was in general multiply periodic. Mention should be made that it is possible to reverse the process. Starting from the recognition that the system motion is multiply periodic, it is possible to introduce the action-angle variables such that the system configuration and the generating function $W'(q, w)$ are multiply periodic in the w's with the period unity, and the Hamiltonian is cyclic in all w's. One can avoid in this manner the necessity of referring to the separation coordinates. For further details see Born, *The Mechanics of the Atom*, Section 15.

It has been emphasized that the system configuration is conditionally periodic only if the frequencies ν_i are not rational fractions of each other. Otherwise the configuration repeats after a sufficiently long time and would therefore be simply periodic. The formal condition for the commensurability of all the frequencies is that there exist $n - 1$ relations of the form

$$\sum_{i=1}^{n} j_i \nu_i = 0, \tag{9-52}$$

where the j_i's are integers. By solving these equations we can then express any ν_i as a rational fraction of any of the other frequencies. When there are only m relations of the form (9-52) between the fundamental frequencies, then the system is said to be *m-fold degenerate*. If m is equal to $n - 1$, so that the motion is simply periodic, then the system is said to be *completely degenerate*. Thus, whenever *the orbit of the system point is closed, the motion will be completely degenerate.**

The simplest examples of degeneracy occur when two or more of the frequencies are equal. If two of the force constants in a three-dimensional harmonic oscillator are equal, then the corresponding frequencies are identical and the system is singly degenerate. In an isotropic linear oscillator the force constants are the same along all directions, all frequencies are equal, and the system is completely degenerate.

Whenever degeneracy is present the fundamental frequencies are no longer independent and the periodic motion of the system can be described by less than the full complement of n frequencies. Indeed, the m conditions of degeneracy can be used to reduce the number of frequencies to $n - m$, and the system motion is said to be $n - m$-fold periodic. The reduction of the frequencies may be most elegantly performed by means of a point

---

* There is an interesting connection between degeneracy and the coordinates in which the Hamilton-Jacobi equation is separable. It can be shown that the path of the system point for a nondegenerate system completely fills a limited region of both configuration and phase space (cf. Born, op. cit., Appendix 1). Now the motion in any one of the separation coordinates is simply periodic, and has been shown to be independent of the motion of the other coordinates. Hence the path of the system point as a whole must be limited by the surfaces of constant q_i, p_i which mark the bounds of the oscillatory motion of the separation variables. (The argument is easily extended to rotation by limiting all angles to the region 0 to 2π.) These surfaces therefore define the volume in space which is densely filled by the system point orbit. It obviously follows that the separation of variables in nondegenerate systems must be unique; the Hamilton-Jacobi equation cannot be separated in two different coordinate systems (aside from trivial variations such as change of scale). The possibility of separating the motion in more than one set of coordinates thus provides sure evidence that the system is degenerate.

transformation of the action-angle variables. It is convenient to summarize the m degeneracy conditions in the form

$$\sum_{i=1}^{n} j_{ki}\nu_i = 0, \qquad k = 1, \ldots m. \tag{9–53}$$

Consider now a point transformation from (w, J) to (w', J') defined by the generating function (cf. Eq. (8–19)):

$$F_2 = \sum_{k=1}^{m} \sum_{i=1}^{n} J'_k j_{ki} w_i + \sum_{k=m+1}^{n} J'_k w_k. \tag{9–54}$$

The transformed coordinates are

$$\begin{aligned} w'_k &= \sum_{i=1}^{n} j_{ki} w_i, && k = 1, \ldots m, \\ w'_k &= w_k, && k = m + 1 \ldots n. \end{aligned} \tag{9–55}$$

Correspondingly, the new frequencies are

$$\begin{aligned} \nu'_k = \dot{w}'_k &= \sum_{i=1}^{n} j_{ki}\nu_i = 0 && k = 1, \ldots m, \\ &= \nu_k && k = m + 1, \ldots n. \end{aligned} \tag{9–56}$$

Thus in the transformed coordinates m of the frequencies are zero and we are left with a set of $n - m$ independent frequencies. It is obvious that the new w'_k may also be termed as angle variables in the sense that the system configuration is multiply periodic in the w'_k coordinates with the fundamental period unity. The corresponding constant action variables are given as the solution of the n equations of transformation

$$J_i = \sum_{k=1}^{m} J'_k j_{ki} + \sum_{k=m+1}^{n} J'_k \delta_{ki}. \tag{9–57}$$

The zero frequencies correspond to constant factors in the Fourier expansion. These are, of course, also present in the original Fourier series in terms of the ν's, Eq. (9–49), occurring whenever the indices j_i are such that degeneracy conditions are satisfied. Since

$$\nu'_i = \frac{\partial H}{\partial J'_i},$$

the Hamiltonian must be independent of the action variables J'_i whose corresponding frequencies vanish. In a completely degenerate system the Hamiltonian can therefore be made to depend on only one of the action variables.

The problem of the bound motion of a particle in an inverse square law central force illustrates many of the phenomena involved in degeneracy. A discussion of this problem also affords an opportunity to show how the action-angle technique is applied to specific systems, and to indicate the connections with Bohr's quantum mechanics. Accordingly, the next section is devoted to a detailed treatment of the Kepler problem in terms of action-angle variables.

9-7 The Kepler problem in action-angle variables. To exhibit all of the properties of the solution we shall examine the motion in space, rather than make use of our a priori knowledge that the orbit lies in a plane. In terms of spherical polar coordinates the kinetic energy is

$$T = \frac{m}{2}\left(\dot{r}^2 + r^2\dot{\theta}^2 + r^2 \sin^2\theta\, \dot{\phi}^2\right), \tag{9-58}$$

where θ is the colatitude and ϕ the azimuth angle about the polar axis. The canonical momenta are therefore

$$p_r = m\dot{r}, \qquad p_\theta = mr^2\dot{\theta}, \qquad p_\phi = mr^2 \sin^2\theta\, \dot{\phi}. \tag{9-59}$$

With these coordinates and momenta the Hamiltonian for the problem becomes

$$H = \frac{1}{2m}\left(p_r^2 + \frac{p_\theta^2}{r^2} + \frac{p_\phi^2}{r^2 \sin^2\theta}\right) - \frac{k}{r}. \tag{9-60}$$

The orbit will be limited, and therefore periodic, only for an attractive force, corresponding to k positive. From the form of the Hamiltonian, the Hamilton-Jacobi equation for W can be immediately written as:

$$\frac{1}{2m}\left[\left(\frac{\partial W}{\partial r}\right)^2 + \frac{1}{r^2}\left(\frac{\partial W}{\partial \theta}\right)^2 + \frac{1}{r^2 \sin^2\theta}\left(\frac{\partial W}{\partial \phi}\right)^2\right] - \frac{k}{r} = \alpha_1 = E, \tag{9-61}$$

where the constant α_1 has, been specifically identified as the total energy. The variables in Eq. (9-61) may be separated by assuming a solution of the form

$$W = W_r(r) + W_\theta(\theta) + W_\phi(\phi). \tag{9-62}$$

Upon substituting this trial solution in Eq. (9-61) it will be noted that the dependence upon ϕ can be involved only in the last term in the brackets. If the equation is to hold identically for all ϕ it must be true that

$$\frac{\partial W_\phi}{\partial \phi} = \alpha_\phi, \tag{9-63a}$$

one of the constants of integration. The Hamilton-Jacobi equation now reduces to

$$\frac{1}{2m}\left[\left(\frac{\partial W_r}{\partial r}\right)^2 + \frac{1}{r^2}\left\{\left(\frac{\partial W_\theta}{\partial \theta}\right)^2 + \frac{\alpha_\phi^2}{\sin^2\theta}\right\}\right] - \frac{k}{r} = E.$$

Again, the term in the curly brackets involves θ only and must therefore be equal to a constant:

$$\left(\frac{\partial W_\theta}{\partial \theta}\right)^2 + \frac{\alpha_\phi^2}{\sin^2\theta} = \alpha_\theta^2. \tag{9–63b}$$

The Hamilton-Jacobi equation now contains only the dependence upon r:

$$\left(\frac{\partial W_r}{\partial r}\right)^2 + \frac{\alpha_\theta^2}{r^2} = 2m\left(E + \frac{k}{r}\right). \tag{9–63c}$$

Each of the three separate equations (9–63) corresponds to a conservation theorem of the motion. The first states that p_ϕ, the component of the angular momentum along the fixed polar axis, is a constant of the motion. Eq. (9–63b) can be put in the form

$$p_\theta^2 + \frac{p_\phi^2}{\sin^2\theta} = \alpha_\theta^2.$$

Now, in plane polar coordinates the Hamiltonian is

$$H = \frac{1}{2m}\left[p_r^2 + \frac{p^2}{r^2}\right] - \frac{k}{r},$$

where p is the magnitude of the total angular momentum. Comparison with the Hamiltonian in Eq. (9–60) shows that α_θ must be identified with p, and Eq. (9–63b) therefore states the conservation of the total angular momentum. Finally, Eq. (9–63c) corresponds to the conservation of energy.

Eqs. (9–63) can be immediately integrated to furnish the form of the generating function. However, this step is actually of little interest, since we are primarily concerned with obtaining the action-angle variables. There are three action variables, defined as:

$$J_1 \equiv J_\phi = \oint p_\phi \, d\phi = \oint \frac{\partial W_\phi}{\partial \phi}\, d\phi, \tag{9–64a}$$

$$J_2 \equiv J_\theta = \oint p_\theta \, d\theta = \oint \frac{\partial W_\theta}{\partial \theta}\, d\theta, \tag{9–64b}$$

$$J_3 \equiv J_r = \oint p_r \, dr = \oint \frac{\partial W_r}{\partial r}\, dr. \tag{9–64c}$$

By Eqs. (9–63) these three defining integrals may be written:

$$J_\phi = \oint \alpha_\phi \, d\phi, \tag{9–65a}$$

$$J_\theta = \oint \sqrt{\alpha_\theta^2 - \frac{\alpha_\phi^2}{\sin^2\theta}} \, d\theta, \tag{9–65b}$$

$$J_r = \oint \sqrt{2mE + \frac{2mk}{r} - \frac{\alpha_\theta^2}{r^2}} \, dr. \tag{9–65c}$$

The first integral is trivial; ϕ goes through 2π radians in a complete revolution and therefore

$$J_\phi = 2\pi\alpha_\phi = 2\pi p_\phi. \tag{9–66}$$

This result could have been predicted beforehand, for ϕ is a cyclic coordinate in H, and Eq. (9–66) is merely a special case of Eq. (9–34′) for the action variables corresponding to cyclic coordinates. Integration of Eq. (9–65b) poses no special mathematical difficulty, and is performed most simply by a procedure suggested by J. H. Van Vleck. It will be recalled that when the defining equations for the generalized coordinates do not involve time explicitly then

$$2T = \sum_i p_i \dot{q}_i,$$

(cf. Section 2–6). Expressing the kinetic energy in both spherical and plane polar coordinates, it follows that

$$p_r\dot{r} + p_\theta\dot{\theta} + p_\phi\dot{\phi} = p_r\dot{r} + p\dot{\psi},$$

where ψ is the plane azimuth angle of the particle in its orbit. Hence $p_\theta \, d\theta$ in Eq. (9–64b) can be replaced by $p\, d\psi - p_\phi \, d\phi$, and the action variable becomes

$$J_\theta = \oint p \, d\psi - \oint p_\phi \, d\phi.$$

As θ goes through a complete cycle of libration ϕ and ψ vary by 2π and the integrals reduce to

$$J_\theta = 2\pi(p - p_\phi) = 2\pi(\alpha_\theta - \alpha_\phi). \tag{9–67}$$

The last integral, for J_r, can now be written as

$$J_r = \oint \sqrt{2mE + \frac{2mk}{r} - \frac{(J_\theta + J_\phi)^2}{4\pi^2 r^2}} \, dr. \tag{9–68}$$

After performing the integration, this equation can be solved for the energy $E \equiv H$ in terms of the three action variables J_ϕ, J_θ, J_r. It will be noted that J_ϕ and J_θ can occur in E only in the combination $J_\theta + J_\phi$, and

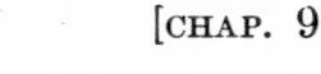

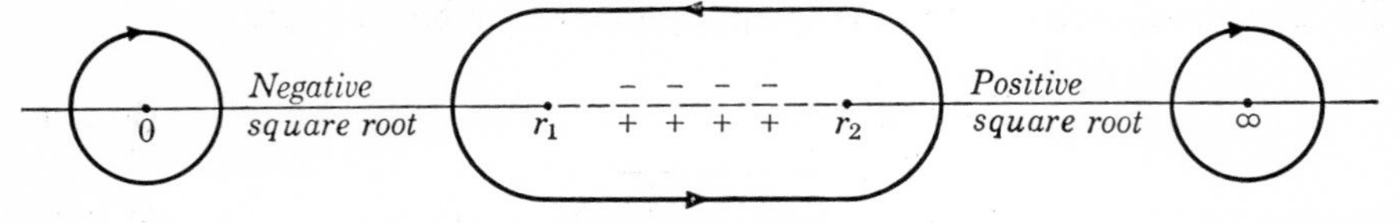

FIG. 9–3. The complex r plane in the neighborhood of the real axis; showing the paths of integration occurring in the evaluation of J_r.

hence the corresponding frequencies ν_ϕ and ν_θ must be equal, indicating a degeneracy. This result has not involved the inverse square law nature of the central force; *any motion produced by a central force is at least singly degenerate.*

The integral involved in Eq. (9–68) can be evaluated by elementary means, but the integration is most elegantly and quickly performed using the method of residues, a procedure first employed by Sommerfeld. For the benefit of those familiar with this technique we shall outline the steps involved in integrating Eq. (9–68). Bound motion can, of course, occur only when E is negative (cf. Section 3–3), and since the integrand is equal to $p_r = m\dot{r}$, the limits of the motion are defined by the roots r_1 and r_2 of the expression in the square root sign. If r_1 is the inner bound, as in Fig. 3–6, a complete cycle of r involves going from r_1 to r_2 and then back again to r_1. On the outward half of the journey, from r_1 to r_2, p_r is positive and we must take the positive square root. However, on the return trip to r_1, p_r is negative and the square root must likewise be negative. The integration thus involves both branches of a double-valued function, with r_1 and r_2 as the branch points. Consequently the complex plane can be represented as one of the sheets of a Riemann surface, slit along the real axis from r_1 to r_2 (as indicated in Fig. 9–3).

Since the path of integration encloses the line between the branch points, the method of residues cannot be applied directly. However, we may also consider the path as enclosing all the rest of the complex plane, the direction of integration now being in the reverse (clockwise) direction.* The integrand is single-valued in this region and there is now no bar to the application of the method of residues. Only two singular points are present, namely, the origin and infinity, and the integration path can be distorted

* To visualize this change in viewpoint it is convenient to think of the complex plane as projected stereographically onto the surface of a sphere, with the origin at the south pole, and the point ∞ at the north pole. The real axis becomes a meridian circle joining the two poles. Any closed integration path on the sphere divides the surface of the sphere into two areas. The path may then be considered as enclosing either of the two areas, depending upon the direction of the integration.

into two clockwise-described circles enclosing these two points. Now, the sign in front of the square root in the integrand must be negative for the region along the real axis below r_1, as can be seen by examining the behavior of the function in the neighborhood of r_1. If the integrand is represented as

$$-\sqrt{A+\frac{2B}{r}-\frac{C}{r^2}},$$

the residue at the origin is

$$R_0 = -\sqrt{-C}.$$

Above r_2 the sign of the square root on the real axis is found to be positive, and the residue is obtained by the standard technique of changing the variable of integration to $z = r^{-1}$:

$$-\oint \frac{1}{z^2}\sqrt{A+2Bz-Cz^2}\,dz.$$

Expansion about $z = 0$ now furnishes the residue

$$R_\infty = -\frac{B}{\sqrt{A}}.$$

The total integral is $-2\pi i$ times the sum of the residues:

$$J_r = 2\pi i\left(\sqrt{-C}+\frac{B}{\sqrt{A}}\right),$$

or, upon substituting the coefficients A, B, and C:

$$J_r = -(J_\theta + J_\phi) + \pi k\sqrt{\frac{2m}{-E}}. \tag{9-69}$$

Eq. (9-69) supplies the functional dependence of H upon the action variables, for solving for E we have

$$H \equiv E = -\frac{2\pi^2 mk^2}{(J_r + J_\theta + J_\phi)^2}. \tag{9-70}$$

It will be noted that, as predicted, J_θ and J_ϕ occur only in the combination $J_\theta + J_\phi$. More than that, all three of the action variables appear only in the form $J_r + J_\theta + J_\phi$. Hence all of the frequencies are equal; *the motion is completely degenerate.* This result could also have been predicted beforehand, for we know that with an inverse square law of force the orbit is closed for negative energies. With a closed orbit, the motion is simply periodic and therefore completely degenerate. If the central force contained an r^{-3} term, such as is provided by relativistic corrections, then the orbit is no longer closed but is in the form of a precessing ellipse. One of the degeneracies will be removed in this case, but the motion is still singly degenerate, since $\nu_\theta = \nu_\phi$ for all central forces. The one frequency for the motion is given by

$$\nu = \frac{\partial H}{\partial J_r} = \frac{\partial H}{\partial J_\theta} = \frac{\partial H}{\partial J_\phi} = \frac{4\pi^2 mk^2}{(J_r + J_\theta + J_\phi)^3}.$$

If we evaluate the sum of the J's in terms of the energy from Eq. (9–70) the period of the orbit is

$$\tau = \pi k \sqrt{\frac{m}{-2E^3}}. \tag{9–71}$$

This formula for the period agrees with Kepler's third law, Eq. (3–54), if it is remembered that the semimajor axis a is equal to $-k/2E$.

The degenerate frequencies may be eliminated by a contact transformation to a new set of action-angle variables, following the procedure outlined in the previous section. Expressing the degeneracy conditions as:

$$\nu_\phi - \nu_\theta = 0, \qquad \nu_\theta - \nu_r = 0,$$

the appropriate generating function is

$$F = (w_\phi - w_\theta)J_1' + (w_\theta - w_r)J_2' + w_r J_3'. \tag{9–72}$$

The new angle variables are

$$\begin{aligned} w_1' &= w_\phi - w_\theta, \\ w_2' &= w_\theta - w_r, \\ w_3' &= w_r, \end{aligned} \tag{9–73}$$

and, as planned, two of the new frequencies, ν_1' and ν_2', are zero. We can obtain the new action variables from the transformation equations:

$$\begin{aligned} J_\phi &= J_1', \\ J_\theta &= J_2' - J_1', \\ J_r &= J_3' - J_2', \end{aligned}$$

which yields the relations

$$\begin{aligned} J_1' &= J_\phi, \\ J_2' &= J_\phi + J_\theta, \\ J_3' &= J_\phi + J_\theta + J_r. \end{aligned} \tag{9–74}$$

In terms of these transformed variables the Hamiltonian appears as

$$H = -\frac{2\pi^2 mk^2}{J_3'^2}, \tag{9–75}$$

a form involving only that action variable for which the corresponding frequency is different from zero.

The values of the angle variables, and their physical significance in relation to the orbit, are obtained from the equations of transformation

$$w_i = \frac{\partial W}{\partial J_i}. \tag{9-76}$$

Integration of Eqs. (9-63) supplies W in terms of the constants of the motion p, p_ϕ, and E, and hence in terms of the action variables. This function may then be substituted in Eqs. (9-76) to relate the w's to the constants of motion. In practice the integrations often become quite lengthy. Fortunately, some qualitative considerations are sufficient to indicate the significance of the constant angle coordinates w_1' and w_2'. The action variable J_1' is 2π times the component of the angular momentum along the polar axis, and hence its conjugate angle variable must be some fixed angle of rotation about this axis. The angle of the line of nodes (the line formed by the intersection of the plane of the orbit with the equatorial plane) is such an angle, and w_1' can therefore differ from it only by a constant. (The value of this constant must be obtained from the explicit integration.) Similarly, $J_2' = J_\theta + J_\phi = 2\pi p$, and is thus proportional to the total angular momentum. The conjugate angle w_2' is therefore related to a fixed angle in the plane of the orbit, as, for example, the angle between the perihelion and the line of nodes. It will also be noted that the ratio J_1'/J_2' must be the cosine of the angle between the polar axis and the angular momentum vector. These three quantities, w_1', w_2', and J_1'/J_2' are in effect the Euler angles which determine the orientation of the orbit in space.

The method of action-angle variables probably does not strike one as the quickest way to solve the Kepler problem, and the practical usefulness of the technique may well be questioned. We have seen that the (w', J') set of variables serves to fix the location of the orbit in space, and it is also possible to express the size and shape of the orbit (i.e., semimajor axis and ellipticity) in terms of the J_i''s. These variables are therefore particularly appropriate for the astronomical study of planetary orbits, where the (w', J') set of action-angle variables is known as the *Delaunay elements* of the orbit. When there are only two bodies involved in the motion these elements (with the exception of w_3') are strict constants of the motion. However, if there are small perturbing forces present, due for example to other planets or satellites, the motion can often be represented by a slow variation of these elements with time. The action-angle variables provide a powerful tool for the study of the effects of such perturbations.

For a long time action-angle variables remained an esoteric technique of classical mechanics used only by astronomers. The situation changed rapidly with the advent of Bohr's quantum theory of the atom, for it was found that the quantum conditions could be stated most simply in terms of the action variables. In classical mechanics the action variables possess a continuous range of values, but this is no longer the case in quantum mechanics. The quantum conditions of Sommerfeld and Wilson required that the motion be limited to such orbits for which the "proper" action variables had discrete values which were integral multiples of h, the quantum of action. (By proper action variables are meant those J's whose frequencies are nondegenerate and different from zero. J_3', for example, is a proper action variable.) As Sommerfeld stated, the method of action-angle variables then provided "a royal road to quantization." One had only to solve the problem in classical mechanics using action-angle variables, and the motion could be immediately quantized by replacing the J's with integral multiples of Planck's constant h.

As an example of this procedure it may be noted that the quantized energy levels for a hydrogenic atom follow at once from Eq. (9–75) if k is set equal to Ze^2 and J_3' is replaced by nh:

$$E = -\frac{2\pi^2 m Z^2 e^4}{n^2 h^2}. \tag{9–77}$$

Here the integer n is known as the *principal quantum number*, and is the sole quantum number for the completely degenerate system. The degeneracy will be partly removed if relativity corrections are introduced, producing a precession of the perihelion in the plane of the orbit. The angle variable w_2' which measures the position of this perihelion then varies with time and the conjugate action variable becomes a "proper" variable and must also be quantized:

$$J_2' = kh,$$

where k is the *orbital* quantum number. Since both ν_3' and ν_2' are different from zero, the energy must depend on both J_3' and J_2', i.e., on n and k. We thus obtain the well-known relativistic fine structure of the hydrogen levels. The degeneracy can be completely removed by introducing a constant magnetic field along the arbitrary polar axis. The plane of the orbit then executes a Larmor precession about the polar axis, producing a uniform increase of the angle variable w_1' with time. J_1' therefore becomes a proper action variable in the presence of a magnetic field and must likewise obey quantum conditions:

$$J_1' = mh,$$

m being the *magnetic* quantum number. The energy now depends on all three quantum numbers, and removal of the degeneracy in this manner thus results in the Zeeman splitting of the atomic levels.*

During the heyday of the old quantum theory the action-angle technique naturally received much attention, and for a time was the daily tool of the research theoretical physicist. Once past the simple hydrogen atom, the problems became progressively more complicated to solve even classically, and it became necessary to treat many of the additional forces as small perturbing elements. Again the physicist turned to the astronomer, this time to borrow the technique of perturbation calculations in classical mechanics. While there are many points of resemblance between classical perturbation theory and the perturbation methods of wave mechanics, the classical techniques are far more involved than their quantum counterparts, especially where degeneracy occurs.

It soon became apparent, however, that the difficulties were not merely mathematical; the Bohr quantum theory was simply not an accurate picture of nature. As is well known, the impasse was broken with the almost simultaneous introduction of wave and matrix mechanics. Techniques for solving quantum problems were entirely different in these new theories, and interest in action-angle variables waned abruptly. By now they have once more been returned to the exclusive domain of astronomy (and texts on classical mechanics). Yet some of the concepts have remained — such as degeneracy and the relation of degeneracy to the separation coordinates — and are still part of the fabric of quantum mechanics.

Strangely enough, the root of the newer wave mechanics also arose out of Hamilton-Jacobi theory. If the Poisson bracket formulation of classical mechanics serves as a point of departure to matrix mechanics, the germ of wave mechanics is contained in the connection between Hamilton-Jacobi theory and geometrical optics. It is to the study of this connection that we now turn our attention.

9–8 Hamilton-Jacobi theory, geometrical optics, and wave mechanics. We shall consider only those systems for which the Hamiltonian is a constant of the motion and is identical with the total energy. Hamilton's principal and characteristic functions are then related according to the equation

$$S(q, P, t) = W(q, P) - Et. \tag{9–78}$$

* The splitting so obtained represents only the normal Zeeman effect. The correct abnormal Zeeman effect can, of course, be calculated only by including the effects of "spin."

Since the characteristic function is independent of time, the surfaces of constant W in configuration space have fixed locations. A surface characterized by a constant value of S must coincide at a given time with some particular surface of constant W. However, the value of W corresponding to a definite value of S changes with time in accordance with Eq. (9–78). Thus at $t = 0$ the surfaces $S = a$ and $S = b$ coincide with the surfaces for which $W = a$ and $W = b$, respectively (cf. Fig. 9–4). At a time dt later the surface $S = a$ now coincides with the surface for which $W = a + E\,dt$, and similarly $S = b$ is located at the surface $W = b + E\,dt$. In effect, in a time dt the surface $S = a$ has moved from $W = a$ to $W = a + E\,dt$. The motion of the surface in time is similar to the propagation of a wave front, such as, for example, that of a shock wave, across space. The surfaces of constant S may thus be considered as *wave fronts propagating in configuration space.*

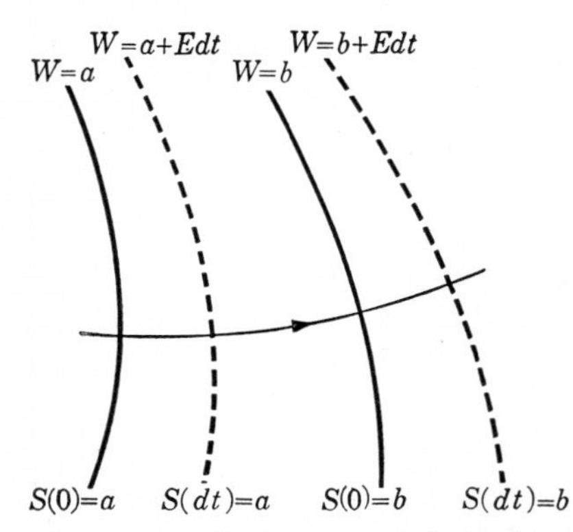

FIG. 9–4. The motion of the surfaces of constant S in configuration space.

Since the constant S surfaces in general change their shape in the course of time, the wave velocity, i.e., the velocity with which the surfaces move, will not be uniform for all points on the surfaces. However, it is possible to calculate the value of the wave velocity at any given point. For convenience we shall consider a system consisting of only one particle and take the cartesian position coordinates as the generalized coordinates. Configuration space then reduces to ordinary three-dimensional space, which greatly simplifies the geometry of the problem. The wave velocity at a particular point on a surface of constant S is given by the perpendicular distance the wave front moves in an infinitesimal time dt, divided by the time interval dt. If the infinitesimal distance normal to the surface is denoted by ds then the wave velocity is

$$u = \frac{ds}{dt}. \tag{9–79}$$

Now in the time dt the S surface travels from a surface W to a new surface on which the value of the characteristic function is $W + dW$, where

$$dW = E\,dt.$$

The change dW is also related to the normal distance ds according to the formula

$$dW = |\nabla W|\,ds, \tag{9–80}$$

so that

$$u = \frac{ds}{dt} = \frac{E}{|\nabla W|}. \tag{9-81}$$

The magnitude of the gradient of W is furnished by the Hamilton-Jacobi equation which has the form

$$\frac{1}{2m}\left[\left(\frac{\partial W}{\partial x}\right)^2 + \left(\frac{\partial W}{\partial y}\right)^2 + \left(\frac{\partial W}{\partial z}\right)^2\right] + V = E, \tag{9-82}$$

or, equivalently,

$$(\nabla W)^2 = 2m(E - V). \tag{9-83}$$

Hence the wave velocity is

$$u = \frac{E}{\sqrt{2m(E - V)}}. \tag{9-84}$$

Eq. (9-84) may be expressed in a number of variant forms. The difference between E and V is simply the kinetic energy T, so that

$$u = \frac{E}{\sqrt{2mT}}. \tag{9-85}$$

For the one-particle system under consideration, $2mT = m^2v^2 = p^2$, and Eq. (9-85) can also be written as

$$u = \frac{E}{p} = \frac{E}{mv}. \tag{9-85'}$$

Eq. (9-85') states that the velocity of a point on a surface of constant S is inversely proportional to the spatial velocity of the particle whose motion is being described by S. It is but a simple step to show that the trajectories of the particle must always be normal to the surfaces of constant S. The direction of the trajectory at any given point in space is determined by the direction of the momentum $\mathbf{p}$. By Eqs. (9-21), however,

$$\mathbf{p} = \nabla W, \tag{9-86}$$

and the gradient of W determines the normal to the surfaces of constant S or W. Any family of surfaces of constant W thus creates a set of trajectories of possible motion which are always normal to the surfaces. As a particle moves along one of the trajectories the surfaces of S generating the motion will also travel through space, but the two motions do not keep step. In fact, when the particle slows down the surfaces move faster, and vice versa.

In these considerations we have specialized on a system of one particle for ease in discussion. But most of the results hold for many-particle systems if we speak in terms of a configuration space whose metric tensor is such that an element of arc length $d\rho$ is given by

$$d\rho^2 = 2T\,dt^2$$

(cf. Eq. 7–42). Instead of the actual trajectory of the particle in space we deal with the path of the system point in configuration space. The wave velocity of the S surfaces is then found to be *

$$u = \frac{E}{\sqrt{2(E-V)}} = \frac{E}{\sqrt{2T}}, \tag{9–84′}$$

which is closely analogous to Eq. (9–84). It will be remembered from Section 7–5 that the velocity of the system point in configuration space is proportional to $\sqrt{T}$, so that the reciprocal relation between the wave and system velocities is preserved. Likewise, the possible system trajectories are again found to be normal to the surfaces of constant S. The transition to a many-particle system thus introduces no new physical results, and to simplify the mathematics we shall continue to confine the discussion to one-particle systems.

The surfaces of constant S have been characterized as wave fronts because they propagate in space in the same manner as wave surfaces of constant phase. We have even gone so far as to compute the wave velocity. But nothing has yet been said about the nature or origin of these waves whose fronts are surfaces of constant S. The most striking features of all wave motion result from their periodicity, and there has been no indication so far of the frequency and wave length spectra of the waves associated with S. To throw light on these questions, let us examine some of the properties of a well-known wave motion — that of light waves.

The scalar wave equation of optics is

$$\nabla^2\phi - \frac{n^2}{c^2}\frac{d^2\phi}{dt^2} = 0, \tag{9–87}$$

where ϕ is a scalar quantity such as the scalar electromagnetic potential, c is the velocity of light in vacuo, and n is the index of refraction equal to the ratio of c to the velocity of light. In general, n depends upon the medium and will be a function of position in space. If n is constant, Eq. (9–87) is satisfied by a plane wave solution of the form

$$\phi = \phi_0 e^{i(\mathbf{k}\cdot\mathbf{r}-\omega t)}, \tag{9–88}$$

* For a discussion of the motion of the S surfaces in configuration space see L. Brillouin, *Les Tenseurs en Mécanique et en Élasticité*, Chapter VIII.

where the wave number k and the frequency ω are connected by the relation

$$k = \frac{2\pi}{\lambda} = \frac{n\omega}{c}. \tag{9-89}$$

Taking the direction of $\mathbf{k}$ for simplicity as being along the z-axis, the plane wave solution can also be written

$$\phi = \phi_0 e^{ik_0(nz - ct)}, \tag{9-90}$$

where k_0 is the wave number in the vacuum. We shall be interested, however, in the case of *geometrical optics*, where n is not exactly constant but varies slowly in space. The plane wave is then no longer a solution of the wave equation (9-87); the variation of the index of refraction with position will distort and bend the wave. Since n is assumed to change only gradually in space, we seek a solution resembling the plane wave as closely as possible:

$$\phi = e^{A(\mathbf{r}) + ik_0(L(\mathbf{r}) - ct)}. \tag{9-91}$$

The quantities A and L are taken as functions of position to be determined and are both considered as real. A therefore is a measure of the amplitude of the wave. If n were constant L would reduce to nz and in consequence is called the *optical path length* or phase of the wave. It is also frequently referred to as the *eikonal*. Successive applications of the gradient operator to the solution ϕ result in the relations

$$\nabla\phi = \phi\,\nabla(A + ik_0L),$$
$$\nabla^2\phi = \phi[\nabla^2(A + ik_0L) + (\nabla(A + ik_0L))^2],$$

or

$$\nabla^2\phi = \phi[\nabla^2A + ik_0\,\nabla^2L + (\nabla A)^2 - k_0^2(\nabla L)^2 + 2ik_0\,\nabla A \cdot \nabla L].$$

The wave equation now becomes

$$ik_0[2\,\nabla A \cdot \nabla L + \nabla^2L]\phi + [\nabla^2A + (\nabla A)^2 - k_0^2(\nabla L)^2 + n^2k_0^2]\phi = 0. \tag{9-92}$$

Since both A and L are real, the equation holds only if the two expressions in the square brackets separately vanish:

$$\nabla^2A + (\nabla A)^2 + k_0^2(n^2 - (\nabla L)^2) = 0, \tag{9-93a}$$
$$\nabla^2L + 2\,\nabla A \cdot \nabla L = 0. \tag{9-93b}$$

So far no approximation has been made; both equations are rigorous. We can now introduce the assumption that n varies only slowly with distance; in particular, that n does not change greatly over distances of the order of the wave length. Effectively, this means that the wave length is small compared to the dimension of any change in the medium, which is the assumption of geometrical optics. The term in $k_0^2 = 4\pi^2/\lambda_0^2$ in Eq. (9–93a) is therefore the prominent one, and the equation reduces to the simple form

$$(\nabla L)^2 = n^2. \tag{9–94}$$

Eq. (9–94) is known as the *eikonal equation of geometrical optics.* The surfaces of constant L determined by this equation are the surfaces of constant optical phase and thus define the wave fronts. The ray trajectories are everywhere perpendicular to the wave fronts and hence are also determined by Eq. (9–94).

We need not digress further into geometrical optics, for it will be seen that the eikonal equation (9–94) is identical in form with the mechanical Hamilton-Jacobi equation for W, (9–83). The characteristic function W plays the same role as the eikonal L and $[2m(E - V)]^{\frac{1}{2}}$ serves as the index of refraction. The Hamilton-Jacobi equation thus tells us that *classical mechanics corresponds to the geometrical optics limit of a wave motion* in which the light rays orthogonal to the wave fronts correspond to the particle trajectories orthogonal to the surfaces of constant S. It is now clear why Huygens' wave theory and Newton's light corpuscles were able to account equally well for the phenomena of reflection and refraction, for both theories of geometrical optics are formally identical. The resemblance of the principle of least action to Fermat's principle of geometrical optics is also explained. According to Eq. (7–40) the least action principle appeared as

$$\Delta \int \sqrt{2mT}\, ds = 0.$$

The integrand is seen to be proportional to the index of refraction for the corresponding wave motion, or inversely proportional to the wave velocity. Hence the principle of least action may also be written in the forms

$$\Delta \int n\, ds = \Delta \int \frac{ds}{u} = 0, \tag{9–95}$$

which are two well-known variations of Fermat's principle for the trajectories of light rays.

We have still not established the frequencies and wave lengths of the waves associated with classical motion. All that has been determined is that the wave length must be very much smaller than the spatial extensions of the forces and potentials. Further than this we cannot go within the realm of classical mechanics. As a species of geometrical optics, classical mechanics is precisely the field in which phenomena depending on the wave length (interference, diffraction, etc.) cannot occur. There is a duality of particle and wave even in classical mechanics, but the particle is the senior partner, and the wave aspect has no opportunity to display its unique characteristics.

We can speculate, nonetheless, on the form of the wave equation for which the Hamilton-Jacobi equation represents the short wave length limit. The similarity of the eikonal equation (9–94) with the Hamilton-Jacobi equation (9–83) does not imply the equivalence of L with W; it is necessary merely that the two quantities be proportional to each other. We shall see that the constant of proportionality is a measure of the magnitude of the wave length. If W corresponds to L, then $S = W - Et$ must be proportional to the total phase of the light wave described by Eq. (9–91):

$$k_0(L - ct) = 2\pi\left(\frac{L}{\lambda_0} - \nu t\right).$$

Hence the particle energy E and the wave frequency ν must be proportional, and we shall denote the constant ratio of the two quantities by the symbol h:

$$E = h\nu. \tag{9–96}$$

The wave length and the frequency are connected by the relation

$$\lambda\nu = u,$$

so that, by Eq. (9–85′), λ is given by

$$\lambda = \frac{u}{\nu} = \frac{E}{E}\bigg|\frac{p}{h},$$

or

$$\lambda = \frac{h}{p}. \tag{9–97}$$

The optical wave equation (9–87) can also be written

$$\nabla^2\phi - \frac{1}{u^2}\frac{d^2\phi}{dt^2} = 0,$$

where u is the wave velocity in the medium of index n. If the time dependence $e^{-i\omega t}$ be substituted in this equation, we obtain

$$\nabla^2\phi + \frac{4\pi^2}{\lambda^2}\phi = 0, \tag{9–98}$$

which is the time-independent wave equation. Corresponding to the wave amplitude ϕ in optics there will be some quantity ψ in the wave theory of mechanics which must satisfy an equation of the same form as

(9–98). But now λ is h/p, where p is $\sqrt{2m(E - V)}$. Hence the wave equation for which W represents the eikonal must be given by

$$\nabla^2\psi + \frac{8\pi^2 m}{h^2}(E - V)\psi = 0. \tag{9–99}$$

Eq. (9–99) will be recognized as the Schrödinger equation of wave mechanics.

It is seen from Eq. (9–97) that the size of the wave length depends on the magnitude of the constant of proportionality h. The smaller h, the smaller the wave length and the better the approximation to geometrical optics. Now, the equivalence of the Hamilton-Jacobi and eikonal equations was first realized by Hamilton in 1834; the corresponding wave equation was first derived by de Broglie and Schrödinger in 1926. It has been stated that had Hamilton only gone a little further, he would have discovered the Schrödinger equation. This is not so; he lacked the experimental authority for the jump. In Hamilton's day classical mechanics was considered to be rigorously true, and there was no justification in experiment for considering it as an approximation to a broader theory. In other words, Hamilton had no reason to believe that the value of h was at all different from zero. The recognition that classical mechanics was only a geometrical optics *approximation* to a wave theory could come only when effects depending on the particle wave length were discovered — as in the interference experiments of Davisson and Germer. Only then could physical reality be ascribed to h, which is, of course, the famous Planck's constant.*

Nevertheless, it can now be seen that classical mechanics contains within it the seeds of the quantum theory, and that the Hamilton-Jacobi formulation is particularly suited to show how to generalize from classical to wave mechanics. To go further into these subjects would take us beyond the scope of this book, which might well be titled "The Geometrical Optics of Wave Mechanics"!

SUGGESTED REFERENCES

M. Born, *The Mechanics of the Atom.* Most of the references listed in the previous chapter deal with Hamilton-Jacobi theory in more or less detail. Born's book is outstanding in the wealth of material it presents on the applications of Hamilton-Jacobi theory and action-angle variables. Its discussions on multiply peri-

* A similar situation occurred in the development of the wave theory of light. Until the phenomena of interference and diffraction were experimentally observed in light there was no reason to prefer Huygens' wave theory over Newton's corpuscular rays.

odic motion and perturbation theory are undoubtedly the most complete in the English language. The reader should be cautious, however, in accepting the statements about atomic structure. Most of them are out of date.

A. SOMMERFELD, *Atomic Structure and Spectral Lines.* The exposition of Hamilton-Jacobi theory and action-angle variables which are scattered through the text and appendices of this book is considerably less detailed than in Born. Probably for that reason they are often more readable. Especially noteworthy is the discussion of the connection between the number of systems of separation coordinates and the degeneracy of the motion. The evaluation of the integrals occurring in the Kepler problem by means of the theory of residues is explained in an appendix (and is also given in Born's book).

J. H. VAN VLECK, *Quantum Principles and Line Spectra.* The chapter of this work entitled "Mathematical Techniques" provides a quick survey of Hamilton-Jacobi theory and action-angle variables, with an introduction into perturbation theory. Most of the rest of the book is of historical interest only. The caution applied to Born's book is equally valid here, and holds to a lesser extent for Sommerfeld's volume.

E. FUES, *Störungsrechnung* in Vol. V of the *Handbuch der Physik.* The preceding article in this volume of the *Handbuch*, "Hamilton-Jacobische Theorie" by L. Nordheim and E. Fues, actually reaches Hamilton-Jacobi theory only in its concluding sections. Besides some generalities on the foundations of the theory there is a brief discussion of the conditions for separability. The Fues article picks up from there, and is much more thorough in its exposition. The nature of multiply periodic motion is carefully explained, action-angle variables are introduced, and the Kepler problem is done in detail (except for the complex integration). More than half of the article is devoted to perturbation theory, whose full complexity as compared with quantum perturbation is revealed in the very detailed treatment.

P. FRANK, *Die Differential Gleichungen allgemeiner Mechanischer Systeme* in Vol. 2 of *Die Differential- und Integralgleichungen der Mechanik und Physik*, by Frank and von Mises. This chapter presents a compact treatment of all of mechanics on a fairly advanced level. The portions of interest for the present chapter are Sections 4 to 7. Especially valuable is the very general treatment of separable systems given in Section 5 which is based on the work of Staeckel. There is some material on the connection between Hamilton-Jacobi theory and geometrical optics. The preceding chapter is in fact a detailed exposition of geometrical optics on the basis of the one-particle Hamilton-Jacobi equation.

C. CARATHÉODORY, *Variationsrechnung.* Mention has been made of the connection between Hamiltonian theory and the general mathematical problem of first-order partial differential equations. The Hamilton-Jacobi equation plays a vital role in this connection. A detailed treatment of these matters will be found in Carathéodory's treatise, along with the so-called theory of "characteristics."

C. L. CHARLIER, *Die Mechanik des Himmels.* The position of Hamilton-Jacobi theory in astronomical mechanics is presented in this standard treatise. Chapter 2 considers multiply periodic motion, while Chapters 6 and 7 discuss perturbation theory and its application to the three-body problem. See also H. Poincaré, *Les Methodes Nouvelles de Mécanique Céleste*, for a more thoroughgoing approach from the viewpoint of the mathematician.

L. BRILLOUIN, *Les Tenseurs en Mécanique et en Élasticité.* The motion of the surfaces of constant S in configuration space is presented in detail in Chapter VIII, and the connections linking classical mechanics, geometrical optics, and wave mechanics are thoroughly discussed in Chapter IX.

EXERCISES

1. In the text the Hamilton-Jacobi equation for S was obtained by seeking a contact transformation from the canonical coordinates (q, p) to the constants (α, β). Conversely, if $S(q_i, \alpha_i, t)$ is any complete solution of the Hamilton-Jacobi equation (9–3), show that the set of variables (q_i, p_i) defined by Eqs. (9–6, 7) are canonical variables, i.e., that they satisfy Hamilton's equations.

2. Solve the problem of the motion of a point projectile in space under the influence of a uniform gravitational force, using the Hamilton-Jacobi method. Find both the equation of the trajectory and the dependence of the coordinates on time.

3. Set up the problem of the heavy symmetrical top, with one point fixed, in the Hamilton-Jacobi method, and obtain the formal solution to the motion as given by Eq. (5–56).

4. Find the frequencies of a three-dimensional harmonic oscillator with unequal force constants, using the method of action-angle variables.

5. (a) Show that if the amplitude of oscillation is small the energy of a simple pendulum is given by

$$E = J\nu.$$

(b) Consider a simple pendulum formed by a point weight on a string which passes vertically through a hole in a horizontal board. Suppose the length of the pendulum is decreased by slowly pulling the string up through the hole. The rate of decrease of length is to be extremely slow compared with the frequency of swing, so that one can still speak of a period of oscillation for any given length. Compute the change in the energy of the pendulum, as the length is shortened, from the work done in pulling against the tension of the string. Show as a result that the action variable $J = E/\nu$ is constant throughout the process. A change in the external parameters of the system at a rate slow compared to the intrinsic frequencies is known as an *adiabatic variation* and the action variable of the pendulum is therefore an *adiabatic invariant.* It is possible to prove in general the adiabatic invariance of the action variables for any system in which degeneracy does not occur. The action variables are thus stable constants of the system under the influence of slowly varying external conditions. Now, the quantum state of a system is also an adiabatic invariant; a slow change of the external parameters does not induce transitions from one state to another. We have here another indication of the suitability of the action variables for describing the quantization of the system states.

6. (a) In the harmonic oscillator of Exercise 4 allow all the frequencies to become equal (isotropic oscillator) so that the motion is completely degenerate.

Transform to the "proper" action-angle variables, expressing the energy in terms of only one of the action variables.

(b) Solve the problem of the isotropic oscillator in action-angle variables using spherical polar coordinates. Transform again to proper action-angle variables and compare with the result of part (a). Are the two sets of proper variables the same? What are their physical significances? This problem illustrates the feasibility of separating a degenerate motion in more than one set of coordinates. The nondegenerate oscillator can be separated only in cartesian coordinates, not in polar coordinates.

7. The motion of a degenerate plane harmonic oscillator can be separated in any cartesian coordinate system. Obtain the relations between the two sets of action-angle variables corresponding to two cartesian systems of axes making an angle θ with each other. Note that the transformation between the two sets is *not* the orthogonal transformation of the rotation.

8. Evaluate the integral appearing in the expression for J_r in the Kepler problem, Eq. (9–68), by elementary methods (using standard tables of integrals, such as Pierce's).

9. By integrating each of the separated Hamilton-Jacobi equations of the Kepler problem, obtain W as the sum of three integrals From the relations

$$w_i' = \frac{\partial W}{\partial J_i'},$$

similarly obtain integral expressions for the three angle variables. Show that w_1', except for an additive constant of integration, is the azimuth of the line of nodes, and that w_2' is the angular position in the plane of the orbit of the perihelion from the line of nodes, again to within an additive constant. In interpreting the integrals it will be found convenient to replace the ratio J_1'/J_2' by $\cos\alpha$, where α is the angle of the plane of the orbit from the polar axis.

10. The equation of the orbit for the Kepler problem can be evaluated in terms of the action-angle variables from Eq. (9–29b), expressing $\alpha_1 = E$ and $\alpha_\phi = l$ as functions of J_2' and J_3'. (Note the change in the meaning of the angle ϕ.) Perform the integration to obtain the orbit equation, and show that the semimajor axis a and the eccentricity ϵ are given by

$$a = \frac{J_3'^2}{4\pi^2 mk}, \qquad \epsilon = \sqrt{1 - \left(\frac{J_2'}{J_3'}\right)^2}.$$

11. Set up the problem of the relativistic Kepler motion in action-angle variables, using the Hamiltonian in the form given by Eq. (7–20). Show in particular that the total energy (including rest mass) is given by

$$\frac{E}{mc^2} = \frac{1}{\sqrt{1 + \dfrac{4\pi^2 k^2}{[(J_3' - J_2')c + \sqrt{J_2'^2 c^2 - 4\pi^2 k^2}]^2}}}.$$

Note that the degeneracy has been partly lifted, because the orbit is no longer closed, but is still confined to a plane. In the limit as c approaches infinity show that this reduces to Eq. (9–75).

CHAPTER 10

SMALL OSCILLATIONS

In the previous chapter multiply periodic systems were considered, where the motion could be represented in a multiple Fourier expansion in terms of the fundamental frequencies ν_i and all their harmonics and combination frequencies. An important special case of multiply periodic motion occurs with oscillations of such small amplitudes that only the fundamental frequencies are excited and none of the harmonics. The general method of action-angle variables is not particularly appropriate for such motion, and more elementary, though highly specialized, techniques have been devised. The theory of small oscillations finds widespread physical applications in acoustics, molecular spectra, and coupled circuits. It also prepares the way for the discussion of the mechanics of continuous systems and fields, to be given in the next chapter. We shall be concerned here primarily with small oscillations about positions of stable equilibrium, although it is also possible to treat small oscillations about stable motion.

10–1 Formulation of the problem. We consider conservative systems in which the potential energy is a function of position only. It will be assumed that the transformation equations defining the generalized coordinates of the system, $q_1 \ldots q_n$, do not involve the time explicitly. Thus, time-dependent constraints are to be excluded. The system is said to be in *equilibrium* when the generalized forces acting on the system vanish:

$$Q_i = \left(\frac{\partial V}{\partial q_i}\right)_0 = 0. \tag{10–1}$$

The potential energy therefore has an extremum at the equilibrium configuration of the system, $q_{01}, q_{02} \ldots q_{0n}$. If the configuration is initially at the equilibrium position, with zero initial velocities $\dot{q}_i$, then the system will continue in equilibrium indefinitely. Examples of the equilibrium of mechanical systems are legion — a pendulum at rest, a suspension galvanometer at its zero position, an egg standing on end.

An equilibrium position is classified as *stable* if a small disturbance of the system from equilibrium results only in small bounded motion about the rest position. The equilibrium is *unstable* if an infinitesimal disturbance produces unbounded motion. A pendulum at rest is in stable equilibrium, but the egg standing on end is an obvious illustration of unstable equilibrium. It can be readily seen that when the extremum of V is a minimum the equilibrium must be stable. Suppose the system is disturbed from the equilibrium by an increase in energy dE above the equilibrium energy. If V is a minimum at equilibrium any deviation from this position will produce an increase in V. By the conservation of energy the velocities must then decrease and eventually come to zero, indicating bound motion. On the other hand, if V decreases as the result of some departure from equilibrium, the kinetic energy and the velocities increase indefinitely, corresponding to unstable motion. The same conclusion may be arrived at graphically by examining the shape of the potential energy curve, as shown symbolically in Fig. 10–1. A more rigorous mathematical proof that stable equilibrium requires a minimum in V will be given in the course of the discussion.

We shall be interested in the motion of the system within the immediate neighborhood of a configuration of stable equilibrium. Since the departures from equilibrium are to be small, all functions may be expanded in a Taylor series about the equilibrium, retaining only the lowest order terms. The deviations of the generalized coordinates from equilibrium will be denoted by η_i:

$$q_i = q_{0i} + \eta_i, \tag{10–2}$$

and these may be taken as the new generalized coordinates of the motion. Expanding the potential energy about q_{0i}, we obtain

$$V(q_1 \ldots q_n) = V(q_{01} \ldots q_{0n}) + \sum_i \left(\frac{\partial V}{\partial q_i}\right)_0 \eta_i + \frac{1}{2} \sum_{i,j} \left(\frac{\partial^2 V}{\partial q_i\, \partial q_j}\right)_0 \eta_i \eta_j + \cdots \tag{10–3}$$

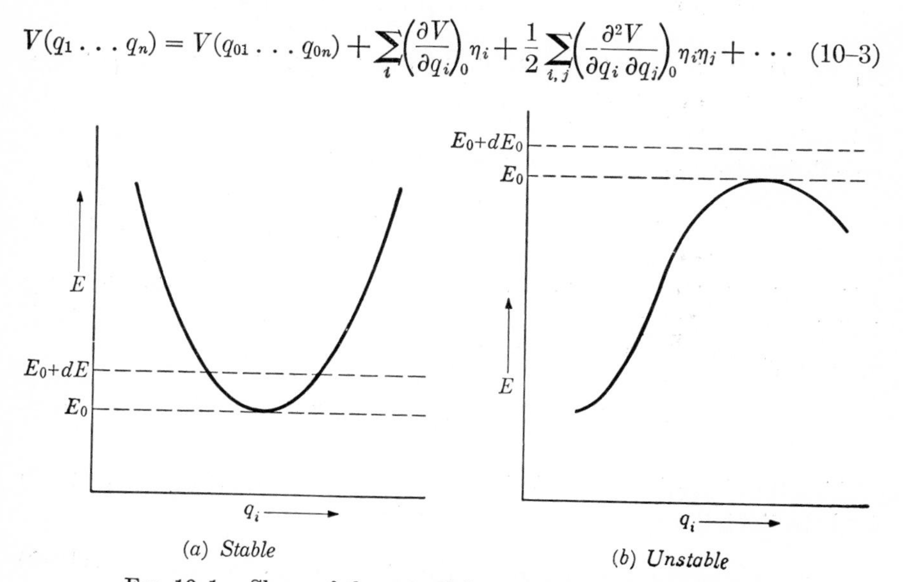

(a) *Stable* (b) *Unstable*

Fig. 10–1. Shape of the potential energy curve at equilibrium.

The terms linear in η_i vanish automatically in consequence of the equilibrium conditions (10–1). The first term in the series is the potential energy of the equilibrium position, and by shifting the arbitrary zero of potential to coincide with the equilibrium potential, this term may also be made to vanish. We are therefore left with the quadratic terms as the first approximation to V:

$$V = \frac{1}{2}\sum_{i,j}\left(\frac{\partial^2 V}{\partial q_i\,\partial q_j}\right)_0 \eta_i\eta_j = \frac{1}{2}\sum_{i,j} V_{ij}\eta_i\eta_j, \tag{10–4}$$

where the second derivatives of V have been designated by the constants V_{ij} depending only on the equilibrium values of the q_i's. It is obvious from their definition that the V_{ij}'s are symmetrical, i.e., that $V_{ij} = V_{ji}$.

A similar series expansion can be obtained for the kinetic energy. Since the generalized coordinates do not involve the time explicitly, the kinetic energy is a homogeneous quadratic function of the velocities (cf. Eq. (1–62)):

$$T = \frac{1}{2}\sum_{i,j} m_{ij}\dot{q}_i\dot{q}_j = \frac{1}{2}\sum_{i,j} m_{ij}\dot{\eta}_i\dot{\eta}_j. \tag{10–5}$$

The coefficients m_{ij} are in general functions of the coordinates q_i, but they may be expanded in a Taylor series about the equilibrium configuration:

$$m_{ij}(q_1 \ldots q_n) = m_{ij}(q_{01} \ldots q_{0n}) + \sum_k \left(\frac{\partial m_{ij}}{\partial q_k}\right)_0 \eta_k + \cdots$$

As Eq. (10–5) is already quadratic in the $\dot{\eta}_i$'s, the lowest nonvanishing approximation to T is obtained by dropping all but the first term in the expansions of m_{ij}. Denoting the constant values of the m_{ij} functions at equilibrium by T_{ij}, we can therefore write the kinetic energy as

$$T = \frac{1}{2}\sum_{i,j} T_{ij}\dot{\eta}_i\dot{\eta}_j. \tag{10–6}$$

It is again obvious that the constants T_{ij} must be symmetric, since the individual terms in Eq. (10–6) are unaffected by an interchange of indices. From Eqs. (10–4) and (10–6) the Lagrangian is given by

$$L = \frac{1}{2}\sum_{i,j}(T_{ij}\dot{\eta}_i\dot{\eta}_j - V_{ij}\eta_i\eta_j). \tag{10–7}$$

Taking the η's as the generalized coordinates, the Lagrangian of Eq. (10–7) leads to the following n equations of motion:

$$\sum_j T_{ij}\ddot{\eta}_j + \sum_j V_{ij}\eta_j = 0, \tag{10–8}$$

where explicit use has been made of the symmetry property of the V_{ij} and T_{ij} coefficients. Each of the equations (10–8) will involve, in general, all of the coordinates η_i, and it is this set of simultaneous differential equations which must be solved to obtain the motion near the equilibrium.

10–2 The eigenvalue equation and the principal axis transformation. The equations of motion (10–8) are linear differential equations with constant coefficients, of a form familiar from electrical circuit theory. We are therefore led to try an oscillatory solution of the form

$$\eta_i = Ca_i e^{-i\omega t}. \tag{10–9}$$

Here Ca_i gives the complex amplitude of the oscillation for each coordinate η_i, the factor C being introduced for convenience as a scale factor, the same for all coordinates. It is understood, of course, that it is the real part of Eq. (10–9) which is to correspond to the actual motion. Substitution of the trial solution (10–9) into the equations of motion leads to the following equations for the amplitude factors:

$$\sum_j (V_{ij}a_j - \omega^2 T_{ij}a_j) = 0. \tag{10–10}$$

Eqs. (10–10) constitute n linear homogeneous equations for the a_i's, and consequently can have a solution only if the determinant of the coefficients vanishes:

$$\begin{vmatrix} V_{11} - \omega^2 T_{11} & V_{12} - \omega^2 T_{12} & \cdots\cdots \\ V_{21} - \omega^2 T_{21} & V_{22} - \omega^2 T_{22} & \\ V_{31} - \omega^2 T_{31} & & \\ \vdots & & \end{vmatrix} = 0. \tag{10–11}$$

This determinantal condition is in effect an algebraic equation of the nth degree for ω^2, and the roots of the determinant provide the frequencies for which Eq. (10–9) represents a correct solution to the equations of the motion. For each of these values of ω^2 the equations (10–10) may be solved for the amplitudes of a_i, or more precisely, for $n - 1$ of the amplitudes in terms of the remaining a_i.

All this has a familiar ring, and we may obtain the proper mathematical perspective by briefly considering a simple variant of the general problem. Suppose the appropriate generalized coordinates were the cartesian coordinates of the system particles. The kinetic energy then contains only the squares of the velocity components. By introducing generalized coordinates which are the cartesian components multiplied by the square root of the particle mass, the kinetic energy can be put in the form

$$T = \frac{1}{2}\sum_i \dot{\eta}_i^2, \tag{10-12}$$

so that in this case $T_{ij} = \delta_{ij}$. If ω^2 is denoted by λ, the homogeneous equations (10–10) simplify to

$$\sum_j V_{ij}a_j = \lambda a_i. \tag{10-13}$$

But this is precisely the formulation of the eigenvalue problem familiar to us from Chapters 4 and 5 (cf. Eq. (5–22)); the only difference is that the vector space has n dimensions rather than three. Considering V_{ij} as an element of an $n \times n$ matrix V and a_i as the component of an n-dimensional vector $\mathbf{a}$, Eq. (10–13) can be put in the form

$$\mathsf{V}\mathbf{a} = \lambda\mathbf{a},$$

which resembles the eigenvalue equation (4–74). Under these conditions the determinantal equation (10–11) similarly reduces to the secular equation for the eigenvalues λ.

Since V is symmetrical and real, the corresponding eigenvalues are real (cf. Section 5–4). If the n sets of the a_i's corresponding to the n eigenvalues are formed into a matrix A then, as in Section 4–6, A must diagonalize V by means of a similarity transformation. Further, the n eigenvectors $\mathbf{a}$ are orthogonal to each other (Section 5–4) and the diagonalizing matrix A must therefore be orthogonal.

These conclusions are valid beyond the special case in which T_{ij} is diagonal; similar results can be proved for the general problem. Eqs. (10–10) do represent a type of eigenvalue equation, for writing T_{ij} as an element of the matrix T, the equations may be written

$$\mathsf{V}\mathbf{a} = \lambda\mathsf{T}\mathbf{a}. \tag{10-14}$$

Here the effect of V on the eigenvector $\mathbf{a}$ is not merely to reproduce the vector times the factor λ, as in the ordinary eigenvalue problem. Instead, the eigenvector is such that V acting on $\mathbf{a}$ produces a multiple of the result of T acting on $\mathbf{a}$. We shall show that the eigenvalues λ for which Eq. (10–14) can be satisfied are all real in consequence of the hermitean property of T and V, and, in fact, must be positive. It will also be shown that the eigenvectors $\mathbf{a}$ are orthogonal — in a sense. In addition, the matrix of the eigenvectors, A, diagonalizes *both* T and V, the former to the unit matrix $\mathsf{1}$ and the latter to a matrix whose diagonal elements are the eigenvalues λ.

Proceeding as in Section 5–4, let a_{jk} represent the jth component of the kth eigenvector. A typical one of the equations (10–14) for the eigenvalue λ_k can be written

$$\sum_j V_{ij}a_{jk} = \lambda_k \sum_j T_{ij}a_{jk}. \tag{10–15}$$

The complex conjugate of the similar equation for λ_l has the form

$$\sum_i V_{ij}a_{il}^* = \lambda_l^* \sum_i T_{ij}a_{il}^*. \tag{10–16}$$

Multiply Eq. (10–16) by a_{jk}, sum over j, and subtract the resulting equation from the similar product of Eq. (10–15) with a_{il}^* summed over i. The left-hand side of the difference equation vanishes, leaving only

$$0 = (\lambda_k - \lambda_l^*) \sum_{i,j} T_{ij}a_{jk}a_{il}^*. \tag{10–17}$$

Consider first the special form of Eq. (10–17) when $l = k$:

$$(\lambda_k - \lambda_k^*) \sum_{i,j} T_{ij}a_{jk}a_{ik}^* = 0. \tag{10–18}$$

The sum over i and j will now be shown to be real, and, in fact, positive definite. To verify this statement, separate a_{jk} into its real and imaginary components:

$$a_{jk} = \alpha_{jk} + i\beta_{jk}.$$

The summation can then be written:

$$\sum_{i,j} T_{ij}a_{jk}a_{ik}^* = \sum_{i,j} T_{ij}\alpha_{jk}\alpha_{ik} + \sum_{i,j} T_{ij}\beta_{jk}\beta_{ik} + i\sum_{i,j} T_{ij}(\beta_{jk}\alpha_{ik} - \beta_{ik}\alpha_{jk}).$$

The imaginary term vanishes in consequence of the symmetry of T_{ij}, for an interchange of the dummy indices i and j changes the sign of the summation. Hence the sum is real. Further, it is seen from the definition of the coefficients T_{ij}, Eq. (10–6), that the two real sums are twice the kinetic energy when the velocities $\dot{\eta}_i$ have the values α_{ik} and β_{ik} respectively. Now, a kinetic energy by its nature must be positive definite for real velocities, and therefore the summation in Eq. (10–18) cannot be zero. It follows that the eigenvalues λ_k must be real.

Since the eigenvalues are real, the ratio of the eigenvector components a_{jk} determined by Eqs. (10–15) must all be real. There is still some indeterminateness, of course, since the value of a particular one of the a_{jk}'s can still be chosen at will without violating Eqs. (10–15). We can require, however, that this component shall be real, and the reality of λ_k then insures the reality of all the other components. (Any complex phase factor in the amplitude of the oscillation will be thrown into the factor C, Eq. (10–9).) Multiply now Eq. (10–15) by a_{ik} and sum over i:

$$\sum_{i,j} V_{ij}a_{ik}a_{jk} = \lambda_k \sum_{i,j} T_{ij}a_{ik}a_{jk},$$

an equation which can be solved for λ_k:

$$\lambda_k = \frac{\sum V_{ij}a_{ik}a_{jk}}{\sum T_{ij}a_{ik}a_{jk}}. \tag{10–19}$$

The denominator of this expression is equal to twice the kinetic energy for velocities a_{ik}, and since the eigenvectors are real the sum must be positive definite. Similarly, the numerator is the potential energy for coordinates a_{ik}, and the condition that V be a minimum at equilibrium requires that the sum must be positive or zero. Neither numerator nor denominator can be negative, and the denominator cannot be zero, hence λ is always finite and positive. (It may, however, be zero.) It will be remembered that λ stands for ω^2, so that positive λ corresponds to real frequencies of oscillation. Were the potential not a strict minimum, the numerator in Eq. (10–19) might be negative, giving rise to imaginary frequencies which would produce an unbounded exponential increase of the η_i with time. Such motion would obviously be unstable, and we have here the promised mathematical proof that a minimum of the potential is required for stable motion.

Let us return for the moment to Eq. (10–17) which, in view of the reality of the eigenvalues and eigenvectors, can be written

$$(\lambda_k - \lambda_l)\sum_{i,j} T_{ij}a_{il}a_{jk} = 0. \tag{10–17′}$$

If all the roots of the secular equation are distinct, then Eq. (10–17′) can hold only if the summation vanishes for l not equal to k:

$$\sum_{i,j} T_{ij}a_{il}a_{jk} = 0. \qquad l \neq k \tag{10–20a}$$

It has been remarked several times that the values of the a_{jk}'s are not completely fixed by the eigenvalue equations (10–10). We can remove this indeterminacy by requiring further that

$$\sum_{i,j} T_{ij}a_{ik}a_{jk} = 1. \tag{10–20b}$$

There are n such equations (10–20b) and they uniquely fix the one arbitrary component of each of the n eigenvectors $\mathbf{a}_k$.* The two equations (10–20a and b) can be combined into one equation of the form

$$\sum_{i,j} T_{ij} a_{il} a_{jk} = \delta_{lk}. \tag{10–21}$$

If two or more of the roots are repeated, the argument leading to Eq. (10–20a) falls through for $\lambda_l = \lambda_k$. We shall reserve a discussion of this exceptional case for a later time. It will suffice for the present to state that a set of a_{jk} coefficients can always be found which satisfies both the eigenvalue equations (10–10) and Eq. (10–20a).

The condition (10–21) can be written in matrix notation as

$$\tilde{\mathsf{A}}\mathsf{T}\mathsf{A} = 1. \tag{10–21'}$$

On the other hand, a matrix A is said to be orthogonal (cf. Eq. (4–36)) if

$$\tilde{\mathsf{A}}\mathsf{A} = 1. \tag{10–22}$$

The two conditions are somewhat similar in form, but Eq. (10–21′) differs from the orthogonality requirement by the presence of the matrix T in the middle. It will be seen, however, that Eq. (10–21′) does represent an orthogonality condition, but in a space which is not in general cartesian. The ordinary orthogonality requirement states in effect that each of the vectors $\mathbf{a}_k$ is of unit magnitude:

$$\mathbf{a}_k \cdot \mathbf{a}_k = \sum_j a_{jk}^2 = 1,$$

and that any two of the vectors are perpendicular:

$$\mathbf{a}_l \cdot \mathbf{a}_k = \sum_j a_{jl} a_{jk} = 0. \qquad j \neq k$$

Now in a noncartesian space with inclined axes, in which the elements of the metric tensor T are constants independent of the coordinates, the length of a vector $\mathbf{a}_k$ is given by (cf. Eq. (7–42)):

$$\mathbf{a}_k \cdot \mathbf{a}_k = \sum_{i,j} T_{ij} a_{ik} a_{jk}.$$

---

* Eq. (10–20b) may be put in a form which explicitly shows that it suffices to remove the indeterminacy in the a_{jk}'s. Suppose that it is the magnitude of a_{1k} which is to be evaluated; the ratio of all the other a_{jk}'s to a_{1k} is obtained from Eqs. (10–10). Then Eq. (10–20b) can be written as

$$\sum_{i,j} T_{ij} \frac{a_{ik}}{a_{1k}} \frac{a_{jk}}{a_{1k}} = \frac{1}{a_{1k}^2}.$$

The left-hand side is completely determined from the eigenvalue equations and may be evaluated directly to provide a_{1k}.

The dot product of any two vectors $\mathbf{a}_l$, $\mathbf{a}_k$ in such a space is, correspondingly,

$$\mathbf{a}_l \cdot \mathbf{a}_k = \sum_{i,j} T_{ij} a_{il} a_{jk}.$$

Comparison with Eqs. (10–20) shows that (10–20b) states the vectors $\mathbf{a}_k$ are of unit length, while (10–20a) implies that any two different vectors $\mathbf{a}_l$, $\mathbf{a}_k$ are perpendicular. Eq. (10–21′) is accordingly the *orthogonality condition for the matrix* A *in the configuration space whose metric tensor is* T. In a cartesian space the metric tensor is the unit tensor $\mathsf{1}$ and the condition (10–21′) then reduces to the ordinary orthogonality requirements.

In Chapter 4 the *similarity* transformation of a matrix C by a matrix B was defined by the equation (cf. Eq. (4–41)):

$$\mathsf{C}' = \mathsf{BCB}^{-1}.$$

We now introduce the related concept of the *congruent* transformation of C by A according to the relation

$$\mathsf{C}' = \tilde{\mathsf{A}}\mathsf{CA}. \tag{10–23}$$

If A is orthogonal, so that $\tilde{\mathsf{A}} = \mathsf{A}^{-1}$, there is no essential difference between the two types of transformation (as may be seen by denoting A^{-1} by the matrix E). Eq. (10–21′) can therefore be read as the statement that A transforms T by a congruent transformation into a diagonal matrix, in particular into the unit matrix.

If a diagonal matrix $\boldsymbol{\lambda}$ with elements $\lambda_{lk} = \lambda_k \delta_{lk}$ be introduced, the eigenvalue equations (10–15) may be written

$$\sum_j V_{ij} a_{jk} = \sum_{j,l} T_{ij} a_{jl} \lambda_{lk},$$

which becomes in matrix notation:

$$\mathsf{VA} = \mathsf{TA}\boldsymbol{\lambda}. \tag{10–24}$$

Multiplying by $\tilde{\mathsf{A}}$ from the left, Eq. (10–24) takes the form

$$\tilde{\mathsf{A}}\mathsf{VA} = \tilde{\mathsf{A}}\mathsf{TA}\boldsymbol{\lambda},$$

which, by Eq. (10–21′) reduces to:

$$\tilde{\mathsf{A}}\mathsf{VA} = \boldsymbol{\lambda}. \tag{10–25}$$

Our final equation (10–25) states that a congruent transformation of V by A changes it into a diagonal matrix whose elements are the eigenvalues λ_k.

The matrix A thus simultaneously diagonalizes both T and V. Remembering the interpretation of T as a metric tensor in configuration space, we can give the following meaning to the diagonalization process. A is the matrix of a linear transformation from a system of *inclined* axes to *cartesian orthogonal* axes, as evidenced by the fact that the transformed metric tensor is $\mathbf{1}$. At the same time, the new axes are the perpendicular *principal axes* of V, so that the matrix V is diagonal in the transformed coordinate system. The entire process of obtaining the fundamental frequencies of small oscillation is thus a particular type of *principal axis transformation,* such as was discussed in Chapter 5.

It remains only to consider the case of multiple roots to the secular equation, a situation which is more annoying in the mathematical theory than it is in practice. If one or more of the roots are repeated, it is found that the number of independent equations among the eigenvalues is insufficient to determine even the ratio of the eigenvector components. Thus, if the eigenvalue λ is a double root, any two of the components a_j may be chosen arbitrarily, the rest being fixed by the eigenvalue equations. To illustrate, let us consider a two-dimensional system, in which the secular equation appears as

$$\begin{vmatrix} V_{11} - \lambda T_{11} & V_{12} - \lambda T_{12} \\ V_{12} - \lambda T_{12} & V_{22} - \lambda T_{22} \end{vmatrix} = 0,$$

or

$$(V_{12} - \lambda T_{12})^2 - (V_{11} - \lambda T_{11})(V_{22} - \lambda T_{22}) = 0.$$

Suppose now that the matrix elements are such that

$$\frac{V_{12}}{T_{12}} = \frac{V_{11}}{T_{11}} = \frac{V_{22}}{T_{22}} = \lambda_0. \tag{10-26}$$

Then the secular equation can be written

$$(T_{12}^2 - T_{11}T_{22})(\lambda_0 - \lambda)^2 = 0,$$

indicating that λ_0 is a double root of the secular equation. But the eigenvalue equations (10-10) for this root are

$$\begin{aligned} (V_{11} - \lambda_0 T_{11})a_1 + (V_{12} - \lambda_0 T_{12})a_2 &= 0, \\ (V_{12} - \lambda_0 T_{12})a_1 + (V_{22} - \lambda_0 T_{22})a_2 &= 0, \end{aligned}$$

and by virtue of the conditions (10-26) all of the coefficients of the a's vanish identically. Any set of values for the two a's will then satisfy the eigenvalue equations. Even with the normalization requirement (10-20b) there will thus be a single infinity of eigenvectors corresponding to a double root of the secular equation, a double infinity for a triple root, and so on.

In general, any pair of eigenvectors randomly chosen out of the infinite set of allowed vectors will not be orthogonal. Nevertheless, it is always possible to construct a pair of allowed vectors which are orthogonal, and these can be used to form the orthogonal matrix A. Consider for simplicity the procedure to be followed for a double root. Let $\mathbf{a}'_k$ and $\mathbf{a}'_l$ be any two allowable eigenvectors for a given double root λ, and further let $\mathbf{a}'_k$ be normalized so as to satisfy Eq. (10–20b). Any linear combination of $\mathbf{a}'_k$ and $\mathbf{a}'_l$ will also be an eigenvector for the root λ. We therefore seek to construct a vector $\mathbf{a}_l$:

$$\mathbf{a}_l = c_1\mathbf{a}'_k + c_2\mathbf{a}'_l, \tag{10–27}$$

where c_1 and c_2 are constants such that $\mathbf{a}_l$ is orthogonal to $\mathbf{a}'_k$. In component form Eq. (10–27) can be written

$$a_{il} = c_1a'_{ik} + c_2a'_{il}. \tag{10–27$'$}$$

Multiply Eq. (10–27′) by $T_{ij}a'_{jk}$ and sum over i and j:

$$\sum_{i,j} T_{ij}a_{il}a'_{jk} = c_1 + c_2\sum_{i,j} T_{ij}a'_{il}a'_{jk}.$$

To satisfy the orthogonality condition (10–20a) the left-hand side of this equation must vanish, which can be insured by setting

$$\frac{c_1}{c_2} = -\sum_{i,j} T_{ij}a'_{il}a'_{jk}.$$

Another equation for the ratio c_1/c_2 is supplied by requiring that $\mathbf{a}_l$ satisfy the normalization condition (10–20b). Together the two equations fix the coefficients c_1 and c_2, and therefore the vector $\mathbf{a}_l$. Both $\mathbf{a}_l$ and $\mathbf{a}_k \equiv \mathbf{a}'_k$ are automatically orthogonal to the eigenvectors of the other distinct eigenvalues, for then the argument based on Eq. (10–17′) remains valid. Hence we have a set of n eigenvectors $\mathbf{a}_j$ whose components form the matrix A satisfying Eq. (10–21′).

A similar procedure is followed for a root of higher multiplicity. If λ is an m-fold root, then orthogonal normalized eigenvectors are formed out of linear combinations of any of the m corresponding eigenvectors $\mathbf{a}'_1 \ldots \mathbf{a}'_m$. The first of the "ortho-normal" eigenvectors $\mathbf{a}_1$ is then chosen as a multiple of $\mathbf{a}'_1$; $\mathbf{a}_2$ is taken as a linear combination of $\mathbf{a}'_1$ and $\mathbf{a}'_2$; and so on. In this manner the number of constants to be determined is equal to the sum of the first m integers, or $\frac{1}{2}m(m+1)$. The normalization requirements provide m conditions, while there are $\frac{1}{2}m(m-1)$ orthogonality conditions, and together these are just enough to fix the constants uniquely.

This process of constructing orthogonalized eigenvectors in the case of multiple roots is completely analogous to the method of constructing a sequence of orthogonal functions out of any arbitrary set of functions. Phrased in geometrical language, it is also seen to be identical with the procedure followed in Chapter 5 for multiple eigenvalues of the inertia tensor. For example, the added indeterminacy in the eigenvector components for a double root means that all of the vectors in a *plane* are eigenvectors. We merely choose any two perpendicular directions in the plane as being the new principal axes, with the eigenvectors in $\mathbf{A}$ as unit vectors along these axes.

Multiple roots of the secular equation are often referred to as degenerate frequencies. It should be noted that the term has a different meaning here than in the preceding chapter. Two distinct frequencies, even if commensurable, are not considered here as degenerate, although they would be so classified in Chapter 9.

10-3 Frequencies of free vibration, and normal coordinates.

The somewhat lengthy arguments of the preceding section demonstrate that the equations of motion will be satisfied by an oscillatory solution of the form (10-9) not merely for one frequency but in general for a set of n frequencies ω_k. A complete solution of the equations of motion therefore involves a superposition of oscillations with all the allowed frequencies. Thus, if the system is displaced slightly from equilibrium and then released, the system performs small oscillations about the equilibrium with the frequencies $\omega_1 \ldots \omega_n$. The solutions of the secular equation are therefore often designated as the frequencies of *free vibration* or as the *resonant frequencies* of the system.

The general solution of the equations of motion may now be written as

$$\eta_i = \sum_k C_k a_{ik} e^{-i\omega_k t}, \tag{10-28}$$

there being a complex scale factor C_k for each resonant frequency. It might be objected that for each solution λ_k of the secular equation there are two resonant frequencies $+\omega_k$ and $-\omega_k$. The eigenvector $\mathbf{a}_k$ would be the same for the two frequencies, but the scale factors C_k^+ and C_k^- could conceivably be different. On this basis the general solution should appear as

$$\eta_i = \sum_k a_{ik}(C_k^+ e^{+i\omega_k t} + C_k^- e^{-i\omega_k t}). \tag{10-29}$$

It is to be remembered, however, that the actual motion is the real part of the complex solution, and the real part of either (10-28) or (10-29) can be written in the form

$$\eta_i = \sum_k f_k a_{ik} \cos(\omega_k t + \delta_k), \tag{10-30}$$

where the amplitude f_k and the phase δ_k are determined from the initial conditions. Either of the solutions (10–28, 29) will therefore represent the actual motion, and the former, of course, is the more convenient.

The orthogonality properties of A greatly facilitate the determination of the scale factors C_k in terms of the initial conditions. At $t = 0$ the real part of Eq. (10–28) reduces to

$$\eta_i(0) = \sum_k \mathrm{Re}\, C_k a_{ik}, \tag{10–31}$$

where Re stands for "real part of." Similarly, the initial value of the velocities is obtained as

$$\dot{\eta}_i(0) = \sum_k \mathrm{Im}\, C_k a_{ik}\omega_k, \tag{10–32}$$

where Im C_k denotes the imaginary part of C_k. From these $2n$ equations the real and imaginary parts of the n constants C_k may be evaluated. Tc solve Eq. (10–31), for example, multiply on both sides by $T_{ij}a_{jl}$ and sum over i and j:

$$\begin{aligned}\sum_{i,j} T_{ij}\eta_i(0)a_{jl} &= \sum_{i,j,k} \mathrm{Re}\, C_k T_{ij} a_{ik} a_{jl} \\ &= \sum_k \mathrm{Re}\, C_k\, \delta_{kl},\end{aligned}$$

by Eq. (10–21). Performing the summation over k, we obtain

$$\mathrm{Re}\, C_l = \sum_{i,j} T_{ij}\eta_i(0)a_{jl}. \tag{10–33}$$

Similarly, the imaginary part of the scale factors is given by

$$\mathrm{Im}\, C_l = -\frac{1}{\omega_l}\sum_{i,j} T_{ij}\dot{\eta}_i(0)a_{jl}. \tag{10–34}$$

Eqs. (10–33, 34) thus permit the direct computation of the complex factors C_l in terms of the initial conditions and the matrices T and A.

The complete solution for η_i, Eq. (10–28), is an example of a multiple periodic motion as discussed in Chapter 9. It is true that it is a particularly simple type of multiple periodicity, for each term of the multiple Fourier expansion involves only one of the fundamental frequencies, and all harmonic terms are absent. But even with this simplification, the motion is only conditionally periodic, for unless the resonant frequencies are commensurable η_i will never repeat its initial value. Hence the coordinates η_i are not in general the separation coordinates of the problem, each of which is simply periodic. We can obtain such a set of periodic coordinates, however, by a point transformation from the η_i's.

We define a new set of coordinates ζ_j related to the original coordinates η_i by the equations

$$\eta_i = \sum_j a_{ij}\zeta_j. \tag{10–35}$$

If η_i and ζ_j be represented as the elements of single-column matrices $\boldsymbol{\eta}$ and $\boldsymbol{\zeta}$ respectively, the defining equations (10–35) appear as

$$\boldsymbol{\eta} = \mathsf{A}\boldsymbol{\zeta}. \tag{10–36}$$

In consequence of the properties of the matrix A the potential and kinetic energies take on simple forms when expressed in the new coordinates. The potential energy (10–4):

$$V = \frac{1}{2}\sum_{i,j}\eta_i V_{ij}\eta_j,$$

can be written matrix-wise as

$$V = \frac{1}{2}\,\tilde{\boldsymbol{\eta}}\mathsf{V}\boldsymbol{\eta}. \tag{10–37}$$

Similarly, the kinetic energy (10–6) can be written as a matrix product

$$T = \frac{1}{2}\,\tilde{\dot{\boldsymbol{\eta}}}\mathsf{T}\dot{\boldsymbol{\eta}}. \tag{10–38}$$

Now, the transposed matrix $\tilde{\boldsymbol{\eta}}$ (which is a one-row matrix) is related to $\tilde{\boldsymbol{\zeta}}$ by the equation

$$\tilde{\boldsymbol{\eta}} = \widetilde{\mathsf{A}\boldsymbol{\zeta}} = \tilde{\boldsymbol{\zeta}}\tilde{\mathsf{A}},$$

(cf. Exercise 2, Chapter 4) and the potential energy therefore appears as

$$V = \frac{1}{2}\,\tilde{\boldsymbol{\zeta}}\tilde{\mathsf{A}}\mathsf{V}\mathsf{A}\boldsymbol{\zeta}.$$

But A diagonalizes V by a congruent transformation, cf. Eq. (10–25), and the potential reduces simply to

$$V = \frac{1}{2}\,\tilde{\boldsymbol{\zeta}}\boldsymbol{\lambda}\boldsymbol{\zeta}. \tag{10–39}$$

Written explicitly, Eq. (10–39) appears as

$$V = \frac{1}{2}\sum_k \omega_k^2\zeta_k^2. \tag{10–40}$$

The kinetic energy has an even simpler form in the new coordinates. Since the velocities transform as the coordinates, T can be written

$$T = \frac{1}{2}\,\tilde{\dot{\boldsymbol{\zeta}}}\tilde{\mathsf{A}}\mathsf{T}\mathsf{A}\dot{\boldsymbol{\zeta}},$$

and by virtue of the "orthogonal" property of A, Eq. (10–21′), this reduces to

$$T = \frac{1}{2}\,\tilde{\dot{\boldsymbol{\zeta}}}\dot{\boldsymbol{\zeta}}. \tag{10–41}$$

In terms of the new velocities, the kinetic energy is therefore

$$T = \frac{1}{2}\sum_k \dot{\zeta}_k^2. \tag{10–42}$$

Equations (10–40 and 42) state that in the new coordinates both the potential and kinetic energies are sums of squares only, without any cross terms. Of course, this result is simply another way of saying that A produces a principal axis transformation. It will be remembered that the principal axis transformation of the inertia tensor was specifically designed to reduce the moment of inertia to a sum of squares; the new axes being the principal axes of the inertia ellipsoid. Here the kinetic and potential energies are also quadratic forms (as was the moment of inertia) and both are diagonalized by A. For this reason the principal axis transformation employed here is a particular example of the well-known algebraic process of the *simultaneous diagonalization of two quadratic forms.*

The equations of motion share in the simplification produced by the new coordinates. The new Lagrangian is

$$L = \frac{1}{2}\sum_k (\dot{\zeta}_k^2 - \omega_k^2\zeta_k^2), \tag{10–43}$$

so that the Lagrange equations for ζ_k are

$$\ddot{\zeta}_k + \omega_k^2\zeta_k = 0. \tag{10–44}$$

Eqs. (10–44) have the immediate solutions

$$\zeta_k = C_k e^{-i\omega_k t}, \tag{10–45}$$

which could have been seen, of course, directly from Eq. (10–28). Each of the new coordinates is thus a simply periodic function involving only *one* of the resonant frequencies. It is therefore customary to call the ζ's the *normal coordinates* of the system. Obviously, the normal coordinates are also the separation coordinates for the problem and $\omega_k t/2\pi$ is the kth angle variable.

Each normal coordinate corresponds to a vibration of the system with only one frequency, and these component oscillations are spoken of as the *normal modes of vibration.* All of the particles in each mode vibrate with the same frequency and with the same phase; * the relative amplitudes being determined by the matrix elements a_{ik}. The complete motion is then built up out of the sum of the normal modes weighted with appropriate amplitude and phase factors contained in the C_k's.

* Particles may be exactly out of phase if the a's have opposite sign.

Harmonics of the fundamental frequencies are absent in the complete motion essentially because of the stipulation that the amplitude of oscillation be small. We were then allowed to represent the potential as a quadratic form, which is characteristic of simple harmonic motion. The normal coordinate transformation emphasizes this point, for the Lagrangian in the normal coordinates (10–43) is seen to be the sum of Lagrangians for harmonic oscillators of frequencies ω_k. We can thus consider the complete motion for small oscillations as being obtained by exciting the various harmonic oscillators with different intensities and phases.*

10–4 Free vibrations of a linear triatomic molecule. To illustrate the technique for obtaining the resonant frequencies and normal modes, we shall consider in detail a model based on a linear symmetrical triatomic molecule. In the equilibrium configuration of the molecule two atoms of mass m are symmetrically located on each side of an atom of mass M (cf. Fig. 10–2). All three atoms are on one straight line, the equilibrium distances apart being denoted by b. For simplicity we shall first consider only vibrations along the line of the molecule, and the actual complicated interatomic potential will be approximated by two springs of force constant k joining the three atoms. There are three obvious coordinates marking the lineal position of the three atoms. In these coordinates the potential energy is

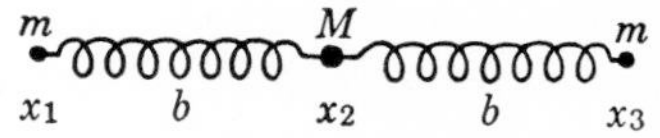

m M m x_1 b x_2 b x_3

FIG. 10–2. Model of a linear symmetrical triatomic molecule.

$$V = \frac{k}{2}(x_2 - x_1 - b)^2 + \frac{k}{2}(x_3 - x_2 - b)^2.$$

We now introduce coordinates relative to the equilibrium positions:

$$\eta_i = x_i - x_{0i},$$

where

$$x_{02} - x_{01} = b = x_{03} - x_{02}.$$

The potential energy then reduces to

$$V = \frac{k}{2}(\eta_2 - \eta_1)^2 + \frac{k}{2}(\eta_3 - \eta_2)^2,$$

---

* It might be mentioned for future reference that the same sort of picture appears in the quantization of the electromagnetic field. The frequencies of the harmonic oscillators are identified with the photon frequencies, and the amplitudes of excitation become the discrete quantized "occupation numbers"—the number of photons of each frequency.

or

$$V = \frac{k}{2}(\eta_1^2 + 2\eta_2^2 + \eta_3^2 - 2\eta_1\eta_2 - 2\eta_2\eta_3). \tag{10–46}$$

Hence the V matrix has the form

$$\mathsf{V} = \begin{pmatrix} k & -k & 0 \\ -k & 2k & -k \\ 0 & -k & k \end{pmatrix}. \tag{10–47}$$

The kinetic energy has an even simpler form:

$$T = \frac{m}{2}(\dot{\eta}_1^2 + \dot{\eta}_3^2) + \frac{M}{2}\dot{\eta}_2^2, \tag{10–48}$$

so that the T matrix is diagonal:

$$\mathsf{T} = \begin{pmatrix} m & 0 & 0 \\ 0 & M & 0 \\ 0 & 0 & m \end{pmatrix}. \tag{10–49}$$

Combining these two matrices, the secular equation appears as

$$|\mathsf{V} - \omega^2\mathsf{T}| = \begin{vmatrix} k - \omega^2 m & -k & 0 \\ -k & 2k - \omega^2 M & -k \\ 0 & -k & k - \omega^2 m \end{vmatrix} = 0. \tag{10–50}$$

Direct evaluation of the determinant leads to the cubic equation

$$\omega^2(k - \omega^2 m)(k(M + 2m) - \omega^2 Mm) = 0, \tag{10–51}$$

with the obvious solutions

$$\omega_1 = 0, \qquad \omega_2 = \sqrt{\frac{k}{m}}, \qquad \omega_3 = \sqrt{\frac{k}{m}\left(1 + \frac{2m}{M}\right)}. \tag{10–52}$$

The first eigenvalue, $\omega_1 = 0$, may appear somewhat surprising and even alarming at first sight. Such a solution does not correspond to an oscillatory motion at all, for the equation of motion for the corresponding normal coordinate is

$$\ddot{\zeta}_1 = 0,$$

which produces a uniform translational motion. But this is precisely the key to the difficulty. The vanishing frequency arises from the fact that the molecule may be translated rigidly along its axis without any change in the potential energy.* Since the restoring force against such motion is

* Such cases, where motion can occur about an equilibrium configuration without disturbing the equilibrium, are referred to as *labile* or *indifferent* equilibrium.

zero, the effective "frequency" must also vanish. We have made the assumption that the molecule has three degrees of freedom for vibrational motion, whereas in reality one of them is a rigid body degree of freedom.

A number of interesting points can be discussed in connection with a vanishing resonant frequency. It is seen from Eq. (10–19) that a zero value of ω can occur only when the potential energy is positive but is not positive definite, i.e., it can vanish even when not all the η_i's are zero. An examination of V, Eq. (10–46), shows that it is not positive definite and that, in fact, V does vanish when all the η's are equal (uniform translation).

Since the zero frequency found here is of no consequence for the vibration frequencies of interest, it is often desirable to phrase the problem so that the root is eliminated from the outset. We can do this here most simply by imposing the condition or constraint that the center of mass remain stationary at the origin:

$$m(x_1 + x_3) + Mx_2 = 0. \qquad (10\text{–}53)$$

Eq. (10–53) can then be used to eliminate one of the coordinates from V and T, reducing the problem to one of two degrees of freedom (cf. Exercise 2, this chapter).

The restriction of the motion to be along the molecular axis allows only one possible type of uniform rigid body motion. However, if the more general problem of vibrations in all three directions is considered, the number of rigid body degrees of freedom will be increased, in general, to six. The molecule may then translate uniformly along the three axes or perform uniform rotations about the axes. Hence in any general system of n degrees of freedom there will be six vanishing frequencies and only $n - 6$ true vibration frequencies. Again, the reduction in the number of degrees of freedom can be performed beforehand by imposing the conservation of linear and angular momentum upon the coordinates.

In addition to rigid body motion, zero resonant frequencies may also arise when the potential is such that both the first *and* second derivatives of V vanish at equilibrium. Small oscillations may still be possible in this case if the fourth derivatives do not also vanish (the third derivatives must vanish for a stable equilibrium), but the vibrations will not be simple harmonic. Such a situation therefore constitutes a breakdown of the customary method of small oscillations, but fortunately it is not of frequent occurrence.

Returning now to the examination of the resonant frequencies, ω_2 will be recognized as the well-known frequency of oscillation for a mass m suspended by a spring of force constant k. We are led to expect, therefore that only the end atoms partake in this vibration; the center molecule remains stationary. It is only in the third mode of vibration, ω_3, that the mass M can participate in the oscillatory motion. These predictions are verified by examining the eigenvectors for the three normal modes.

The components a_{ij} are determined for each frequency by the equations:

$$\begin{aligned}(k - \omega_j^2 m)a_{1j} \qquad - ka_{2j} \qquad\qquad &= 0,\\ -ka_{1j} + (2k - \omega_j^2 M)a_{2j} \qquad - ka_{3j} &= 0,\\ - ka_{2j} + (k - \omega_j^2 m)a_{3j} &= 0,\end{aligned} \tag{10–54}$$

along with the normalization condition:

$$m(a_{1j}^2 + a_{3j}^2) + Ma_{2j}^2 = 1. \tag{10–55}$$

For $\omega_1 = 0$, it follows immediately from the first and third of Eqs. (10–54) that all three coefficients are equal: $a_{11} = a_{21} = a_{31}$. This, of course, is exactly what was expected from the translational nature of the motion (cf. Fig. 10–3a). The normalization condition then fixes the value of a_{1j} so that:

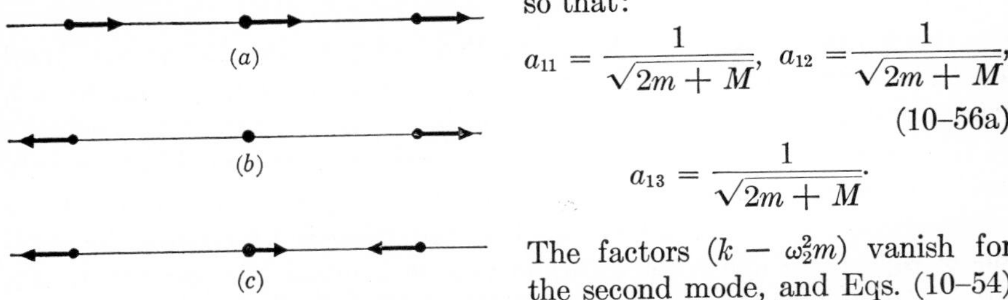

FIG. 10–3. Longitudinal normal modes of the linear symmetric triatomic molecule.

$$a_{11} = \frac{1}{\sqrt{2m + M}}, \quad a_{12} = \frac{1}{\sqrt{2m + M}}, \tag{10–56a}$$

$$a_{13} = \frac{1}{\sqrt{2m + M}}.$$

The factors $(k - \omega_2^2 m)$ vanish for the second mode, and Eqs. (10–54) show immediately that $a_{22} = 0$ (as predicted) and $a_{12} = -a_{32}$. The numerical value of these quantities is then determined by Eq. (10–55):

$$a_{12} = \frac{1}{\sqrt{2m}}, \qquad a_{22} = 0, \qquad a_{32} = -\frac{1}{\sqrt{2m}}. \tag{10–56b}$$

In this mode the center atom is at rest, while the two outer ones vibrate exactly out of phase (as they must in order to conserve linear momentum), cf. Fig. 10–3b. Finally, when $\omega = \omega_3$ it can be seen from the first and third of Eqs. (10–54) that a_{13} and a_{33} must be equal. The rest of the calculation for this mode is not quite as simple as for the others, and it will be sufficient to state the final result:

$$a_{13} = \frac{1}{\sqrt{2m\left(1 + \frac{2m}{M}\right)}}, \quad a_{23} = \frac{-2}{\sqrt{2M\left(2 + \frac{M}{m}\right)}}, \quad a_{33} = \frac{1}{\sqrt{2m\left(1 + \frac{2m}{M}\right)}}. \tag{10–56c}$$

Here the two outer atoms vibrate with the same amplitude, while the inner one oscillates out of phase with them and has a different amplitude, cf. Fig. 10-3c. Any general longitudinal vibration of the molecule which does not involve a rigid translation will be some linear combination of the normal modes ω_2 and ω_3. The amplitudes of the normal modes, and their phases relative to each other, will of course be determined by the initial conditions.

We have spoken so far only of vibrations along the axis; in the actual molecule there will also be normal modes of vibration perpendicular to the axis. The complete set of normal modes is naturally more difficult to determine than merely the longitudinal modes, for the general motion in all directions corresponds to nine degrees of freedom. While the procedure is straightforward, the algebra rapidly becomes quite complicated, and it is not feasible to present the detailed calculation here. However, it is possible to give a qualitative discussion on the basis of general principles, and most of the conclusions of the complete solution can be predicted beforehand.

The general problem will have a number of zero resonant frequencies corresponding to the possibility of rigid body motion. For the linear molecule there will be three degrees of freedom for rigid translation, but rigid rotation can account for only *two* degrees of freedom. Rotation about the axis of the molecule is obviously meaningless and will not appear as a mode of rigid body motion. We are therefore left with four true modes of vibration. Two of these are the longitudinal modes which have already been examined, so that there can only be two modes of vibration perpendicular to the axis. However, the symmetry of the molecule about its axis shows that these two modes of perpendicular vibration must be degenerate. There is nothing to distinguish a vibration in the y direction from a vibration in the z direction, and the two frequencies must be equal. The additional indeterminacy of the eigenvectors of a degenerate mode appears here in that all directions perpendicular to the molecular axis are alike. Any two orthogonal axes in the plane normal to the molecule may be chosen as the directions of the degenerate modes of vibration. The complete motion of the atoms normal to the molecular axis will depend on the amplitudes and relative phases of the two degenerate modes. If both are excited, and they are exactly in phase, then the atoms will move on a straight line passing through the equilibrium configuration. But if they are out of phase, the composite motion is an elliptical Lissajous figure, exactly as in a two-dimensional isotropic oscillator. The two modes then represent a rotation, rather than a vibration.

It is obvious from the symmetry of the molecules that the amplitudes of the end atoms must be identical. The complete calculation shows that the end atoms also travel in the same direction along the Lissajous figure. Hence the center atom must revolve in the opposite direction, in order to conserve angular momentum. Fig. 10–4 illustrates the motion when the degenerate modes are ninety degrees out of phase.

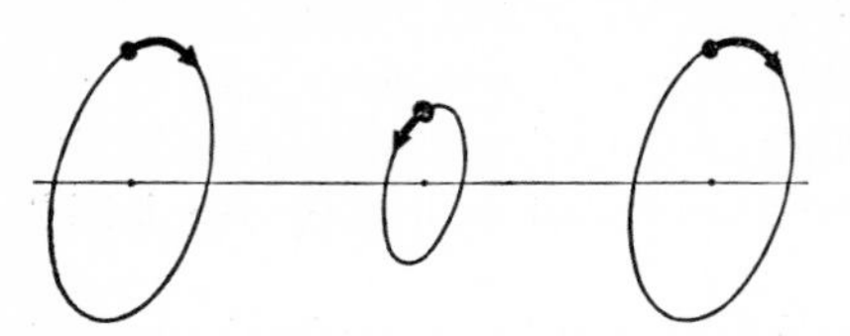

FIG. 10–4. Degenerate modes of the symmetrical triatomic molecule.

10–5 Forced vibrations and the effect of dissipative forces. Free vibrations occur when the system is displaced initially from the equilibrium configuration and is then allowed to oscillate by itself. Very often, however, the system is set into oscillation by an external driving force which continues to act on the system after $t = 0$. The frequency of such a *forced oscillation* is then determined by the frequency of the driving force and not by the resonant frequencies. Nevertheless, the normal modes are of great importance in obtaining the *amplitudes* of the forced vibration, and the problem is greatly simplified by use of the normal coordinates obtained from the free modes.

If F_j is the generalized force corresponding to the coordinate η_j, then by Eq. (1–46) the generalized force Q_i for the normal coordinate ζ_i is

$$Q_i = \sum_j a_{ji} F_j. \tag{10–57}$$

The equations of motion when expressed in normal coordinates now become

$$\ddot{\zeta}_i + \omega_i^2 \zeta_i = Q_i. \tag{10–58}$$

Equations (10–58) are a set of n inhomogeneous differential equations which can be solved only when we know the dependence of Q_i on time. While the solution will not be as simple as in the free case, note that the normal coordinates preserve their advantage of separating the variables, and each equation involves only a single coordinate.

Frequently the driving force varies sinusoidally with time. In an acoustic problem, for example, the driving force might arise from the pressure of a sound wave impinging on the system, and Q_i then has the same frequency as the sound wave. Or, if the system is a polyatomic molecule, a sinusoidal driving force is present if the molecule is illuminated by a monochromatic light beam. Each atom in the molecule is then subject to an electric force whose frequency is that of the incident light. In all such systems the force Q_i can be written

$$Q_i = Q_{0i} \cos(\omega t + \delta_i), \tag{10–59}$$

where ω is the angular frequency of the external force. The equations of motion now appear as

$$\ddot{\zeta}_i + \omega_i^2 \zeta_i = Q_{0i} \cos(\omega t + \delta_i). \tag{10–60}$$

A complete solution of Eq. (10–60) consists of the general solution to the homogeneous equation (i.e., the free modes of vibrations) plus a particular solution to the inhomogeneous equation. By a proper choice of initial conditions, the superimposed free vibrations can be made to vanish,* centering our interest on the particular solution of Eqs. (10–60) which will obviously have the form

$$\zeta_i = B_i \cos(\omega t + \delta_i). \tag{10–61}$$

Here the amplitudes B_i are determined by substituting the solution in Eqs. (10–60):

$$B_i = \frac{Q_{0i}}{\omega_i^2 - \omega^2}. \tag{10–62}$$

The complete motion is then

$$\eta_j = \sum_i a_{ji} \zeta_i = \sum_i \frac{a_{ji} Q_{0i} \cos(\omega t + \delta_i)}{\omega_i^2 - \omega^2}. \tag{10–63}$$

Thus the vibration of each particle is again composed of linear combinations of the normal modes, but now each normal oscillation occurs at the frequency of the driving force.

Two factors determine the extent to which each normal mode is excited. One is the amplitude of the generalized driving force, Q_{0i}. If the force on each particle has no component in the direction of vibration of some particular normal mode, then obviously the generalized force corresponding to the mode will vanish and Q_{0i} will be zero. *An external force can excite a normal mode only if it tends to move the particles in the same direction as in the given mode.* The second factor is the closeness of the driving frequency to the free frequency of the mode. In consequence of the denominators in Eq. (10–63) the closer ω approaches to any ω_i the stronger will that mode be excited relative to the other modes. Indeed, Eq. (10–63) apparently predicts infinite amplitude when the driving frequency agrees exactly with one of the ω_i's — the familiar phenomenon of resonance. Actually, of course, Eq. (10–63) presupposes only small deviations from equilibrium,

* The free vibrations are essentially the transients generated by the application of the driving forces. If we consider the system to be initially in an equilibrium configuration, and then slowly build up the driving forces from zero, these transients will not appear. Alternatively, dissipative forces can be assumed present (see pages following) which will damp out the free vibrations.

and the amplitude of the mode remains finite at resonance. Note that the oscillations are in phase with the driving force when the frequency is less than the resonant frequency, but that there is a phase change of π in going through the resonance.

Our discussion has been unrealistic in that the absence of dissipative or frictional forces has been assumed. In most physical systems these forces, when present, are proportional to the particle velocities, and can therefore be derived from a dissipation function $\mathfrak{F}$ (cf. Section 1–5). Let us first consider the effects of frictional forces on the free modes of vibration.

From its definition, $\mathfrak{F}$ must be a homogeneous quadratic function of the velocities:

$$\mathfrak{F} = \frac{1}{2}\sum_{i,j} \mathfrak{F}_{ij}\dot{\eta}_i\dot{\eta}_j. \tag{10–64}$$

The coefficients $\mathfrak{F}_{ij}$ are clearly symmetric, $\mathfrak{F}_{ij} = \mathfrak{F}_{ji}$, and in general will be functions of the coordinates. Since we are concerned only with small vibrations about equilibrium, it is sufficient to expand the coefficients about equilibrium and retain only the first, constant term, exactly as was done for the kinetic energy. In future applications of Eq. (10–64) we shall take $\mathfrak{F}_{ij}$ as denoting these constant factors. It will be remembered that $2\mathfrak{F}$ is the rate of energy dissipation due to the frictional forces. The dissipation function $\mathfrak{F}$ therefore can never be negative, and in consequence the coefficients $\mathfrak{F}_{ij}$ must be either positive or zero. The complete set of Lagrange equations of motion now become (cf. Section 1–5):

$$\sum_j T_{ij}\ddot{\eta}_j + \sum_j \mathfrak{F}_{ij}\dot{\eta}_j + \sum_j V_{ij}\eta_j = 0. \tag{10–65}$$

It sometimes happens that the principal axis transformation which diagonalizes T and V also diagonalizes $\mathfrak{F}$. This will be the case, for example, if the frictional force is proportional both to the particle's velocity *and* its mass. In such exceptional circumstances the equations of motion in terms of the normal coordinates are

$$\ddot{\zeta}_i + \mathfrak{F}_i\dot{\zeta}_i + \omega_i^2\zeta_i = 0, \tag{10–66}$$

where the $\mathfrak{F}_i$'s are the positive coefficients in the diagonalized form of $\mathfrak{F}$. Being a set of linear differential equations with constant coefficients. Eqs. (10–66) may be solved by functions of the form

$$\zeta_i = C_i e^{-i\omega_i' t},$$

where ω_i' satisfies the quadratic equation

$$\omega_i'^2 + i\omega_i'\mathfrak{F}_i - \omega_i^2 = 0. \tag{10–67}$$

Eq. (10-67) has the two solutions

$$\omega_i' = \pm\sqrt{\omega_i^2 - \frac{\mathfrak{F}_i^2}{4}} - i\frac{\mathfrak{F}_i}{2}. \tag{10-68}$$

The motion is therefore not a pure oscillation, for ω' is complex. It is seen from Eq. (10-68) that the imaginary part of ω_i' results in a factor $e^{-\mathfrak{F}_i t/2}$, and by reason of the positive nature of the $\mathfrak{F}_i$'s this is always an exponentially decreasing function of time. The presence of a damping factor due to the friction is hardly unexpected. As the particles vibrate, they do work against the frictional forces, and the energy of the system (and hence the vibration amplitudes) must decrease with time. The real part of Eq. (10-68) corresponds to the oscillatory factor in the motion, and it will be noted that the presence of friction also affects the frequency of the vibration. However, if the dissipation is small, the squared term in $\mathfrak{F}_i$ may be neglected, and the frequency of oscillation reduces to the friction-free value. The complete motion is then simply an exponential damping of the free modes of vibration:

$$\zeta_i = C_i e^{-\mathfrak{F}_i t/2} e^{-i\omega_i t}. \tag{10-69}$$

If the dissipation function cannot be diagonalized along with T and V, the solution is much more difficult to obtain. The general nature of the solution remains pretty much the same, however: an exponential damping factor times an oscillatory exponential function. Suppose we seek a solution to Eqs. (10-65) of the form

$$\eta_j = Ca_j e^{-i\omega t} = Ca_j e^{-\kappa t} e^{-2\pi i\nu t}. \tag{10-70}$$

With this solution Eqs. (10-65) become a set of simultaneous linear equations:

$$\sum_j V_{ij}a_j - i\omega\sum_j \mathfrak{F}_{ij}a_j - \omega^2\sum_j T_{ij}a_j = 0, \tag{10-71}$$

or, equivalently, writing ω as $i\gamma$:

$$\sum_j V_{ij}a_j + \gamma\sum_j \mathfrak{F}_{ij}a_j + \gamma^2\sum_j T_{ij}a_j = 0. \tag{10-71$'$}$$

Eqs. (10-71) or (10-71′) can be solved for the a_j only for certain values of ω or γ, which will in general be complex. Without actually evaluating the corresponding secular equation we can show that the imaginary part of ω (or the real part of γ) must always be negative. Multiply Eq. (10-71′) by a_i^* and sum over i:

$$\sum_{i,j} V_{ij}a_i^*a_j + \gamma\sum_{i,j} \mathfrak{F}_{ij}a_i^*a_j + \gamma^2\sum_{i,j} T_{ij}a_i^*a_j = 0. \tag{10-72}$$

As was shown in Section 10–2, the symmetry of the coefficients of V_{ij}, $\mathfrak{F}_{ij}$, and T_{ij} is sufficient to insure that each of the sums in Eq. (10–72) is real. By writing a_j in terms of its real and imaginary parts we can, in fact, put Eq. (10–72) in the form

$$\sum_{i,j} V_{ij}(\alpha_i\alpha_j + \beta_i\beta_j) + \gamma\sum_{i,j}\mathfrak{F}_{ij}(\alpha_i\alpha_j + \beta_i\beta_j) + \gamma^2\sum_{i,j}T_{ij}(\alpha_i\alpha_j + \beta_i\beta_j) = 0. \tag{10–72′}$$

Now, if γ is a solution of the quadratic equation (10–72′), then its complex conjugate γ^* must also be a root of the equation. The sum of these two roots is twice the real part of γ, and at the same time must be equal to the coefficient of γ in Eq. (10–72′) divided by the coefficient of γ^2:

$$\gamma + \gamma^* = 2\kappa = \frac{\sum \mathfrak{F}_{ij}(\alpha_i\alpha_j + \beta_i\beta_j)}{\sum T_{ij}(\alpha_i\alpha_j + \beta_i\beta_j)}. \tag{10–73}$$

The dissipation function $\mathfrak{F}$ must always be positive, and T is positive definite; hence κ can only be positive. The oscillations of the system may decrease exponentially with time, but they can never increase with time. Note that if $\mathfrak{F}$ is positive definite, κ *must* be different from zero (and positive), and all modes will have an exponential damping factor. The frequencies of oscillation, given by the real part of ω, will of course be affected by the dissipative forces, but the change will be small if the damping is not very large during a period of oscillation.

We may finally consider forced oscillations in the presence of dissipative forces. Representing the variation of the driving force with time by

$$F_j = F_{0j}e^{-i\omega t},$$

where F_{0j} may be complex, the equations of motion are

$$\sum_j V_{ij}\eta_j + \sum_j \mathfrak{F}_{ij}\dot{\eta}_j + \sum_j T_{ij}\ddot{\eta}_j = F_{0i}e^{-i\omega t}. \tag{10–74}$$

If we seek a particular solution to these equations of the form

$$\eta_j = A_j e^{-i\omega t},$$

we obtain the following set of inhomogeneous linear equations for the amplitudes A_j:

$$\sum_j A_j(V_{ij} - i\omega\mathfrak{F}_{ij} - \omega^2 T_{ij}) - F_{0i} = 0. \tag{10–75}$$

The solution to these equations may easily be obtained from Cramer's rule:

$$A_j = \frac{D_j(\omega)}{D(\omega)}, \tag{10–76}$$

where $D(\omega)$ is the determinant of the coefficients of A_j in Eq. (10–75) and $D_j(\omega)$ is the modification in $D(\omega)$ resulting when the jth column is replaced by $F_{01} \ldots F_{0n}$. It is the denominator $D(\omega)$ which is of principal interest to us here, for the resonances arise essentially out of the algebraic form of the denominator. Now, D is the determinant appearing in the secular equation; its roots are the complex frequencies of the free modes of vibration $\omega_1 \ldots \omega_n$. Hence $D(\omega)$ can be represented as

$$D(\omega) = G(\omega - \omega_1)(\omega - \omega_2)(\omega - \omega_3) \cdots (\omega - \omega_n),$$

where G is some constant. Using the product notation, this can be written

$$D(\omega) = G\prod_{i=1}^{n}(2\pi(\nu - \nu_i) + i\kappa_i). \tag{10–77}$$

When we rationalize Eq. (10–76) to separate A_j into its real and imaginary parts, the denominator will be

$$D^*(\omega)D(\omega) = GG^*\prod_{i=1}^{n}\left(4\pi^2(\nu - \nu_i)^2 + \kappa_i^2\right). \tag{10–78}$$

It can be seen that a resonance will occur when the frequency ν coincides with each of the resonant frequencies ν_i, but as a result of the damping constants κ_i the resonance denominators no longer vanish. Each free mode is still most strongly excited when the driving frequency agrees with the free frequency, but now the driving force must do work against friction and the resonance amplitudes are not infinite.

We have discussed the properties of small oscillations solely in terms of mechanical systems. The reader, however, has undoubtedly noticed the similarity with the theory of the oscillations of electrical networks. The equations of motion (10–65) become the circuit equations for n coupled circuits if we read the V_{ij} coefficients as reciprocal capacitances, the $\mathfrak{F}_{ij}$'s as resistances, and the T_{ij}'s as inductances. Driving forces are replaced by generators of frequency ω applied to one or more of the circuits, and the equations of forced vibration (10–74) reduce to the electrical circuit equations (2–39) mentioned in Chapter 2. We have presented here only a fraction of the techniques that have been devised for handling small oscillations, and of the general theorems about the motion. However, to pursue the subject further would take us more into electrical circuit theory than into mechanics.

Instead we shall turn our attention to another outgrowth of small oscillation theory — the methods of handling the oscillations of continuous systems. Historically, the transition from discrete to continuous systems was first devised by Rayleigh and others in order to treat the vibrations of strings, membranes, or bars. More recently it has been recognized that any continuous variation of one or more quantities in space — a field, in other words — could be considered as representing a continuous system. The techniques developed for continuous mechanical systems can therefore be applied, for example, to the electromagnetic field. In their quantized forms these techniques have become of greatest importance to the present-day theoretical physicist in working with the fields of elementary particles currently being invented in great profusion. In the next chapter we shall therefore present a brief introduction to the classical mechanics of continuous systems.

SUGGESTED REFERENCES

H. Margenau and G. M. Murphy, *The Mathematics of Physics and Chemistry.* A brief introduction to small oscillation theory is given in Chapter 9, and the relevant mathematical background in matrix algebra will be found in Chapter 10. The approach differs somewhat from that given in the text, but abundant use is made of matrix algebra.

A. G. Webster, *Dynamics.* Chapter V contains a somewhat old-fashioned treatment of small oscillations that is especially valuable for the discussion of systems containing dissipative forces. Forced oscillations and the transition to continuous systems are also well treated. But most valuable are the notes at the end of the treatise which provide a short course in quadratic forms and their principal axis transformations. An explicit method is given for simultaneously diagonalizing T and V which does not use matrix algebra but which neatly avoids the difficulty of multiple roots.

E. T. Whittaker, *Analytical Dynamics.* Chapter VII is on the theory of vibrations and gives the explicit proof that T and V can be diagonalized together, but the treatment of this point is much clearer in Webster. More valuable are the later sections of the chapter on the effects of constraints, and on vibrations about steady motion, which is discussed in considerable detail. Section 94 of Chapter VIII on vibrations in the presence of dissipative forces is fragmentary and restricted to two degrees of freedom.

M. Bôcher, *Introduction to Higher Algebra.* Chapter XIII is concerned with the diagonalization of quadratic forms, using a method similar to that of Webster and Whittaker. The earlier chapters on the solution of simultaneous linear equations are also of interest.

S. Timoshenko and D. H. Young, *Advanced Dynamics.* This is an engineering text and the general theory is rather rudimentary. However, vibrations in systems of two and three degrees of freedom are considered in great detail and solutions are worked out for many illustrative examples. These relatively simple systems provide a physical feeling for such concepts as normal modes, resonance, etc., which is often lost in the abstractness of a general theory. A number of unusual topics are also treated, such as methods for the approximate solution of the secular equation, and small vibrations about steady motion.

Lord Rayleigh, *Theory of Sound.* One of the classics of physics literature, this treatise contains a wealth of theorems and physical illustrations on all of the aspects of vibration theory. Rayleigh himself was responsible for developing much of the theory, especially the introduction of the dissipation function. His treatment is smooth-flowing and clear, and contains rarely discussed topics, as on the effects of constraints and the stationary properties of the eigenfrequencies. Both Rayleigh and Webster lean heavily on the work of Routh, who in his Adams Prize Essay of 1877 and in his text *Rigid Dynamics* was one of the first to give a systematic discussion of small vibrations.

E. A. Guillemin, *The Mathematics of Circuit Analysis.* This reference is included to indicate the importance of small vibration theory in modern electrical engineering. Considerable attention is paid to quadratic forms and their principal axis transformations. The treatment, which makes abundant use of matrix algebra, is advanced and elegant.

G. Herzberg, *Infrared and Raman Spectra of Polyatomic Molecules.* This treatise provides many illustrations of the application of classical small vibration theory to molecular structure. The practical difficulties in solving the secular equation are very much in evidence when the number of degrees of freedom greatly exceeds two or three. All possible use must then be made of the constants of motion and the symmetry properties of the system to reduce the complexity of the problem. These techniques are considered in detail here, and the reader will find explicit solutions for many molecular models with illustrative diagrams of the various normal modes.

EXERCISES

1. Obtain the normal modes of vibration for the double pendulum shown in Fig. 1–5, assuming equal lengths, but not equal masses. Show that when the lower mass is small compared to the upper one the two resonant frequencies are almost equal. If the pendula are set in motion by pulling the upper mass slightly away from the vertical and then releasing it, show that subsequent motion is such that at regular intervals one pendulum is at rest while the other has its maximum amplitude. This is the familiar phenomenon of "beats."

2. The problem of the linear triatomic molecule can be reduced to one of two degrees of freedom by introducing coordinates $y_1 = x_2 - x_1$, $y_2 = x_3 - x_2$, and eliminating x_2 by requiring that the center of mass remain at rest. Obtain the frequencies of the normal modes in these coordinates and show that they agree with the results of Section 10–4. The distances between the atoms, y_1 and y_2, are known as the *internal coordinates* of the molecule.

3. Obtain the frequencies of longitudinal vibration of the molecule discussed in Section 10–4, except that now the center atom is to be considered bound to the origin by a spring of force constant k. Show that the translational mode disappears under these conditions.

4. The equilibrium configuration of a molecule is represented by three atoms of equal mass at the vertices of a 45° right triangle connected by springs of equal force constant. Obtain the secular determinant for the modes of vibration in the plane and show by rearrangement of the columns that the secular equation has a

triple root $\omega = 0$. Reduce the determinant to one of the third rank and obtain the nonvanishing frequencies of free vibration.

5. Show directly that the equations of motion of the preceding problem are satisfied by (a) a uniform translation of all atoms along the x-axis, (b) a uniform translation along the y-axis, and (c) a uniform rotation about the z-axis.

6. If the generalized driving forces Q_i are not sinusoidal, show that the forced vibrations of the normal coordinates in the absence of damping are given by

$$\zeta_i = \frac{1}{\sqrt{2\pi}} \int_{-\infty}^{+\infty} \frac{G_i(\omega)}{\omega_i^2 - \omega^2} e^{-\iota\omega t}\, d\omega,$$

where $G_i(\omega)$ is the Fourier transform of Q_i defined by

$$Q_i(t) = \frac{1}{\sqrt{2\pi}} \int_{-\infty}^{+\infty} G_i(\omega) e^{-\iota\omega t}\, d\omega.$$

If the dissipation function is simultaneously diagonalized along with T and V, show that the forced vibrations are given by

$$\zeta_i = \frac{1}{\sqrt{2\pi}} \int_{-\infty}^{+\infty} \frac{G_i(\omega)(\omega_i^2 - \omega^2 + i\omega\mathfrak{F}_i)}{(\omega_i^2 - \omega^2)^2 + \omega^2\mathfrak{F}_i^2} e^{-\iota\omega t}\, dt,$$

which has the typical resonance denominator form. These results are simple illustrations of the powerful technique of the *operational calculus* for handling transient vibrations.

7. A particle moves in a circular orbit under the influence of a central force directed from the center of the circle. Examine the motion of the particle if it is slightly perturbed from this equilibrium orbit. For this purpose, introduce the difference coordinates $\rho = r - r_0$, $\phi = \theta - \omega t$, where r_0 is the radius of the circular orbit and ω the equilibrium angular frequency. T and V may be expressed in these coordinates, which are considered as small quantities so that terms higher than quadratic are dropped. Obtain the equations of motion, and find the conditions for stable oscillations. Show that if V is a power law of the form r^{-n+1} stable oscillation is obtained if n is less than 3. Show also that there is always a zero frequency of oscillation, corresponding to displacement into another circular orbit.

CHAPTER 11

INTRODUCTION TO THE LAGRANGIAN AND HAMILTONIAN FORMULATIONS FOR CONTINUOUS SYSTEMS AND FIELDS

All the formulations of mechanics discussed up to this point have been devised for treating systems with a finite or at most a denumerably infinite number of degrees of freedom. There are some mechanical problems, however, which involve continuous systems, as, for example, the problem of a vibrating elastic solid. Here each point of the continuous solid partakes in the oscillations, and the complete motion can only be described by specifying the position coordinates of *all* points. It is not difficult to modify the previous formulations of mechanics so as to handle such problems. The most direct method is to approximate the continuous system by one containing discrete particles, and then examine the change in the equations describing the motion as the continuous limit is approached.

11–1 The transition from a discrete to a continuous system. We shall apply this procedure to an infinitely long elastic rod which can undergo small longitudinal vibrations, i.e., oscillatory displacements of the particles of the rod parallel to the axis of the rod. A system composed of discrete particles which approximates the continuous rod is an infinite chain of

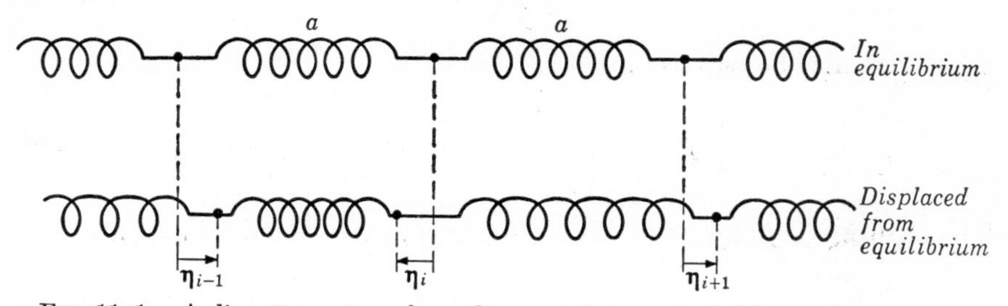

FIG. 11–1. A discrete system of equal mass points connected by springs, as an approximation to a continuous elastic rod.

equal mass points spaced a distance a apart and connected by uniform massless springs having force constants k (cf. Fig. 11–1). It will be assumed that the mass points can move only along the length of the chain. The discrete system will be recognized as an extension of the linear polyatomic molecule discussed in the preceding chapter. We can therefore obtain the equations describing the motion by the customary techniques for small oscillations. Denoting the displacement of the ith particle from its equilibrium position by η_i, the kinetic energy is:

$$T = \frac{1}{2}\sum_i m\dot{\eta}_i^2, \tag{11–1}$$

where m is the mass of each particle. The corresponding potential energy is the sum of the potential energies of each spring as the result of being stretched or compressed from its equilibrium length (cf. Sec. 10–4):

$$V = \frac{1}{2}\sum_i k(\eta_{i+1} - \eta_i)^2. \tag{11–2}$$

That Eq. (11–2) is the correct potential energy may be seen also by calculating directly the force on the ith particle and comparing it with the force predicted from V. The force due to the spring on the right of the particle will be $k(\eta_{i+1}-\eta_i)$, while the spring on the left exerts the force $-k(\eta_i-\eta_{i-1})$, so that the total force is

$$F_i = k(\eta_{i+1} - \eta_i) - k(\eta_i - \eta_{i-1}),$$

which agrees with $F_i = -\dfrac{\partial V}{\partial \eta_i}$ as obtained from Eq. (11–2). Combining Eqs. (11–1) and (11–2), the Lagrangian for the system is

$$L = T - V = \frac{1}{2}\sum_i\left(m\dot{\eta}_i^2 - k(\eta_{i+1} - \eta_i)^2\right), \tag{11–3}$$

which can also be written as

$$L = \frac{1}{2}\sum_i a\left[\frac{m}{a}\dot{\eta}_i^2 - ka\left(\frac{\eta_{i+1} - \eta_i}{a}\right)^2\right] = \sum_i aL_i. \tag{11–4}$$

The resulting Lagrange equations of motion for the coordinates η_i are

$$\frac{m}{a}\ddot{\eta}_i - ka\left(\frac{\eta_{i+1} - \eta_i}{a^2}\right) + ka\left(\frac{\eta_i - \eta_{i-1}}{a^2}\right) = 0. \tag{11–5}$$

The particular form of L in Eq. (11–4), and of the corresponding equations of motion, has been chosen for convenience in going to the limit of a continuous rod as a approaches zero. It is clear that m/a reduces to μ, the mass per unit length of the continuous system, but the limiting value of ka may not be so obvious. For an elastic rod obeying Hooke's law it will be remembered that the extension of the rod *per unit length* is directly proportional to the force or tension exerted on the rod, a relation that can be written as

$$F = Y\xi,$$

where ξ is the elongation per unit length and Y is Young's modulus. Now the extension of a length a of a discrete system, per unit length, will be $\xi = (\eta_{i+1} - \eta_i)/a$. The force necessary to stretch the spring by this amount is

$$F = k(\eta_{i+1} - \eta_i) = ka\left(\frac{\eta_{i+1} - \eta_i}{a}\right),$$

so that ka must correspond to the Young's modulus of the continuous rod. In going from the discrete to the continuous case, the integer index i identifying the particular mass point becomes the continuous position coordinate x; instead of the variable η_i we have $\eta(x)$. Further, the quantity

$$\frac{\eta_{i+1} - \eta_i}{a} = \frac{\eta(x + a) - \eta(x)}{a}$$

occurring in L_i obviously approaches the limit

$$\frac{d\eta}{dx},$$

as a, playing the role of dx, approaches zero. Finally, the summation over a discrete number of particles becomes an integral over x, the length of the rod, and the Lagrangian (11-4) appears as

$$L = \frac{1}{2}\int\left(\mu\dot{\eta}^2 - Y\left(\frac{d\eta}{dx}\right)^2\right)dx. \qquad (11\text{-}6)$$

In the limit as a goes to zero, the last two terms in the equation of motion (11-5) become

$$\underset{a\to 0}{\mathrm{L}} -\frac{Y}{a}\left\{\left(\frac{d\eta}{dx}\right)_x - \left(\frac{d\eta}{dx}\right)_{x-a}\right\},$$

which clearly defines a second derivative of η. Hence the equation of motion for the continuous elastic rod is

$$\mu\frac{d^2\eta}{dt^2} - Y\frac{d^2\eta}{dx^2} = 0. \qquad (11\text{-}7)$$

This simple example is sufficient to illustrate the salient features of the transition from a discrete to a continuous system. The most important fact to grasp is the role played by the position coordinate x. It is *not* a generalized coordinate; it serves merely as a continuous index replacing the discrete i. Just as each value of i corresponded to a different one of the generalized coordinates, η_i, of the system, so here for each value of x there is a generalized coordinate $\eta(x)$. Since η depends also on the continuous variable t, we should perhaps write more accurately $\eta(x, t)$, indicating that x, like t, can be considered as a parameter entering into the Lagrangian. If the continuous system were three-dimensional, rather than one-dimensional as here, the generalized coordinates would be distinguished by three continuous indices x, y, z, and would be written as $\eta(x, y, z, t)$. Eq. (11–6) also shows that the Lagrangian appears as an integral over the continuous index x; in the corresponding three-dimensional case the Lagrangian would have the form

$$L = \iiint \mathfrak{L}\, dx\, dy\, dz, \tag{11–8}$$

where $\mathfrak{L}$ is known as the *Lagrangian density*. For the longitudinal vibrations of the continuous rod the Lagrangian density is

$$\mathfrak{L} = \frac{1}{2}\left\{\mu\left(\frac{\partial\eta}{\partial t}\right)^2 - Y\left(\frac{\partial\eta}{\partial x}\right)^2\right\}, \tag{11–9}$$

corresponding to the continuous limit of the quantity L_i appearing in Eq. (11–4). It is the Lagrangian density, rather than the Lagrangian itself, which will be used to describe the motion of the system.

11–2 The Lagrangian formulation for continuous systems. It will be noted from Eq. (11–9) that $\mathfrak{L}$ for the elastic rod, besides being a function of $\dot{\eta} \equiv \dfrac{\partial\eta}{\partial t}$, also involves a spatial derivative of η, namely, $\dfrac{\partial\eta}{\partial x}$; x and t thus play a similar role as parameters of the Lagrangian density. Of course, $\mathfrak{L}$ in specific cases may be a function of η itself, and may also involve t and x explicitly. In any general three-dimensional continuous system the Lagrangian density will appear as a function of the form

$$\mathfrak{L} = \mathfrak{L}\left(\eta, \frac{\partial\eta}{\partial x}, \frac{\partial\eta}{\partial y}, \frac{\partial\eta}{\partial z}, \frac{\partial\eta}{\partial t}, x, y, z, t\right). \tag{11–10}$$

The Lagrangian played an important role in the mechanics of discrete systems because we could obtain the equations of motion from it. For continuous systems, however, the equations of motion for $\eta(x, t)$ are given directly in terms of the Lagrangian density, $\mathfrak{L}$. Fundamentally, the equations of motion must come from Hamilton's principle, which now takes the form

$$\delta I = \delta \int_1^2\!\!\iiint \mathfrak{L}\, dx\, dy\, dz\, dt = 0. \tag{11–11}$$

The nature of the variation involved in Eq. (11–11) differs only slightly from that previously discussed. There can be no variation of the parameters x, y, z; the virtual displacements constructing the varied paths are for constant xyz as well as for constant t. The variation process affects neither the limits of integration over time, nor the region of volume integration. Just as the variation of η is taken to be zero at the end points t_1 and t_2, so the variation of η on the surface of the volume of integration is also to be zero. The technique of converting the variational principle into an ordinary extremum problem by labelling the varied paths with some parameter α works as well here as in the discrete case. Sufficient experience has been gained by now in handling the δ-variation so that we can dispense with the parameter notation and deal directly with the variations themselves, remembering always that

$$\delta \rightarrow d\alpha \frac{\partial}{\partial \alpha}.$$

Since the Lagrangian density is a function not only of η and $\dot{\eta}$ but also of the spatial derivatives of η, the variation of $\mathfrak{L}$ can be written

$$\delta \mathfrak{L} = \frac{\partial \mathfrak{L}}{\partial \eta} \delta\eta + \frac{\partial \mathfrak{L}}{\partial \dot{\eta}} \delta\dot{\eta} + \sum_{k=1}^{3} \frac{\partial \mathfrak{L}}{\partial\left(\dfrac{\partial \eta}{\partial x_k}\right)} \delta\left(\frac{\partial \eta}{\partial x_k}\right), \tag{11–12}$$

where for convenience the spatial coordinates have been changed from xyz to $x_1x_2x_3$. Hamilton's principle therefore becomes

$$\int_1^2\!\!\iiint \left[\frac{\partial \mathfrak{L}}{\partial \eta} \delta\eta + \frac{\partial \mathfrak{L}}{\partial \dot{\eta}} \delta\dot{\eta} + \sum_{k=1}^{3} \frac{\partial \mathfrak{L}}{\partial\left(\dfrac{\partial \eta}{\partial x_k}\right)} \delta\left(\frac{\partial \eta}{\partial x_k}\right)\right] dx_1\, dx_2\, dx_3\, dt = 0. \tag{11–13}$$

Making use of the same integration by parts occurring in the derivation of the ordinary Lagrange equations, we obtain the relation

$$\int_1^2 \frac{\partial \mathfrak{L}}{\partial \dot{\eta}} \delta\dot{\eta}\, dt = -\int_1^2 \frac{d}{dt}\left(\frac{\partial \mathfrak{L}}{\partial \dot{\eta}}\right) \delta\eta\, dt.$$

The integrals involving the spatial derivatives of η can be manipulated in a similar manner. Interchanging the derivative with respect to x_k and the δ-variation, we have

$$\int \frac{\partial \mathfrak{L}}{\partial\left(\dfrac{\partial \eta}{\partial x_k}\right)} \delta\left(\frac{\partial \eta}{\partial x_k}\right) dx_k = \int \frac{\partial \mathfrak{L}}{\partial\left(\dfrac{\partial \eta}{\partial x_k}\right)} \frac{\partial\, \delta\eta}{\partial x_k}\, dx_k. \tag{11–14}$$

An integration by parts now converts the integral into *

$$\frac{\partial \mathfrak{L}}{\partial \left(\dfrac{\partial \eta}{\partial x_k}\right)} \delta\eta - \int \frac{d}{dx_k}\left(\frac{\partial \mathfrak{L}}{\partial \left(\dfrac{\partial \eta}{\partial x_k}\right)}\right)\delta\eta \, dx_k. \tag{11-15}$$

The integrated term in Eq. (11–15) vanishes, as in the time integral, because the variation process is such that $\delta\eta$ vanishes at the extremities of the spatial integration. There is some difficulty, in principle at least, if the system is infinite in extent. Very often in such systems the disturbance measured by η falls off sufficiently fast so that the integrated term is zero at infinity regardless of the nature of the variation. In any case, the integral can always be taken formally as over a finite region and, after dropping the integrated term, the volume can be allowed to become infinite. Consequently, Hamilton's principle takes on the form:

$$\int_1^2 \iiint \delta\eta \left[\frac{\partial \mathfrak{L}}{\partial \eta} - \frac{d}{dt}\frac{\partial \mathfrak{L}}{\partial \dot{\eta}} - \sum_{k=1}^{3} \frac{d}{dx_k}\left(\frac{\partial \mathfrak{L}}{\partial \left(\dfrac{\partial \eta}{\partial x_k}\right)}\right)\right] dx_1\, dx_2\, dx_3\, dt = 0. \tag{11-16}$$

The integral can vanish identically only if the separate coefficients of the independent variations $\delta\eta(x_1, x_2, x_3, t)$ vanish, yielding the equations of motion:

$$\frac{d}{dt}\frac{\partial \mathfrak{L}}{\partial \dot{\eta}} + \sum_{k=1}^{3} \frac{d}{dx_k}\left(\frac{\partial \mathfrak{L}}{\partial \left(\dfrac{\partial \eta}{\partial x_k}\right)}\right) - \frac{\partial \mathfrak{L}}{\partial \eta} = 0. \tag{11-17}$$

A system of n discrete degrees of freedom will have n Lagrange equations of motion; for the continuous system with an infinite number of degrees of freedom we seem to obtain only one Lagrange equation! It must be remembered, however, that the equation of motion for η_i is a differential equation involving the time only, and in that sense Eq. (11–17) furnishes a separate equation of motion for each value of x_k. The continuous nature of the indices x_k appears in that Eq. (11–17) is a partial differential equation in the four variables $x_1 x_2 x_3$, and t, yielding η as $\eta(x_1, x_2, x_3, t)$.

* The change from a partial derivative with respect to x_k in (11–14) to a total derivative in (11–15) may occasion some difficulty. In the first case a partial derivative is used to indicate that η is a function not only of x_k but of t and the other spatial coordinates. Use of a partial derivative in (11–15) might have implied that only the explicit dependence of $\mathfrak{L}$ on x_k was in question. To emphasize that the derivative must also involve the implicit dependence on x_k through η, we have therefore written a total derivative in (11–15). In any case, the operations actually to be performed are quite clear and unequivocal.

The form of the Lagrangian density and the subsequent equations of motion have been discussed assuming that each point in the system can suffer only one type of displacement, indicated by η. In a more complicated problem, such as the vibration of an elastic solid, a particle will, of course, undergo displacements along all three axes x_1, x_2, and x_3. In such case there will be three types of generalized coordinates, which we may denote by an integer index j: $\eta_j(x_1, x_2, x_3, t)$. In a general problem we must therefore expect to have generalized coordinates characterized by both a discrete index (or set of indices) and the continuous space indices $x_1x_2x_3$. The Lagrangian density $\mathcal{L}$ will be a function of all the generalized coordinates and their space and time derivatives. For each $\eta_j(x_1, x_2, x_3, t)$ there will be an equation of motion of the form:

$$\frac{d}{dt}\frac{\partial \mathcal{L}}{\partial \dot{\eta}_j} + \sum_k \frac{d}{dx_k}\left(\frac{\partial \mathcal{L}}{\partial\left(\frac{\partial \eta_j}{\partial x_k}\right)}\right) - \frac{\partial \mathcal{L}}{\partial \eta_j} = 0, \quad j = 1, 2, \ldots \tag{11–18}$$

Considerable simplification in notation is obtained by introducing a quantity known as a *functional derivative** or *variational derivative*. The functional derivative of the Lagrangian L with respect to η_j is defined as

$$\frac{\delta L}{\delta \eta_j} = \frac{\partial \mathcal{L}}{\partial \eta_j} - \sum_{k=1}^{3} \frac{d}{dx_k} \frac{\partial \mathcal{L}}{\partial\left(\frac{\partial \eta_j}{\partial x_k}\right)}. \tag{11–19}$$

A similar definition holds for the functional derivative of L with respect to $\dot{\eta}_j$, but since $\mathcal{L}$ does not depend on the gradients of $\dot{\eta}_j$, we have simply

$$\frac{\delta L}{\delta \dot{\eta}_j} = \frac{\partial \mathcal{L}}{\partial \dot{\eta}_j}. \tag{11–20}$$

The great advantage of the functional derivative notation is that it enables us in effect to forget the complicating dependence of $\mathcal{L}$ on the spatial derivatives of η. Thus the δ-variation of L can be seen from Eqs. (11–8), (11–12), and (11–15) to be equal to

$$\delta L = \int \sum_j \left(\frac{\partial \mathcal{L}}{\partial \eta_j}\,\delta\eta_j - \sum_k \frac{d}{dx_k}\frac{\partial \mathcal{L}}{\partial\left(\frac{\partial \eta_j}{\partial x_k}\right)}\,\delta\eta_j + \frac{\partial \mathcal{L}}{\partial \dot{\eta}_j}\,\delta\dot{\eta}_j\right) dV \tag{11–21}$$

(where dV is an element of volume). In functional derivative notation, this result becomes simply

* The name is applied because the functional derivative of, say, L with respect to η gives the change in L due to a change in the value of the *function* $\eta(x)$ at a specific space point x, there being no change in the dependence of η on t.

$$\delta L = \int \sum_j \left(\frac{\delta L}{\delta \eta_j} \delta \eta_j + \frac{\delta L}{\delta \dot{\eta}_j} \delta \dot{\eta}_j \right) dV, \tag{11–22}$$

which is exactly what would have been obtained if the δ-variation were applied directly to Eq. (11–8), ignoring the dependence on the gradients of η. The equations of motion (11–18) similarly take on a simple form in terms of functional derivatives:

$$\frac{d}{dt} \frac{\delta L}{\delta \dot{\eta}_j} - \frac{\delta L}{\delta \eta_j} = 0, \tag{11–23}$$

greatly resembling the ordinary Lagrange equations in appearance.

While the functional derivative notation greatly simplifies some of the manipulation of the variational principles, its use tends to obscure the fact that the equations of motion are *partial differential* equations in space and time. It also singles out the time coordinate as different in nature from the spatial coordinates, whereas the derivation actually treats x_k and t in equal fashion as parameters appearing in $\mathfrak{L}$. This equal status reminds us of special relativity; in fact, the entire procedure slips very easily into a Lorentz covariant formulation. The product $dx_1\, dx_2\, dx_3\, dt$ is essentially an element of volume in world space and hence is an invariant under a Lorentz transformation. Hamilton's principle (11–11) is therefore automatically Lorentz invariant providing only that $\mathfrak{L}$ is a world scalar. The equation of motion, (11–17), in covariant notation becomes simply

$$\sum_{\mu=1}^{4} \frac{d}{dx_\mu} \frac{\partial \mathfrak{L}}{\partial \left(\frac{\partial \eta}{\partial x_\mu} \right)} - \frac{\partial \mathfrak{L}}{\partial \eta} = 0, \tag{11–24}$$

which will be relativistically invariant providing $\mathfrak{L}$ and η are world scalars. Similarly, the covariance of the equations of motion (11–18) is assured if $\mathfrak{L}$ is a world scalar and the η_i's have some definite Lorentz transformation properties, e.g., are the components of a four-vector.

As a simple example of the Lagrangian procedure for obtaining the equation of motion, let us return to the longitudinal vibrations of a long elastic rod, treated in the previous section. The Lagrangian density is given by Eq. (11–9) and the various derivatives appearing in Eq. (11–17) are

$$\frac{\partial \mathfrak{L}}{\partial \eta} = 0, \qquad \frac{\partial \mathfrak{L}}{\partial \dot{\eta}} = \mu \dot{\eta}, \qquad \frac{\partial \mathfrak{L}}{\partial \left(\frac{\partial \eta}{\partial x} \right)} = -Y \frac{\partial \eta}{\partial x}.$$

Hence the equation of motion obtained from the Lagrangian density is:

$$\mu \frac{d^2\eta}{dt^2} - Y \frac{d^2\eta}{dx^2} = 0,$$

which agrees with (11–7), found earlier. Eq. (11–7) will be recognized as a one-dimensional wave equation with the propagation velocity:

$$v = \sqrt{\frac{Y}{\mu}}, \tag{11–25}$$

which is the well-known formula for the velocity of longitudinal elastic waves.

11–3 Sound vibrations in gases as an example of the Lagrangian formulation. To illustrate the Lagrangian procedure for handling the motion of continuous mechanical systems, we shall seek the equations of motion for the longitudinal vibrations of a gas. These vibrations, of course, constitute a *sound* field, and what we will obtain will be the wave equation for the propagation of sound. The displacement of the gas will be denoted by the vector $\boldsymbol{\eta}$, with components η_i, $i = 1, 2, 3$. Each point xyz in space will thus have three generalized coordinates associated with it. It will be assumed that the disturbance is always small, so that the pressure P and density μ differ only slightly from their equilibrium values P_0 and μ_0 respectively.

In a discrete system the problem is set up in the Lagrangian formulation by finding the kinetic and potential energies and writing the Lagrangian as the difference of these quantities. Here the Lagrangian we seek is the volume integral of a density $\mathfrak{L}$. The kinetic and potential energies can similarly be obtained as volume integrals of densities $\mathfrak{T}$ and $\mathfrak{V}$ respectively, with the relation

$$\mathfrak{L} = \mathfrak{T} - \mathfrak{V}. \tag{11–26}$$

The kinetic energy density presents no problem; bearing in mind that the displacements from equilibrium are small, we have

$$\mathfrak{T} = \frac{\mu_0}{2}\dot{\boldsymbol{\eta}}^2 = \frac{\mu_0}{2}(\dot{\eta}_1^2 + \dot{\eta}_2^2 + \dot{\eta}_3^2).$$

To obtain the potential energy density is a more difficult task. The potential energy of the gas is a measure of the work the gas can do in expanding against the pressure. Essentially, it arises from what the seventeenth century scientists were fond of calling the "spring" of the gas.

Consider a mass of gas M with equilibrium volume

$$V_0 = \frac{M}{\mu_0} \tag{11-27}$$

sufficiently small that $\mathcal{V}$ is constant over the volume. Then $\mathcal{V}V_0$ represents the potential energy of the quantity of gas. As a result of the sound disturbance, the volume changes from V_0 to $V_0 + \Delta V$. Now, in a change in volume dV the work performed on the system, i.e., the increase in the potential energy, is $-P\,dV$.* Hence the potential energy corresponding to a volume change from V_0 to $V_0 + \Delta V$ is

$$\mathcal{V}V_0 = -\int_{V_0}^{V_0+\Delta V} P\,dV.$$

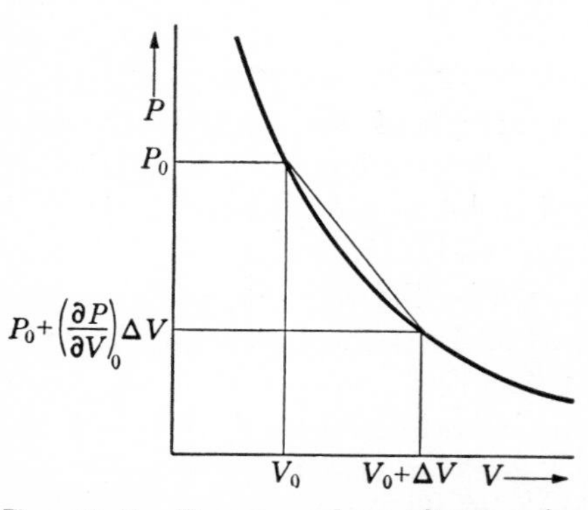

FIG. 11-2. Pressure-volume diagram for a gas.

It might be thought that since ΔV is small, the integral can be approximated by $P_0\,\Delta V$. As we shall see, this term actually does not contribute to the equations of motion. It is therefore necessary to go to the next approximation, in which the curve of P vs. V is replaced by a straight line in the region from V_0 to $V_0 + \Delta V$ (cf. Fig. 11-2):

$$\int_{V_0}^{V_0+\Delta V} P\,dV = P_0\,\Delta V + \frac{1}{2}\left(\frac{\partial P}{\partial V}\right)_0 (\Delta V)^2. \tag{11-28}$$

To evaluate the derivative of P with respect to V, we must digress for a moment into thermodynamics. The first inclination might be to use Boyle's law,

$$PV = C, \tag{11-29}$$

for the relation between the pressure and the volume, and this was the procedure followed by Newton. It leads to the wrong result, however, because (11-29) assumes that the changes in pressure and volume occur

* The customary elementary derivation is as follows. The force exerted on an element of surface dA by the external system is $P\,dA$, pointing inwards. In expanding, the surface moves a distance dx outward along the normal, and the external work done is $-P\,dA\,dx = -P\,dV$.

isothermally. Actually, the vibrations of sound are almost always so rapid that there is no time for conduction to remove the heat developed and equalize the temperatures. The contractions and expansions instead take place *adiabatically*, i.e., without loss of heat. Under these conditions the relation between P and V is

$$PV^{\gamma} = C, \tag{11-30}$$

where γ is the constant ratio of the specific heats at constant pressure and volume.* Hence the desired derivative is

$$\left(\frac{\partial P}{\partial V}\right)_0 = -\frac{\gamma P_0}{V_0}. \tag{11-31}$$

It is convenient to express the change in volume in terms of the associated density change. Since $V = M/\mu$, the change in V is given by

$$\Delta V = -\frac{M}{\mu_0^2}\Delta\mu = -V_0\sigma, \tag{11-32}$$

where the fractional change in the density has been denoted by σ:

$$\mu = \mu_0(1 + \sigma). \tag{11-33}$$

Combining Eqs. (11-27, 28, 31, and 32) the potential energy density appears as

$$\mathcal{V} = P_0\sigma + \frac{\gamma P_0}{2}\sigma^2. \tag{11-34}$$

This is still not in the form useful for the Lagrangian; we have yet to express σ in terms of $\boldsymbol{\eta}$. Consider any finite volume V in space. The mass flowing out of this volume due to the small disturbance from equilibrium is given by

$$\mu_0 \int \boldsymbol{\eta} \cdot d\mathbf{A},$$

evaluated over the surface of the volume. The volume integral of the change in density must be exactly equal to this mass transport:

$$-\mu_0 \int \sigma\, dV = \mu_0 \int \boldsymbol{\eta} \cdot d\mathbf{A}. \tag{11-35}$$

By the divergence theorem the relation (11-35) can be written

$$-\int \sigma\, dV = \int \nabla \cdot \boldsymbol{\eta}\, dV,$$

* For derivation see M. W. Zemansky, *Heat and Thermodynamics*, McGraw-Hill, Chapter VI.

and since the equality holds for any arbitrary volume, we must have *

$$\sigma = -\nabla \cdot \boldsymbol{\eta}. \tag{11–36}$$

With this connection, the final form of the potential energy density is

$$\mathcal{V} = -P_0 \nabla \cdot \boldsymbol{\eta} + \frac{\gamma P_0}{2} (\nabla \cdot \boldsymbol{\eta})^2. \tag{11–37}$$

It can now be seen that the term in $\mathcal{V}$ linear in σ cannot contribute to the total potential energy. By Eq. (11–35) the volume integral of σ is minus the surface integral of $\boldsymbol{\eta}$, and if the surface completely encloses the system this must be zero, i.e., there is no transport of mass out of the system. That this term has a vanishing contribution to L is not yet sufficient reason to omit it from $\mathcal{V}$. Conceivably, the functional behavior of the term might still have an effect on the equation of motion. (It will be remembered that the covariant Hamiltonian of a system may be zero, but the equations of motion, of course, do not vanish.) We shall therefore retain the term for a few more steps.

The complete Lagrangian density can now be written as

$$\mathcal{L} = \frac{1}{2} (\mu_0 \dot{\boldsymbol{\eta}}^2 + 2P_0 \nabla \cdot \boldsymbol{\eta} - \gamma P_0 (\nabla \cdot \boldsymbol{\eta})^2). \tag{11–38}$$

In obtaining the equations of motion, we will need the following derivatives:

$$\frac{\partial \mathcal{L}}{\partial \dot{\eta}_i} = \mu_0 \dot{\eta}_i, \qquad \frac{\partial (\nabla \cdot \boldsymbol{\eta})}{\partial \left(\frac{\partial \eta_i}{\partial x_k}\right)} = \delta_{ik}, \tag{11–39}$$

and

$$\frac{\partial (\nabla \cdot \boldsymbol{\eta})^2}{\partial \left(\frac{\partial \eta_i}{\partial x_k}\right)} = 2(\nabla \cdot \boldsymbol{\eta})\, \delta_{ik}.$$

It follows from Eq. (11–39) that the term $2P_0 \nabla \cdot \boldsymbol{\eta}$ cannot contribute to the equations of motion and can finally be dropped from $\mathcal{L}$:

$$\mathcal{L} = \frac{1}{2} (\mu_0 \dot{\boldsymbol{\eta}}^2 - \gamma P_0 (\nabla \cdot \boldsymbol{\eta})^2). \tag{11–40}$$

---

* Eq. (11–36) may be recognized from its more familiar form,

$$\dot{\mu} = -\nabla \cdot \mu \dot{\boldsymbol{\eta}},$$

as the *equation of continuity* for the gas flow.

The resultant equations of motion are now

$$\mu_0 \frac{\partial^2 \eta_i}{\partial t^2} - \gamma P_0 \frac{\partial(\nabla \cdot \boldsymbol{\eta})}{\partial x_i} = 0, \quad i = 1, 2, 3. \tag{11–41}$$

Eqs. (11–41) may be combined into one vector equation as

$$\mu_0 \frac{\partial^2 \boldsymbol{\eta}}{\partial t^2} - \gamma P_0 \nabla\nabla \cdot \boldsymbol{\eta} = 0. \tag{11–42}$$

In this guise the physical significance of the equation of motion may still not be very apparent, but we can easily reduce it to a familiar form. Take the divergence of Eq. (11–42), remembering that $\nabla \cdot \nabla$ is the Laplacian ∇^2, and $\nabla \cdot \boldsymbol{\eta} = -\sigma$:

$$\nabla^2 \sigma - \frac{\mu_0}{\gamma P_0} \frac{\partial^2 \sigma}{\partial t^2} = 0. \tag{11–43}$$

Eq. (11–43) is now readily recognized as the three-dimensional wave equation, with the velocity

$$v = \sqrt{\frac{\gamma P_0}{\mu_0}}, \tag{11–44}$$

which is the customary expression for the velocity of sound in gases. The Lagrangian density (11–40) thus correctly represents the propagation of sound waves in gases, and we have achieved our goal of describing the sound vibrations by a Lagrangian formulation.

11–4 The Hamiltonian formulation for continuous systems. It is possible to obtain a Hamiltonian formulation for systems with a continuous set of coordinates much as was done in Chapter 7 for discrete systems. To indicate the method of approach, we return briefly to the linear chain of mass points discussed in Section 11–1. Conjugate to each η_i there is a canonical momentum

$$p_i = \frac{\partial L}{\partial \dot{\eta}_i} = a \frac{\partial L_i}{\partial \dot{\eta}_i}. \tag{11–45}$$

The Hamiltonian for the system is therefore

$$H \equiv \sum_i p_i \dot{\eta}_i - L = \sum_i a \frac{\partial L_i}{\partial \dot{\eta}_i} \dot{\eta}_i - L,$$

or

$$H = \sum_i a \left(\frac{\partial L_i}{\partial \dot{\eta}_i} \dot{\eta}_i - L_i \right). \tag{11–46}$$

It will be remembered that in the limit of the continuous rod, when a goes to zero, $L_i \to \mathcal{L}$ and the summation in Eq. (11–46) becomes an integral:

$$H = \int dx \left(\frac{\partial \mathcal{L}}{\partial \dot{\eta}} \dot{\eta} - \mathcal{L} \right). \tag{11–47}$$

The individual canonical momenta p_i, as given by Eq. (11–45), vanish in the continuous limit, but we can define a *momentum density*, π, which remains finite:

$$\pi = \frac{\partial \mathcal{L}}{\partial \dot{\eta}}.$$

Eq. (11–47) is in the form of a space integral over a *Hamiltonian density*, $\mathcal{H}$, defined by

$$\mathcal{H} = \pi \dot{\eta} - \mathcal{L}. \tag{11–48}$$

It is clear, therefore, that in any general three-dimensional system with more than one type of generalized coordinate the total Hamiltonian will be a volume integral of a Hamiltonian density:

$$H = \iiint \mathcal{H}\, dx_1\, dx_2\, dx_3 = \iiint \left(\sum_k \pi_k \dot{\eta}_k - \mathcal{L} \right) dx_1\, dx_2\, dx_3, \tag{11–49}$$

where

$$\pi_i = \frac{\partial \mathcal{L}}{\partial \dot{\eta}_i} = \frac{\delta L}{\delta \dot{\eta}_i}. \tag{11–50}$$

Canonical field equations of motion can be obtained from the Hamiltonian density by following the same procedure used in Section 7–1. We consider $\mathcal{H}$ to be given as a function of the coordinates $\eta_k(x_j, t)$, the canonical momentum densities $\pi_k(x_j, t)$, the spatial derivatives of the coordinates $\frac{\partial \eta_k}{\partial x_j}$, and possibly the time t. An infinitesimal change in H will then be written as

$$dH =$$

$$\iiint \left\{ \sum_k \left[\frac{\partial \mathcal{H}}{\partial \eta_k} d\eta_k + \frac{\partial \mathcal{H}}{\partial \pi_k} d\pi_k + \sum_j \frac{\partial \mathcal{H}}{\partial \left(\frac{\partial \eta_k}{\partial x_j} \right)} d\left(\frac{\partial \eta_k}{\partial x_j} \right) \right] + \frac{\partial \mathcal{H}}{\partial t} dt \right\} dx_1\, dx_2\, dx_3.$$

The integral appearing here roughly resembles the volume integral occurring in Hamilton's principle, Eq. (11–13). As was done in that instance, we perform a parts integration on the integral

$$\int \frac{\partial \mathcal{H}}{\partial \left(\frac{\partial \eta_k}{\partial x_j} \right)} d\left(\frac{\partial \eta_k}{\partial x_j} \right) dx_j.$$

The integrated term can be made to vanish by making the volume of integration larger than the system. Alternatively, the region of integration can be made so large that the values of η and $\mathcal{H}$ on the surface at infinity are small enough to make the integrated term negligible. As a result dH can be written

$$dH = \iiint \left\{ \sum_k \left[\frac{\partial \mathcal{H}}{\partial \eta_k} d\eta_k + \frac{\partial \mathcal{H}}{\partial \pi_k} d\pi_k - \sum_j \frac{d}{dx_j} \left(\frac{\partial \mathcal{H}}{\partial \left(\frac{\partial \eta_k}{\partial x_j} \right)} \right) d\eta_k \right] + \frac{\partial \mathcal{H}}{\partial t} dt \right\} dx_1\, dx_2\, dx_3. \tag{11–51}$$

By making use of the functional derivative as defined in Eq. (11–19) this expression can be reduced to

$$dH = \iiint \left\{ \sum_k \left(\frac{\delta H}{\delta \eta_k} d\eta_k + \frac{\delta H}{\delta \pi_k} d\pi_k \right) + \frac{\partial \mathcal{H}}{\partial t} dt \right\} dx_1\, dx_2\, dx_3, \tag{11–52}$$

since $\mathcal{H}$ is not a function of the spatial derivatives of π_k. Whenever we have to take a differential of a quantity whose density depends upon the spatial derivatives of η or π, a similar parts integration can be performed, and the various derivatives will be grouped so as to form functional derivatives. It is therefore always feasible to express the differentials from the start in terms of functional derivatives, which greatly streamlines the operations.

The differential of H can be expressed in another way, using its definition, Eq. (11–49), in terms of L:

$$dH = \iiint \left\{ \sum_k \left(\pi_k\, d\dot{\eta}_k + \dot{\eta}_k\, d\pi_k - \frac{\delta L}{\delta \eta_k} d\eta_k - \frac{\delta L}{\delta \dot{\eta}_k} d\dot{\eta}_k \right) - \frac{\partial \mathcal{L}}{\partial t} dt \right\} dx_1\, dx_2\, dx_3. \tag{11–53}$$

From the definition of π_k, Eq. (11–50), the first and the last terms in the parentheses cancel, while in the third term

$$\frac{\delta L}{\delta \eta_k} = \frac{d}{dt} \frac{\delta L}{\delta \dot{\eta}_k} = \dot{\pi}_k,$$

from the Lagrange equations of motion (11–23). Hence Eq. (11–53) can be written

$$dH = \iiint \left\{ \sum_k \left(-\dot{\pi}_k\, d\eta_k + \dot{\eta}_k\, d\pi_k \right) - \frac{\partial \mathcal{L}}{\partial t} dt \right\} dx_1\, dx_2\, dx_3. \tag{11–54}$$

Comparison of Eq. (11–54) with Eq. (11–52) results in the set of equations

$$\frac{\delta H}{\delta \eta_k} = -\dot{\pi}_k, \qquad \frac{\delta H}{\delta \pi_k} = \dot{\eta}_k, \tag{11–55}$$

and the identity

$$\frac{\partial \mathcal{H}}{\partial t} = -\frac{\partial \mathcal{L}}{\partial t}.$$

Equations (11–55) are the analog of Hamilton's canonical equations for the continuous system; in terms of the Hamiltonian density they can be written

$$\frac{\partial \mathcal{H}}{\partial \eta_k} - \sum_j \frac{d}{dx_j}\left(\frac{\partial \mathcal{H}}{\partial \left(\frac{\partial \eta_k}{\partial x_j}\right)}\right) = -\dot{\pi}_k, \qquad \frac{\partial \mathcal{H}}{\partial \pi_k} = \dot{\eta}_k. \tag{11–56}$$

In this form they lose much of their symmetry, since $\mathcal{H}$ is not a function of the gradients of π_k.

As a simple example of the Hamiltonian procedure, we can examine the sound vibrations in a gas, as discussed in the previous section. The momentum density is simply

$$\pi_k = \mu_0 \dot{\eta}_k,$$

or, forming a vector momentum density:

$$\boldsymbol{\pi} = \mu_0 \dot{\boldsymbol{\eta}},$$

and the Hamiltonian density is (cf. Eq. (11–40)):

$$\mathcal{H} = \boldsymbol{\pi} \cdot \dot{\boldsymbol{\eta}} - \mathcal{L} = \frac{\boldsymbol{\pi}^2}{2\mu_0} + \frac{P_0 \gamma}{2} (\nabla \cdot \boldsymbol{\eta})^2. \tag{11–57}$$

The Hamiltonian density is thus equal to the sum of the kinetic and potential energy densities:

$$\mathcal{H} = \mathcal{T} + \mathcal{V},$$

and can therefore be identified in this case as an energy density.* The canonical equations obtained from $\mathcal{H}$ are

$$\dot{\eta}_k = \frac{\pi_k}{\mu_0}$$

(which merely repeats the definition of π_k), and

$$-\dot{\pi}_k = -\frac{d}{dx_k}(P_0 \gamma \nabla \cdot \boldsymbol{\eta}),$$

* The term linear in $\nabla \cdot \boldsymbol{\eta}$ has been omitted, for, as we have seen, it does not contribute to the total energy.

and the combination of these two sets of equations is identical with the equations of motion given by Eq. (11–41).

Much of the formal development of the Hamiltonian formulation — conservation theorems, Poisson brackets, etc. — can easily be recast in a form suitable for continuous systems. Thus, the modified Hamilton's principle becomes

$$\delta \int_1^2 \iiint \left\{ \sum_k \pi_k \dot{\eta}_k - \mathcal{H} \right\} dx_1\, dx_2\, dx_3\, dt = 0. \tag{11–58}$$

As an example of the conservation theorems, let us investigate the conditions under which the Hamiltonian is a constant of the motion. The total time derivative of H is

$$\frac{dH}{dt} = \iiint \left\{ \sum_k \left(\frac{\delta H}{\delta \eta_k} \dot{\eta}_k + \frac{\delta H}{\delta \pi_k} \dot{\pi}_k \right) + \frac{\partial \mathcal{H}}{\partial t} \right\} dx_1\, dx_2\, dx_3.$$

But by the equations of motion (11–55), this reduces to

$$\frac{dH}{dt} = \iiint \left\{ \sum_k \left(\frac{\delta H}{\delta \eta_k} \frac{\delta H}{\delta \pi_k} - \frac{\delta H}{\delta \pi_k} \frac{\delta H}{\delta \eta_k} \right) + \frac{\partial \mathcal{H}}{\partial t} \right\} dx_1\, dx_2\, dx_3,$$

so that

$$\frac{dH}{dt} = \iiint \frac{\partial \mathcal{H}}{\partial t}\, dx_1\, dx_2\, dx_3.$$

The total Hamiltonian is thus conserved if $\mathcal{H}$ (or H) is not an explicit function of time.

Similar techniques can be used to obtain an expression for the time derivative of any function G, not depending explicitly on time, which can be represented as the volume integral of a density function $\mathcal{G}$:

$$G = \iiint \mathcal{G}\, dx_1\, dx_2\, dx_3.$$

The total time derivative of G under these conditions is

$$\frac{dG}{dt} = \iiint \sum_k \left(\frac{\delta G}{\delta \eta_k} \dot{\eta}_k + \frac{\delta G}{\delta \pi_k} \dot{\pi}_k \right) dx_1\, dx_2\, dx_3.$$

By the equations of motion, this can be written

$$\frac{dG}{dt} = \iiint \sum_k \left(\frac{\delta G}{\delta \eta_k} \frac{\delta H}{\delta \pi_k} - \frac{\delta G}{\delta \pi_k} \frac{\delta H}{\delta \eta_k} \right) dx_1\, dx_2\, dx_3. \tag{11–59}$$

Now, the integral on the right in Eq. (11–59) is the exact analog of the Poisson bracket of G with H (cf. Eq. (8–42)), the summation over the generalized coordinates appearing as an integral over the continuous indices x_1, x_2, and x_3, and a discrete summation over the index k. We can therefore write

$$\frac{dG}{dt} = [G, H], \tag{11–60}$$

and it is clear that if G depends on time explicitly we obtain

$$\frac{dG}{dt} = [G, H] + \frac{\partial G}{\partial t},$$

in complete agreement with Eq. (8–58). Any integral function that is not an explicit function of time, and whose Poisson bracket with H vanishes, will thus be a constant of the motion. Note that the derivation is valid even when $\mathcal{G}$ is a function of the spatial derivatives of η or π, for this is exactly the situation the functional derivatives were designed to include.

The conservation theorems for integral quantities thus follow as in the ordinary theory. We have the same connection between the constants of the motion and the symmetry properties of the system as was found previously. It must be mentioned, however, that besides these "macroscopic" constants of the motion, there are "microscopic" conservation theorems. These deal directly with the densities rather than with the integrated quantities. One can obtain, for example, theorems which are essentially equations of continuity for the internal flow of energy and both linear and angular momentum. Unfortunately, to discuss such matters would lead us too far afield, and we shall have to refer the interested reader to the works cited at the end of the chapter.

11–5 Description of fields by variational principles. The Lagrangian and Hamiltonian formulations for a continuous set of generalized coordinates were developed in order to treat continuous mechanical systems, such as an elastic solid. But the formulations may also be used, even in the absence of a mechanical system, to describe the equations governing a *field*. Mathematically, a field is no more than one or more independent functions of space and time, and the generalized coordinates $\eta_j(x_1, x_2, x_3, t)$ fit this definition accurately. Indeed, fields occurring in physics often arose historically as the vibrations of some continuous system. The sound "field" in gases, after all, actually refers to the longitudinal vibrations of the gas particles, discussed in Section 11–3. Similarly, the electromagnetic field was long thought of in terms of the elastic vibrations of a mysterious ether. Only in recent times was it realized (to quote an apt phrase of Professor S. L. Quimby) that the ether had no other role than being the subject of the verb "to undulate." We can equally well recognize that the variational principle procedures developed in the previous sections also stand independent of the notion of a continuous mechanical system, and that they will serve to furnish the equations describing any space-time field. Hamilton's principle then becomes, in effect, a convenient abbreviation for the field which on expansion provides the field equations.

Since the Lagrangian density for a field is not associated with a definite mechanical system, it will not necessarily be given as the difference of a kinetic and potential energy density. Instead, we may use any expression for $\mathcal{L}$ which leads to the desired field equations. Thus, the Lagrangian density for the sound field was obtained previously in Section 11-3 by considering the mechanical system in which the sound was propagated. We were then naturally led to describe the field by using the components of the vector displacement $\boldsymbol{\eta}$ as the generalized coordinates. It will be noted, however, that the field is really a scalar one, since the properties of sound can be discussed completely in terms of the single scalar σ, the fractional change in the density. The quantity σ is therefore the natural generalized coordinate to use if we set out to describe the field by a Lagrangian without reference to any mechanical system. $\mathcal{L}$ must then be such that Lagrange's equation of motion is the wave equation in σ, Eq. (11-43). It is easily seen that a Lagrangian density in terms of σ which fulfills this requirement is

$$\mathcal{L} = \frac{1}{2}\left(\frac{\mu_0}{\gamma P_0}\dot{\sigma}^2 - (\nabla\sigma)^2\right) = \frac{\mu_0}{2\gamma P_0}\dot{\sigma}^2 - \frac{1}{2}\sum_k\left(\frac{\partial\sigma}{\partial x_k}\right)^2. \tag{11-61}$$

The various derivatives needed for the equation of motion are

$$\frac{\partial\mathcal{L}}{\partial\sigma} = 0, \qquad \frac{\partial\mathcal{L}}{\partial\dot{\sigma}} = \frac{\mu_0\dot{\sigma}}{\gamma P_0}, \qquad \frac{\partial\mathcal{L}}{\partial\left(\dfrac{\partial\sigma}{\partial x_k}\right)} = -\frac{\partial\sigma}{\partial x_k},$$

so that Lagrange's equation (11-17) becomes

$$\frac{\mu_0}{\gamma P_0}\frac{\partial^2\sigma}{\partial t^2} - \sum_k\frac{\partial^2\sigma}{\partial x_k^2} = 0,$$

which agrees with Eq. (11-43). Note that the Lagrangian density (11-61) is not the same as the density obtained in Section 11-3, Eq. (11-40). Neither of the terms in (11-61) is the mechanical kinetic or potential energy density. But the Lagrangian density does lead to the correct wave equation, and this is all that is required.

Similarly, one could obtain a satisfactory description of the sound field by using a Hamiltonian density

$$\mathcal{H} = \frac{1}{2}\left(\frac{\gamma P_0}{\mu_0}\pi^2 + (\nabla\sigma)^2\right). \tag{11–62}$$

Eq. (11–62) obviously leads to the proper field equation, but it is not the same as (11–53), nor is it an energy density of the mechanical system.

As a more complicated example of tailoring a Lagrangian density to fit a given field, let us consider the electromagnetic field in the absence of material media. In such case, $\mathbf{E} = \mathbf{D}$ and $\mathbf{B} = \mathbf{H}$ in Gaussian units, and Maxwell's equations (1–55) reduce to

$$\nabla \cdot \mathbf{B} = 0, \qquad \nabla \times \mathbf{E} + \frac{1}{c}\frac{\partial \mathbf{B}}{\partial t} = 0, \tag{11–63}$$

$$\nabla \cdot \mathbf{E} = 4\pi\rho, \qquad \nabla \times \mathbf{B} - \frac{1}{c}\frac{\partial \mathbf{E}}{\partial t} = \frac{4\pi\mathbf{j}}{c}. \tag{11–64}$$

The first two equations, it will be remembered, serve mainly to define $\mathbf{E}$ and $\mathbf{B}$ in terms of a scalar and vector potential:

$$\mathbf{E} = -\nabla\phi - \frac{1}{c}\frac{\partial \mathbf{A}}{\partial t}, \tag{1–58}$$

and

$$\mathbf{B} = \nabla \times \mathbf{A}. \tag{1–57}$$

It is the second pair of equations which describe the generation of the fields by external charges and currents, and which may therefore be considered as the desired field equations. The six field components are not very suitable as generalized coordinates, since they are not independent, but can be expressed in terms of the four potential components. We shall therefore use $\mathbf{A}$ and ϕ as the generalized coordinates of the field.

It is now stated that a Lagrangian density which leads to the field equations (11–64) is

$$\mathcal{L} = \frac{E^2 - B^2}{8\pi} - \rho\phi + \frac{\mathbf{j}\cdot\mathbf{A}}{c}, \tag{11–65}$$

where $\mathbf{E}$ and $\mathbf{B}$ are to be expressed in terms of the potentials by means of Eqs. (1–57, 58). To verify this statement, let us first find the equation of motion corresponding to the coordinate ϕ. The scalar potential itself appears only in the term $-\rho\phi$, and the spatial derivatives of ϕ occur only in E^2. (Note that there are no time derivatives of ϕ anywhere.) Hence we have

$$\frac{\partial \mathcal{L}}{\partial \phi} = -\rho$$

and

$$\frac{\partial \mathcal{L}}{\partial\left(\frac{\partial \phi}{\partial x_k}\right)} = \frac{E_k}{4\pi} \frac{\partial E_k}{\partial\left(\frac{\partial \phi}{\partial x_k}\right)} = -\frac{E_k}{4\pi},$$

the last step following from Eq. (1–58). Lagrange's equation for ϕ is therefore

$$\frac{1}{4\pi} \sum_k \frac{\partial E_k}{\partial x_k} - \rho = 0$$

or

$$\nabla \cdot \mathbf{E} = 4\pi\rho,$$

which is the first of Eqs. (11–64).

The evaluation of the equations of motion corresponding to the components of **A** is somewhat lengthier. Consider a typical component of **A**, say A_1. The Lagrangian density contains both spatial and time derivatives of A_1, with the component itself occurring in the term $(\mathbf{j} \cdot \mathbf{A})/c$. The required derivatives of $\mathcal{L}$ are therefore

$$\frac{\partial \mathcal{L}}{\partial A_1} = \frac{j_1}{c},$$

$$\frac{\partial \mathcal{L}}{\partial \dot{A}_1} = \frac{E_1}{4\pi} \frac{\partial E_1}{\partial \dot{A}_1} = -\frac{E_1}{4\pi c},$$

$$\frac{\partial \mathcal{L}}{\partial\left(\frac{\partial A_1}{\partial x_2}\right)} = -\frac{1}{4\pi} B_3 \frac{\partial B_3}{\partial\left(\frac{\partial A_1}{\partial x_2}\right)} = \frac{B_3}{4\pi},$$

and finally,

$$\frac{\partial \mathcal{L}}{\partial\left(\frac{\partial A_1}{\partial x_3}\right)} = -\frac{B_2}{4\pi}.$$

Combining these expressions, Lagrange's equation for A_1 appears as

$$\frac{1}{4\pi}\left(\frac{\partial B_3}{\partial x_2} - \frac{\partial B_2}{\partial x_3}\right) - \frac{1}{4\pi c}\frac{dE_1}{dt} - \frac{j_1}{c} = 0. \tag{11–66}$$

Equation (11–66) is readily recognized as the 1-component of the remaining Maxwell equation

$$(\nabla \times \mathbf{B}) - \frac{1}{c}\frac{d\mathbf{E}}{dt} = \frac{4\pi \mathbf{j}}{c},$$

the other components of which are the equations of motion for A_2 and A_3. Thus the four field equations obtained from the Lagrangian density (11–65) are identical with the Maxwell's equations (11–64).*

The charge and current density are connected, of course, by the relation

$$\mathbf{j} = \rho \mathbf{v}, \tag{11–67}$$

where $\mathbf{v}$ is the velocity of the charges, a function of position in space. The volume integral of the Lagrangian density (11–65) is the total Lagrangian for the electromagnetic field, and by virtue of Eq. (11–67) can be written

$$L = \int \left\{ \frac{E^2 - B^2}{8\pi} - \rho \left(\phi - \frac{\mathbf{v} \cdot \mathbf{A}}{c} \right) \right\} dV. \tag{11–68}$$

It will be noted that the combination of ϕ and $\mathbf{A}$ in the parentheses appears also in the Lagrangian Eq. (1–61) for a charged particle in an electromagnetic field. Indeed, we can show that this part of the field Lagrangian corresponds exactly to the generalized potential part of the particle Lagrangian.

The term "particle" indicates that the mass and charge are concentrated at a point in space. Hence the charge density corresponding to a particle must be zero everywhere except at the position of the particle. There it must be infinite, but in such a manner that the volume integral of the charge density is equal to the total charge on the particle. The mathematical function satisfying these peculiar requirements is a multiple of the well-known volume δ-function introduced by Dirac, and defined by the conditions:

$$\begin{aligned} \delta(\mathbf{r} - \mathbf{r}_1) &= 0, \qquad \mathbf{r} \neq \mathbf{r}_1 \\ \int \delta(\mathbf{r} - \mathbf{r}_1)\, dV &= 1, \end{aligned} \tag{11–69}$$

where $\mathbf{r}_1$ is the position vector of the particle. An obvious property of the δ-function is that if $f(\mathbf{r})$ is any function of position, then

$$\int f(\mathbf{r})\, \delta(\mathbf{r} - \mathbf{r}_1)\, dV = f(\mathbf{r}_1). \tag{11–70}$$

---

* In some respects the electromagnetic field is an unfortunate example, for it presents a number of unique difficulties. It was mentioned that $\mathcal{L}$ does not contain $\dot{\phi}$. Hence there is no canonical momentum conjugate to ϕ, and the Hamiltonian formulation of Section 11–4 falls through. Essentially, the source of the difficulty is that the scalar and vector potentials are not entirely independent quantities, but are connected by the so-called *gauge condition.* The gauge requirement acts as a supplementary condition which can be used, in effect, to eliminate one of the generalized coordinates, so that the field is described only by independent coordinates. For further details, see G. Wentzel, *Introduction to the Quantum Theory of Fields.*

The density function for a group of n particles can therefore be represented as

$$\rho(\mathbf{r}) = \sum_{i=1}^{n} q_i\, \delta(\mathbf{r} - \mathbf{r}_i), \tag{11-71}$$

where q_i is the charge of the ith particle and $\mathbf{r}_i$ its position vector.

Let us now evaluate the integral

$$\int \rho(\mathbf{r}) \left\{ \phi(\mathbf{r}) - \frac{\mathbf{v} \cdot \mathbf{A}(\mathbf{r})}{c} \right\} dV$$

for the charge density (11-71). By Eq. (11-70) the integral is then equal to

$$\sum_i q_i \left(\phi(\mathbf{r}_i) - \frac{\mathbf{v}_i \cdot \mathbf{A}(\mathbf{r}_i)}{c} \right),$$

so that when the charges and currents arise from a system of particles the total Lagrangian for the field becomes

$$L = \int \frac{E^2 - B^2}{8\pi}\, dV - \sum_i q_i \left(\phi_i - \frac{\mathbf{v}_i \cdot \mathbf{A}_i}{c} \right). \tag{11-72}$$

Here the subscripts indicate that ϕ and $\mathbf{A}$ are to be evaluated at the positions of the particles. Comparison with Eq. (1-61) now shows that the summation in (11-72) is *exactly* that part of the Lagrangian for the n particles which produces the electromagnetic forces acting on the particles! We can, in fact, combine the two Lagrangians by adding the kinetic energy of the particles to (11-72):

$$L = \int \frac{E^2 - B^2}{8\pi}\, dV - \sum_i q_i \left(\phi_i - \frac{\mathbf{v}_i \cdot \mathbf{A}_i}{c} \right) + \frac{1}{2} \sum_i m_i v_i^2. \tag{11-73}$$

An alternative form for Eq. (11-73) is

$$L = \int \left\{ \frac{E^2 - B^2}{8\pi} - \sum_i q_i\, \delta(\mathbf{r} - \mathbf{r}_i) \left(\phi - \frac{\mathbf{v}_i \cdot \mathbf{A}}{c} \right) \right\} dV + \frac{1}{2} \sum_i m_i v_i^2. \tag{11-73$'$}$$

In Eq. (11-73) or (11-73′) we have a total Lagrangian which describes both the electromagnetic field on the one hand, and the mechanical motion of the n particles on the other. It is a function of the generalized coordinates ϕ, $\mathbf{A}$ with continuous space indices x_1, x_2, x_3, *and* the particle generalized coordinates $\mathbf{r}_i$ distinguished by a discrete index i. A single Hamilton's principle thus suffices for both systems! Variation with respect to the potentials produces the Maxwell equations of the field, while variation with respect to the particle coordinates results in the particle equations of motion. Note that the first term in Eq. (11–73) represents the Lagrangian for the field in the absence of the charged particles. Similarly, the last term represents the Lagrangian for the particles in the absence of the field. The middle term then furnishes the mutual interaction between particle and field.

Expressing fields in terms of a variational principle formulation thus results in a description that is compact and elegant. One may well ask, however, what practical advantage the formulation has over the direct use of the field equations. The most important applications actually lie outside the domain of classical physics, but may be mentioned here briefly.

First, the Lagrangian formulation provides a convenient technique for inventing new types of fields and investigating their properties. The possible terms in the Lagrangian density are closely limited by the requirements that $\mathcal{L}$ contain only the coordinates and their first spatial and time derivatives, and that $\mathcal{L}$ be a Lorentz invariant. For example, if there is only one type of generalized coordinate η, which must be a world scalar (or pseudoscalar), there are only three possible terms meeting these requirements:

$$\eta, \qquad \sum_\mu \left(\frac{\partial \eta}{\partial x_\mu}\right)^2, \qquad \sum_\mu A_\mu \frac{\partial \eta}{\partial x_\mu},$$

where A_μ is an external world vector (or pseudovector). Any Lagrangian density for a scalar field must therefore be composed of combinations of these terms. One can thus examine many of the general properties of such a scalar field without knowing the physical mechanism producing the field. This is the technique currently used in much of the theoretical work on meson fields.

The second application has to do with the quantization of fields. It has already been remarked that the transition from classical to quantum theory can be stated only in terms of the canonical variables describing a system. Thus it was pointed out that the classical Poisson brackets of the canonical coordinates, or functions of them, correspond to the quantum commutation relations. In effect, we know how to quantize a system only when we can speak of it in mechanical terms. If it is desired to construct a quantum theory of the electromagnetic field, or of any other field, we must first obtain a description of the field in the language of mechanics. The Lagrangian and Hamiltonian formulations presented in this chapter form the basis for such a description.

SUGGESTED REFERENCES

J. C. SLATER AND N. H. FRANK, *Mechanics.* In discussing the mechanics of continuous systems, we have stopped at the equations of motion, and have not considered their solutions in detail. To treat the vibrations of strings, membranes, fluids, or solids would require a separate volume in itself. The text by Slater and Frank devotes almost half the space to such questions, and forms a readable, if somewhat elementary, introduction to the subject. In particular, the transition from the discrete chain to the continuous string is discussed in Chapter VII for *transverse* vibrations.

LORD RAYLEIGH, *The Theory of Sound.* This treatise naturally contains much material on the vibrations of continuous bodies. A discussion of the wave equation for the propagation of sound in gases will be found in Chapter XI, Volume 2, where the question of adiabatic vs. isothermal motion of the gas is examined in great detail.

G. WENTZEL, *Introduction to the Quantum Theory of Fields.* No single reference can be given which contains a detailed and comprehensive treatment of the classical mechanics of fields. Because the classical theory was usually developed as a preliminary to quantization of the field, most discussions are to be found in works on quantum mechanics. The best source is probably the excellent and well written volume of Wentzel, especially Chapter 1. Valuable material is also given in Chapters XIII and XIV of *Quantum Mechanics* by L. I. Schiff. The latter chapter, in particular, is devoted to the electromagnetic field. An earlier reference is Section 9 of the Appendix to W. Heisenberg's *The Physical Principles of the Quantum Theory.* The pioneering work in the theory of fields, and still quite valuable, is the paper by W. Heisenberg and W. H. Pauli, Zeitschr. f. Phys. **56,** 1 (1929).

EXERCISES

1. (a) The transverse vibrations of a stretched string can be approximated by a discrete system consisting of equally spaced mass points located on a weightless string. Show that if the spacing is allowed to go to zero, the Lagrangian approaches the limit

$$L = \frac{1}{2}\int\left[\mu\dot{\eta}^2 - T\left(\frac{\partial\eta}{\partial x}\right)^2\right]dx$$

for the continuous string, where T is the fixed tension. What is the equation of motion if the density μ is a function of position?

(b) Obtain the Lagrangian for the continuous string by finding the kinetic and potential energies corresponding to transverse motion. The potential energy can be obtained from the work done by the tension force in stretching the string in the course of the transverse vibration.

2. Obtain Hamilton's equations of motion for a continuous system from the modified Hamilton's principle (11–58), following the procedure of Section 7–4.

3. Show that if ψ and ψ^* are taken as two independent field variables, the Lagrangian density

$$\mathfrak{L} = \frac{h^2}{8\pi^2 m} \nabla\psi \cdot \nabla\psi^* + V\psi^*\psi + \frac{h}{2\pi i}(\psi^*\dot{\psi} - \psi\dot{\psi}^*),$$

leads to the Schrödinger equation

$$-\frac{h^2}{8\pi^2 m} \nabla^2\psi + V\psi = \frac{ih}{2\pi} \frac{\partial\psi}{\partial t},$$

and its complex conjugate. What are the canonical momenta? Obtain the Hamiltonian density corresponding to $\mathfrak{L}$.

4. Show that

$$G_i = -\int \sum_k \pi_k \frac{\partial \eta_k}{\partial x_i} \, dV$$

is a constant of the motion if the Hamiltonian density is not an explicit function of position. The quantity G_i can be identified as the total linear momentum of the field along the x_i direction. The similarity of this theorem with the usual conservation theorem for linear momentum of discrete systems (cf. Section 8–6) is obvious.

5. A Lagrangian density for the electromagnetic field is given by the relativistic covariant form:

$$\mathfrak{L} = -\frac{1}{16\pi} \sum_{\mu,\nu} \left(\frac{\partial A_\mu}{\partial x_\nu} - \frac{\partial A_\nu}{\partial x_\mu}\right)^2 - \frac{1}{8\pi} \sum_\mu \left(\frac{\partial A_\mu}{\partial x_\mu}\right)^2 + \sum_\mu \frac{j_\mu A_\mu}{c},$$

where A_μ is the potential four-vector and j_μ a four-vector with components $\mathbf{j}$ and $i\rho c$. Show that the Lagrangian density leads directly to the wave equations for the vector potential:

$$\Box^2 A_\mu = \frac{4\pi j_\mu}{c}.$$

Show also that this Lagrangian density is identical with that given in the text, except for the middle term (which has zero value anyhow as a result of the gauge condition).

BIBLIOGRAPHY

Where a reference has been reprinted recently (as in the case of many of the German books), the place and date of reprinting have been added in parentheses. If a given work is listed in the *Suggested References* section of a chapter, this fact is indicated by the appropriate chapter number in parentheses after the reference.

General Treatises on Classical Mechanics

1. Ames, Joseph Sweetman and Murnaghan, Francis D., *Theoretical Mechanics.* Boston: Ginn and Company, 1929.

2. Appell, Paul, *Traité de Mécanique Rationelle* (5 vols.) 2d ed. Paris: Gauthier-Villars, 1902–37.

3. Coe, Carl Jeness, *Theoretical Mechanics.* New York: Macmillan Co., 1938. (Chapter 1)

4. Destouche, Jean Louis, *Principes de la Mécanique Classique.* Paris: Centre National de la Recherche Scientifique, 1948.

5. Lamb, Horace, *Higher Mechanics*, 2d ed. Cambridge: Cambridge University Press, 1929.

6. MacMillan, William Duncan, *Theoretical Mechanics.* New York: McGraw-Hill. Vol. 1: *Statics and the Dynamics of a Particle*, 1927 (Chapter 3). Vol. 3: *Dynamics of Rigid Bodies*, 1936 (Chapter 5).

7. Milne, E. A., *Vectorial Mechanics.* New York: Interscience Publishers, 1948. (Chapters 1, 5)

8. Osgood, William F., *Mechanics.* New York: Macmillan, 1937. (Chapters 1, 2, 4)

9. Schaefer, Clemens, *Einführung in die Theoretische Physik.* Vol. 1: *Mechanik materieller Punkte; Mechanik starrer Körper und Mechanik der Kontinua (Elastizität und Hydrodynamik)*, 3d ed. Berlin: Walter de Gruyter, 1929. (Ann Arbor: J. W. Edwards, 1948.)

10. Slater, John C. and Frank, Nathaniel H., *Mechanics.* New York: McGraw-Hill, 1947. (Chapters 3, 11)

11. Sommerfeld, Arnold, *Vorlesungen über Theoretische Physik.* Vol. 1: *Mechanik*, 4th ed. Wiesbaden: Dieterich'sche Verlagsbuchhandlung, 1949. (Chapters 2, 3, 5)

12. Synge, John L. and Griffith, Byron A., *Principles of Mechanics*, 2d ed. New York: McGraw-Hill, 1949. (Chapters 1, 5)

13. Timoshenko, S. and Young, D. H., *Advanced Dynamics.* New York: McGraw-Hill, 1948. (Chapters 5, 10)

14. WEBSTER, ARTHUR GORDON, *The Dynamics of Particles and of Rigid, Elastic, and Fluid Bodies.* Leipzig: B. G. Teubner, 1904. (New York: Stechert-Hafner, 1920.) (Chapters 5, 7, 10)

15. WHITTAKER, E. T., *A Treatise on the Analytical Dynamics of Particles and Rigid Bodies*, 4th ed. Cambridge: Cambridge University Press, 1937. (New York: Dover Publications, 1944.) (Chapters 1–4, 7, 8, 10)

16. —— *Encyklopädie der Mathematischen Wissenschaften.* Vols. IV_1 and IV_2: *Mechanik, A: Grundlagen der Mechanik; B: Mechanik der Punkte und Starrer Systeme.* Leipzig: B. G. Teubner, 1901–35.

17. —— *Handbuch der Physik.* Vol. V: *Grundlagen der Mechanik der Punkte und Starrer Körper.* Berlin: Julius Springer, 1927. (Chapters 1, 5, 7, 8, 9)

WORKS ON SPECIAL ASPECTS OF CLASSICAL MECHANICS

18. BERGMANN, PETER GABRIEL, *Introduction to the Theory of Relativity.* New York: Prentice-Hall, 1942. (Chapter 6)

19. BYERLY, WILLIAM ELWOOD, *An Introduction to the Use of Generalized Coordinates in Mechanics and Physics.* Boston: Ginn & Co., 1913. (Chapter 2)

20. CHARLIER, CARL LUDWIG, *Die Mechanik des Himmels* (2 vols.) 2d ed. Berlin: Walter de Gruyter, 1927. (Chapter 9)

21. DAVIDSON, MARTIN, *The Gyroscope and Its Applications*, rev. ed. London: Hutchinson, 1947. (Chapter 5)

22. EINSTEIN, ALBERT, *The Meaning of Relativity*, 4th ed. Princeton: Princeton University Press, 1950. (Chapter 6)

23. GRAY, ANDREW, *A Treatise on Gyrostatics and Rotational Motion.* London: Macmillan Co., 1918. (Chapter 5)

24. KASNER, EDWARD, *Differential-Geometric Aspects of Dynamics.* New York: American Mathematical Society, 1913.

25. KLEIN, FELIX, *The Mathematical Theory of the Top.* New York: Scribners, 1897. (Chapter 5)

26. KLEIN, FELIX AND SOMMERFELD, ARNOLD, *Über die Theorie des Kreisels* (4 vols.). Leipzig: B. G. Teubner, 1897–1910. (Chapter 5)

27. LANCZOS, CORNELIUS, *The Variational Principles of Mechanics.* Toronto University of Toronto Press, 1949.

28. MACH, ERNST, *The Science of Mechanics*, 5th English ed. LaSalle, Illinois: Open Court Publishing Co., 1942. (Chapter 1)

29. NEWBOULT, H. O., *Analytical Method in Dynamics.* Oxford: Oxford University Press, 1946. (Chapter 4)

30. OLSON, HARRY F., *Dynamical Analogies.* New York: D. Van Nostrand, 1946. (Chapter 2)

31. PAULI, WOLFGANG, JR., *Relativitätstheorie.* Leipzig: B. G. Teubner, 1921.

32. POINCARÉ, HENRI, *Les Méthodes Nouvelles de la Mécanique Céleste* (3 vols.). Paris: Gauthier-Villars, 1892–99. (Chapter 9)

33. ROUTH, E. J., *The Advanced Part of a Treatise on the Dynamics of a System of Rigid Bodies*, 6th ed. London: Macmillan, 1905.

34. Schaefer, Clemens, *Die Prinzipe der Dynamik.* Berlin: Walter de Gruyter, 1919. (Chapter 7)

35. Thomson, J. J., *Applications of Dynamics to Physics and Chemistry.* London: Macmillan, 1888. (Chapter 2)

36. Wintner, Aurel, *The Analytical Foundations of Celestial Mechanics.* Princeton: Princeton University Press, 1941.

Works in Other Branches of Physics and Mathematics, Containing Material of Interest for Classical Mechanics

37. Becker, R., *Theorie der Elektrizität.* Vol. II: *Elektronentheorie*, 6th ed. Leipzig: B. G. Teubner, 1933. (Ann Arbor: J. W. Edwards, 1946.) (Chapter 6)

38. Bergmann, Peter Gabriel, *Basic Theories of Physics: Mechanics and Electrodynamics.* New York: Prentice-Hall, 1949.

39. Bliss, Gilbert Ames, *Calculus of Variations* (Carus Mathematical Monographs, 1). LaSalle, Illinois: Open Court Publishing Co., 1925. (Chapter 2)

40. Bôcher, Maxime, *Introduction to Higher Algebra.* New York: Macmillan, 1907. (Chapters 4, 10)

41. Born, Max, *The Mechanics of the Atom*, tr. by J. W. Fisher. London: G. Bell and Sons, 1927. (Chapters 8, 9)

42. Born, Max and Jordan, Pascual, *Elementare Quantenmechanik.* Berlin: Julius Springer, 1930. (Ann Arbor: J. W. Edwards, 1946.) (Chapter 8)

43. Brand, Louis, *Vector and Tensor Analysis.* New York: John Wiley & Sons, 1947.

44. Brillouin, Léon, *Les Tenseurs en Mécanique et en Élasticité.* Paris: Masson et cie, 1938. (New York: Dover Publications, 1946.) (Chapters 4, 9)

45. Carathéodory, Constantin, *Variationsrechnung und Partielle Differentialgleichungen Erster Ordnung.* Leipzig: B. G. Teubner, 1935. (Ann Arbor: J. W. Edwards, 1945.) (Chapters 8, 9)

46. Courant, R. and Hilbert D., *Methoden der Mathematischen Physik* (2 vols.) 2d ed. Berlin: Julius Springer, 1931–37. (New York: Interscience Publishers, 1943.) (Chapter 4)

47. Dirac, P. A. M., *The Principles of Quantum Mechanics*, 3d ed. Oxford: Oxford University Press, 1944. (Chapter 8)

48. Epstein, Paul S., *Textbook of Thermodynamics.* New York: John Wiley & Sons, 1937. (Chapter 7)

49. Frank, Philipp and von Mises, Richard, *Die Differential- und Integralgleichungen der Mechanik und Physik* (2 vols.) 2d ed. Braunschweig: F. Vieweg & Sohn, 1930–35. (New York: Mary S. Rosenberg, 1943.) (Chapters 8, 9)

50. Gibbs, J. Willard, *Vector Analysis*, ed. by E. B. Wilson. New York: Scribner, 1901. (New Haven: Yale University Press, 1931.) (Chapter 5)

51. Guillemin, Ernst A., *The Mathematics of Circuit Analysis.* New York: John Wiley & Sons, 1949. (Chapter 10)

52. HEISENBERG, WERNER, *The Physical Principles of the Quantum Theory*, tr. by Carl Eckart and Frank C. Hoyt. Chicago: University of Chicago Press, 1930. (New York: Dover Publications, 1949.) (Chapter 11)

53. HERZBERG, GERHARD, *Infrared and Raman Spectra of Polyatomic Molecules.* New York: D. Van Nostrand, 1945. (Chapters 4, 10)

54. JEFFREYS, H. AND JEFFREYS, BERTHA S., *Methods of Mathematical Physics*, 2d ed. Cambridge: Cambridge University Press, 1950. (Chapter 4)

55. JOOS, GEORG, *Theoretical Physics*, tr. by I. M. Freeman. New York: G. E. Stechert, 1934. (Chapter 1)

56. LAMB, HORACE, *Hydrodynamics*, 6th ed. Cambridge: Cambridge University Press, 1932. (New York: Dover Publications, 1945.)

57. LEVI-CIVITA, TULLIO, *The Absolute Differential Calculus*, tr. by M. Long. London: Blackie & Son, 1929.

58. LINDSAY, ROBERT BRUCE, *Introduction to Physical Statistics.* New York: John Wiley & Sons, 1941. (Chapter 3)

59. LINDSAY, ROBERT BRUCE AND MARGENAU, HENRY, *Foundations of Physics.* New York: John Wiley & Sons, 1936. (Chapters 1, 6, 7)

60. MARGENAU, HENRY AND MURPHY, GEORGE MOSELEY, *The Mathematics of Physics and Chemistry.* New York: D. Van Nostrand, 1943.

61. MORSE, PHILIP M., *Vibration and Sound*, 2d ed. New York: McGraw-Hill, 1948.

62. LORD RAYLEIGH, *The Theory of Sound* (2 vols.) 2d ed. London: Macmillan, 1894–96. (New York: Dover Publications, 1945.) (Chapters 1, 10, 11)

63. SCHIFF, LEONARD I., *Quantum Mechanics.* New York: McGraw-Hill, 1949. (Chapter 11)

64. SOMMERFELD, ARNOLD, *Atomic Structure and Spectral Lines*, tr. by H. L. Brose from 5th German ed. of 1931. New York: E. P. Dutton, 1934. (Chapters 8, 9)

65. SOMMERFELD, ARNOLD, *Vorlesungen über Theoretische Physik.* Bd III: *Elektrodynamik.* Wiesbaden: Dieterich'sche Verlagsbuchhandlung, 1948.

66. TOLMAN, RICHARD C., *The Principles of Statistical Mechanics.* Oxford: Oxford University Press, 1938. (Chapter 8)

67. VAN VLECK, J. H., *Quantum Principles and Line Spectra.* Washington: National Research Council, 1926. (Chapter 9)

68. WENTZEL, GREGOR, *Quantum Theory of Fields*, tr. by C. Houtermans and J. W. Jauch. New York: Interscience Publishers, 1949. (Chapter 11)

69. WIGNER, EUGEN, *Gruppentheorie und ihre Anwendungen auf die Quantenmechanik der Atomspektren.* Braunschweig: F. Vieweg & Sohn, 1931. (Ann Arbor: J. W. Edwards, 1944.)

70. WILLS, A. P., *Vector Analysis, with an Introduction to Tensor Analysis.* New York: Prentice-Hall, 1931. (Chapter 5)

71. ZEMANSKY, MARK W., *Heat and Thermodynamics*, 2d ed. New York: McGraw-Hill, 1943.

INDEX OF SYMBOLS

In choosing the various symbols, certain general principles have been followed whenever possible. Vectors have been denoted by bold-faced Roman characters, while tensors of second rank or higher, and matrices, are represented by bold-faced sans serif letters. When a vector is specifically considered as a tensor of the first rank or as a one-column (or row) matrix, the sans serif type face is used. For Greek symbols, the same bold face is used for vectors, matrices, and tensors.

A dot above a letter invariably denotes differentiation with respect to time. Primes are frequently used to denote quantities which have been subjected to a transformation of some kind. In Chapter 4 primes on coordinates refer to body sets of axes, as distinguished from the unprimed space sets of axes, but after Chapter 4 this convention, having served its purpose, is no longer followed.

In discussing canonical transformations, lower case letters are used for the original variables, and capitals for the transformed variables. Subscripts 0 frequently denote initial or equilibrium values. As is customary, complex conjugates are denoted by an asterisk.

This index of symbols is not intended to be complete; it lists only the important symbols, and those which may possibly give rise to ambiguity.

INDEX OF GREEK SYMBOLS

INDEX

ABCDE698765